Engineering Materials and Processes

Series Editor

Professor Brian Derby, Professor of Materials Science
Manchester Materials Science Centre, Grosvenor Street, Manchester, M1 7HS, UK

Other titles published in this series:

Fusion Bonding of Polymer Composites
C. Ageorges and L. Ye

Composite Materials
D.D.L. Chung

Titanium
G. Lütjering and J.C. Williams

Corrosion of Metals
H. Kaesche

Corrosion and Protection
E. Bardal

Intelligent Macromolecules for Smart Devices
L. Dai

Microstructure of Steels and Cast Irons
M. Durand-Charre

Phase Diagrams and Heterogeneous Equilibria
B. Predel, M. Hoch and M. Pool

Computational Mechanics of Composite Materials
M. Kamiński

Gallium Nitride Processing for Electronics, Sensors and Spintronics
S.J. Pearton, C.R. Abernathy and F. Ren

Materials for Information Technology
E. Zschech, C. Whelan and T. Mikolajick

Nigel Sammes (Ed.)

Fuel Cell Technology

Reaching Towards Commercialization

With 139 Figures

Springer

Professor Nigel Sammes, BSc, PhD, MBA
Herman F. Coors Distinguished Professor of Ceramic Engineering
Department of Metallurgical and Materials Engineering
Colorado School of Mines
1500 Illinois Street
Golden, Colorado 80401
USA

British Library Cataloguing in Publication Data
Fuel cell technology: reaching towards commercialization.
- (Engineering materials and processes)
1. Fuel cells
I. Sammes, Nigel M.
621.3'12429

Engineering Materials and Processes ISSN 1619-0181
e-ISBN 1-84628-207-1
ISBN-13: 978-1-84996-983-3 e-ISBN-13: 978-1-84628-207-2

Softcover reprint of the hardcover 1st edition 2006

9 8 7 6 5 4 3 2

Springer Science+Business Media
springer.com

Preface

Fuel cells are electrochemical devices that convert the energy of a fuel (such as hydrogen, natural gas or other hydrocarbon-based fuels) directly into electricity. All fuel cells consist of an electrolyte layer in contact with an anode and a cathode on either side. The oxidation reaction occurs on the anode side of the fuel cell, while reduction takes place on the cathode. Single fuel cells are usually connected either in series, or in parallel, to form a stack, which is capable of producing several watts to many kW of power, depending on the requirements.

The fuel cell stack requires a number of other components to complete the system. These components are usually termed the balance-of-plant (BoP) and consist of the fuel cell processing section, the power section (the components surrounding the stack itself) and the power conditioning and control units. Fuel processing is required to produce a hydrogen-rich gas (and possibly to desulfurize the gas), while the power-conditioning unit is there to convert variable DC to controlled AC current, with a specific frequency, active and reactive power. It also acts as feedback to control the fuel flow to the stack.

This book is a consequence of many years of work performed by the faculty at the University of Connecticut, in most of the areas described above. As is apparent, the development of fuel cells is a multi-disciplinary activity, and thus most of the Engineering and Science departments at the University are involved in fuel cell research and development. Chapters 1 and 2 study the materials and design implications of solid oxide fuel cells (SOFC) and polymer electrolyte membrane fuel cells (PEM) respectively. Chapter 3 looks at fuel processing of a number of different hydrocarbon-based fuels, from both a chemical and engineering approach, with emphasis on the production of a hydrogen-rich fuel. The potential for internal reforming is also discussed. Chapters 4 to 6 study the modeling aspects of fuel cells from a micro (electrochemical and transport aspects, Chapter 4) to accelerated testing (Chapter 5) to macro modeling of the stack as a whole (Chapter 6). Chapter 7 looks at work being performed on new materials for sealing and for increasing the surface area of electrodes; the aerogel material. Chapter 8 studies the control and electrical conversion aspects of fuel cells. Finally, Chapter 8 looks at some very new systems based on microbial fuel cells.

Acknowledgments

The editor acknowledges the support of all the faculty, at the University of Connecticut, who contributed to this book. Their hard work over the years in fuel cell research and development will, hopefully, be acknowledged by the fuel cell community.

I would also like to thank Mr Jakub Pusz, PhD student, for helping rationalize the chapters and ensuring that the book got to the publishers on time.

I would also like to thank my colleagues without whom none of this work could have been realized in the first place.

Finally, I would like to thank my family for their continued support in my ongoing mission to modestly try and help fuel cells become a commercial reality.

Nigel Sammes
Storrs, CT
USA

Contents

List of Contributors

A. F. M. Anwar
Department of Electrical and
Computer Engineering
University of Connecticut
371 Fairfield Rd.,
Storrs, CT 06269

F. Barbir
Connecticut Global Fuel Cell Center
University of Connecticut
44 Weaver Rd,
Storrs, CT 06269

R. Bove
Industrial Engineering Department
University of Perugia
Via Duranti 67,
06125 Perugia,
Italy

X. Dong
Connecticut Global Fuel Cell Center
University of Connecticut
44 Weaver Rd,
Storrs, CT 06269

C. Erkey
Department of Chemical
Engineering
University of Connecticut
191 Auditorium
Storrs, CT 06269

E. W. Faraclas
Department of Electrical and
Computer Engineering
University of Connecticut
371 Fairfield Rd.,
Storrs, CT 06269

S. Haji
Department of Chemical
Engineering
University of Connecticut
191 Auditorium
Storrs, CT 06269

H. Hara
Connecticut Global Fuel Cell Center
University of Connecticut
44 Weaver Rd,
Storrs, CT 06269

X. Huang
Connecticut Global Fuel Cell Center
University of Connecticut
44 Weaver Rd,
Storrs, CT 06269

S. S. Islam
Department of Electrical and
Computer Engineering
University of Connecticut
371 Fairfield Rd.,
Storrs, CT 06269

K. A. Malinger
Department of Chemistry
University of Connecticut
55 North Eagleville
Storrs, CT 06269

K. Noll
Molecular and Cell Biology
University of Connecticut
91 North Eagleville Rd,
Storrs, CT 06269

R. Pitchumani
Department of Mechanical
Engineering
University of Connecticut
191 Auditorium Rd.
Storrs, CT 06269

J. Pusz
Connecticut Global Fuel Cell Center
University of Connecticut
44 Weaver Rd,
Storrs, CT 06269

K. Reifsnider
Department of Mechanical
Engineering
University of Connecticut
191 Auditorium Rd.
Storrs, CT 06269

N. M. Sammes
Department of Mechanical
Engineering
University of Connecticut
191 Auditorium Rd.
Storrs, CT 06269

A. Smirnova
Connecticut Global Fuel Cell Center
University of Connecticut
44 Weaver Rd,
Storrs, CT 06269

S. L. Suib
Department of Chemistry
University of Connecticut
55 North Eagleville
Storrs, CT 06269

J. Tang
Department of Mechanical
Engineering
University of Connecticut
191 Auditorium Rd.
Storrs, CT 06269

X. Xie
Department of Mechanical
Engineering
University of Connecticut
191 Auditorium Rd.
Storrs, CT 06269

F. Yang
Department of Mechanical
Engineering
University of Connecticut
191 Auditorium Rd.
Storrs, CT 06269

1

Solid Oxide Fuel Cells

Nigel M. Sammes, Roberto Bove and Jakub Pusz

1.1 Introduction

A Solid Oxide Fuel Cell (SOFC) is typically composed of two porous electrodes, interposed between an electrolyte made of a particular solid oxide ceramic material. The system originates from the work of Nernst in the nineteenth century. In his patent [1], Nernst proposed that a solid electrolyte could be made to electrically conduct, using a heater; the system then "glowed" by the passage of an electric current. The systems originally studied by Nernst were based on simple metal oxides. In 1937, Bauer and Preis [2] operated the first ceramic fuel cell at 1000°C, showing that the so-called "Nernst Mass" (85% zirconia and 15% yttria), and other zirconia-based materials present a reasonable ionic conduction at high temperature (600–1000° C). These works were really the prelude to the modern SOFC.

In Figure 1.1, a schematic representation of an SOFC is depicted.

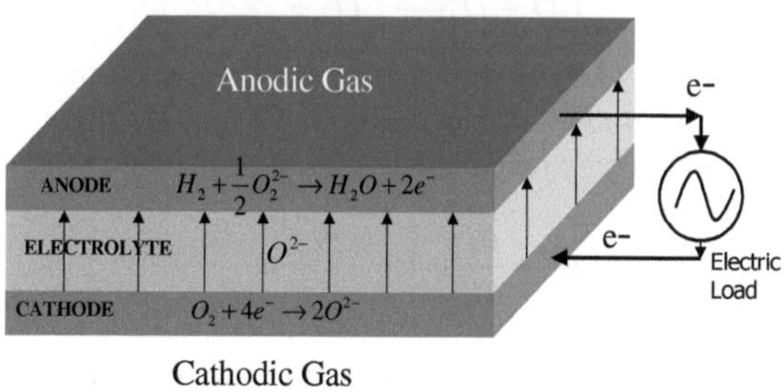

Figure 1.1. Schematic representation of a Solid Oxide Fuel Cell

At the cathode the reduction of oxygen occurs via

$$O_2 + 4e^- \rightarrow 2O^{2-} \tag{1.1}$$

The dense structure of the electrolyte does not allow the passage of the cathodic gas through it, while the high ionic conductivity and the high electrical resistance allow only O^{2-} ions to migrate from the cathode to the anode. At the anode, O^{2-} reacts with hydrogen producing water:

$$H_2 + O^{2-} \rightarrow H_2O + 2e^- \tag{1.2}$$

As reaction [1.2] occurs, electrons are released at the anode and migrate from the cathode through an external electric circuit, thus generating an electric current.

The overall reaction occurring in the cell is:

$$H_2 + \frac{1}{2}O_2 \rightarrow H_2O \tag{1.3}$$

Compared to fuel cells with proton conductive electrolyte, *i.e.* Proton Exchange Membrane Fuel Cells (PEM), Alkaline Fuel Cells (AFC) and Phosphoric Acid Fuel Cells (PAFC), for SOFC, water is produced at the anode.

1.2 Operation and Performance

Due to the high operating temperature, no precious metals (such as platinum used in low temperature fuel cells) are needed as the catalyst, thus carbon monoxide does not represent a harmful substance for the anode, in fact, CO represents additional fuel for the fuel cell. The presence of water, occuring via reaction (1.3) or externally provided, makes the following reaction (shift) occur:

$$CO + H_2O \leftrightarrow H_2 + CO_2 \tag{1.4}$$

thus providing additional hydrogen. Direct oxidation of carbon monoxide is also well established [3]:

$$CO + O^{2-} \rightarrow CO_2 + 2e^- \tag{1.5}$$

Whichever reaction, (1.4) or (1.5), is considered, each mole of carbon monoxide produces the same amount of electricity that hydrogen does.

The open circuit voltage can be calculated through the Nernst equation:

$$E = E^O + \frac{RT}{4F}\ln P_{O_2} + \frac{RT}{2F}\ln\frac{P_{H_2}}{P_{H_2O}} \tag{1.6}$$

Where E^O is the reversible voltage at the standard pressure and the operating temperature, *i.e.*:

$$E^O = \frac{\Delta G^0}{2F} \tag{1.7}$$

where ΔG^0 is the Gibbs free energy variation of reaction (1.3).

When the operating temperature is increased, ΔG^0 decreases, thus reducing the open circuit voltage. An increase of the temperature, however, leads to a reduction of the electrodes polarization and an increase in the electrolyte conductivity, and, consequently, an increase in the performance.

Another advantage inherent to the high operating temperature is the possibility of internally reforming hydrocarbons. The steam reforming reaction of a general hydrocarbon can be written as:

$$C_xH_yO_z + (2x - z)H_2O \rightarrow xCO_2 + \left(\frac{y}{2} + 2x - z\right)H_2 \quad (1.8)$$

Reaction (1.8) is endothermic, thus thermal energy is required. If reaction (1.8) occurs inside the anode section, the thermal energy requirements are directly recovered from reaction (1.3), which is exothermic. Moreover, nickel, which is a typical anode material, presents good catalytic activity for reaction (1.8) in the SOFC temperature range.

No thermal losses are associated with the heat transfer from the fuel cell section to the reforming reactor.

Reduced costs for the fuel processing unit. When internal reforming is conducted, in fact, only a small part of the fuel must be externally reformed, thus the cost of the Balance of Plant (BoP) can be substantially reduced.

As reaction (1.3) proceeds, if water is not removed, the reversible voltage decreases (equation (1.6)), due to the increased partial pressure of water and reduced partial pressure of hydrogen. The consumption of water through reaction (1.8) reduces this phenomenon.

On the other hand, the disadvantages of internal reforming are the increased temperature gradient inside the cell [4–5] and possible carbon deposition [5]. In order to avoid these problems, as mentioned above, part of the fuel is externally pre-reformed before entering the anode. The pre-reforming reaction can be achieved using different hydrocarbon conversion techniques, such as autothermal reforming, steam reforming, or catalytic partial oxidation. Bove and Sammes [6] have analyzed the effect of different fuel processing on the performance of SOFC system.

1.3 SOFC Materials

Due to the high operating temperature, and to the all-solid state nature of the components, the thermal expansion of each material composing the stack must be as close as possible, in order to avoid mechanical fracture and material delamination. Other desired characteristics include chemical stability under the relative operational conditions, high electrical conductivity for the electrodes and electrical interconnects, high ionic conductivity and almost zero electrical conductivity for the electrolyte, and low cost. The choice of materials is usually the result of a compromise between the above characteristics, and, at the present, a variety of materials are being investigated. There are, however, some materials that are considered 'traditional' for SOFC, due to the large number of applications in the last few years. Table 1.1 summarizes these materials.

Table 1.1. Traditional materials for SOFC

Component composition		Specific conductivity (S/m) at SOFC running temperature	Conductivity depends upon
Anode	*Ni/YSZ cermet*	*400 - 1000*	*Particle size ratio* *Ni content*
Cathode	$Sr_xLa_{1-x}MnO_{3-\delta}$	*6 - 60*	*Cathode* *Porosity* *Sr content*
Electrolyte	Y_2O_3-ZrO_2	*10 - 15*	*Electrolyte Density*

In the 'typical' modern SOFC, the electrolyte is based on yttria-stabilized zirconia (YSZ), traditionally fully cubic, having a relatively high ionic conductivity at temperatures above 700°C, with negligible electronic conductivity (required in an SOFC). The material is stable in both oxidizing and reducing environments, and can be fabricated in many forms, using ceramic processing and thin-film technologies. The requirement for a dense YSZ is critical as it is imperative that the fuel and air gases are not physically (and thus chemically) mixed. The porous cathode (traditionally doped $LaMnO_3$) must be electronically conducting, must have good catalytic activity towards oxygen reduction, and must be stable under oxidizing environments. The anode, must also be electronically conducting, must be stable under reducing environments and must be catalytically active towards the oxidation reaction of the fuel of choice. Traditionally, hydrogen is used as the fuel, thus Ni exploits the electrocatalyst functions. Ni, however, has a very high thermal expansion coefficient and tends to sinter (thus lowering the active sites available for oxidation, and, consequently, lowering the overall efficiency of the fuel cell) with time at temperature and under electrical load. For this reason, YSZ is added to form a Ni/YSZ composite to reduce the sinterability and also to lower the coefficient of thermal expansion closer to that of the electrolyte. YSZ in the composite matrix can also act to increase the number of active sites (the triple phase boundary) by allowing some oxygen ion conduction within the anode structure [7].

1.3.1 Anode

Although in the early development of SOFCs precious metals like platinum and gold were considered, due to their low instability, the most widely used material for the anode is currently a composite of Ni/8YSZ (*i.e.* Ni with 8% mole of YSZ). As mentioned above, Ni/YSZ provides the anode with high electrical conductivity, an adequate ionic conductivity, and a high activity for the electrochemical reactions and reforming. Moreover, the mechanical characteristics enable the anode to be the supporting structure of the entire cell, thus realizing anode supported cells [8–12].

Ni/YSZ, however, is not without drawbacks. When hydrocarbons are used as the fuel, carbon deposition may occur [13–14] and sulfur (that is usually present in natural gas) can cause a loss of performance even in concentrations below 50 ppb [15]. Another characteristic of the Ni/YSZ cermet anode is that it is prepared with NiO that reduces to Ni. However, if the anode is exposed to air at high temperature, Ni re-oxidizes [16], thus it must be kept in a reducing environment all the time. In order to overcome this issue, a variety of other materials are being investigated, including chromite/titanate based perovskites [17], $Y_{0.2}Ti_{0.18}Zr_{0.62}O_{1.9\pm\delta}$ (YZT) [17], ceria based oxide materials [18–20], $La_xSr_{1-x}VO_{3-\delta}$ [21], and $La_{0.8}Sr_{0.2}Cr_{0.97}V_{0.03}O_3$ [22]. Trofimenko *et al.* [23] compared the performance of four different anodes, made of $Ni/Zr_{0.862}Y_{0.138}O_{1.931}$ (Ni/8YSZ) cermet, $Ni/Ce_{0.8}Gd_{0.2}O_{1.9}$ (Ni/CGO) cermet, $La_{0.47}Ca_{0.4}Ti_{0.8}Cr_{0.2}O_3$ (LCaTC) perovskite and $La_{0.47}Ca_{0.4}Ti_{0.8}Cr_{0.2}O_3/Zr_{0.862}Y_{0.138}O_{1.931}$ (LCaTC/8YSZ) at different temperature and oxygen partial pressure and found that the LCaTC and LCaTC/8YSZ present a too low electrochemical activity, compared to Ni/8YSZ and Ni/CGO anodes. Fouquet *et al.* [24] studied the behavior of Ni/8YSZ and Ni/CGO operating on methane and found similar electrical performance at 800°C.

1.3.2 Electrolyte

Yttria-stabilized zirconia (YSZ) is currently the most widely employed material for the electrolyte manufacture. After the studies of Nernst [1] and Bauer and Preis [2], Brown Boveri in Germany and Westinghouse in the United States [25] have demonstrated that high power density could be reached in SOFC when YSZ is used as the electrolyte. YSZ presents good ionic conductivity and negligible electronic conductivity at high temperature (above 700°C), it is chemical stable at the SOFC operating conditions, and presents a thermal expansion that is compatible with the anode and the cathode. As can be observed from Figure 1.2, the ionic conductivity presents a strong dependence on the operating temperature. The first developed SOFC typically operated around 1000°C [3, 7, 26]. At this temperature the ohmic loss associated with a current density of 250 mA/cm^2 across a 200 μm electrolyte is about 50 mV [3].

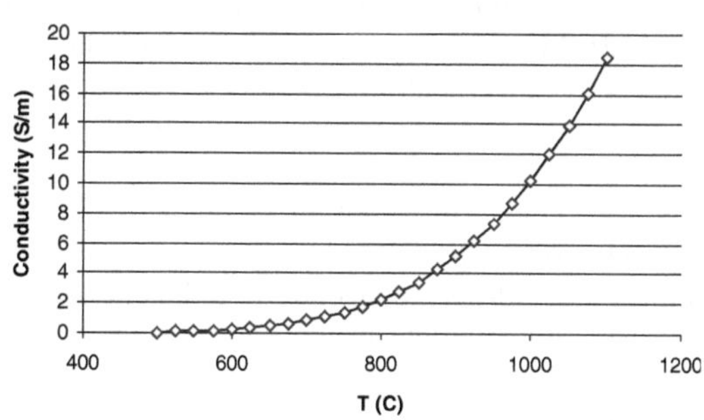

Figure 1.2. Conductivity of YSZ versus temperature

Although Siemens-Westinghouse [27], Rolls-Royce [28] and Mitsubishi Heavy Industries-Chubu Electric Power Company [29] seek to keep on operating their systems at 1000°C, there is a common effort in the scientific community to reduce the operating temperature below 800°C [3, 7–9, 11–12, 30–38]. The benefits associated with a reduced temperature are mainly connected to the possibility of replacing the expensive ceramic interconnect materials with metallic interconnects, such as ferritic stainless steels [3]. Other advantages are related to the higher chemical stability and the reduced difference between the thermal expansion of different components.

As can be observed from Figure 1.2, however, the temperature reduction causes a relevant reduction of the ionic conductivity. In order to obtain the same performance of high temperature SOFC, the electrolyte thickness of the so called Intermediate Temperature SOFC (ITSOFC) needs to be drastically reduced. Due to the improvement of the mechanical characteristics of the anode in recent years, most of the developers have replaced the electrolyte supported structure with the anode supported configuration, thus reducing the electrolyte thickness to several microns (typically around 10μm).

Besides the effort in reducing the electrolyte thickness, alternative materials are being investigated. Studies on scandia stabilized zirconia (ScSZ) [39–43] show that this has the highest ionic conductivity of all the doped zirconium oxide systems (approximately twice that of YSZ). The cubic phase exists above 8.5 mol% Sc_2O_3, and above 10 mol% a rhombohedral (β-phase) co-exists with the cubic. The main drawback of ScSZ are represented by the high costs [42] and the performance degradation with time. Experimental tests on the degradation of the zirconia-based electrolyte, reported by Müller *et al.* [43], showed that, apart from 10YSZ and 10 ScCeSZ, the performance reduction is about 50% after 1000 hours operation.

Doped-CeO_2 materials, like gadolinium doped ceria (GDC) and samarium doped ceria (SDC), have been found to have a very high ionic conductivity but they are prone to reduction in reducing atmospheres (Ce^{4+} becomes Ce^{3+}), as described in [44].

Another alternative material for ITSOFC is $LaGaO_3$. Since 1994, when the work of Ishihara *et al.* [45] was published, different dopants have been proposed. Of particular interest is the lanthanum gallate with the strontium doping on the A site of the perovskite and magnesium on the B-site (LSGM), proposed by Goodenough [46]. The issues for LSGM remain the mis-matching of the thermal expansion, mechanical strength, chemical compatibilities and high cost [3].

For an extensive review of the ionic conductive electrolyte, the reader is referred to [25].

All the above mentioned electrolytes are ionic conductive, and, as a result, water is produced in the anode. When an SOFC is operating on pure hydrogen, the production of water dilutes the hydrogen concentration, thus the open circuit voltage is reduced (3.6). If proton conductive electrolytes are considered, instead, the reduction of hydrogen partial pressure due to reaction (3.3) is much less (in the case that the cell operates on pure hydrogen, the hydrogen molar fraction is constantly equal to 1). The result is that SOFC based on proton conductive electrolyte can operate at higher fuel utilization, compared to ionic conductive SOFC. Since 1981, when Iwata *et al.* published their work, reporting the proton conductivity of $SrCeO_3$ for steam electrolysis [47], several studies have been published, proposing doped $SrCeO_3$ and other oxides as the SOFC electrolyte [48–54]. However, many studies are still needed to verify the potential of these materials.

1.3.3 Cathode

As shown in Table 1.1., the typical material for the cathode is the strontium doped $LaMnO_3$ (LSM), even if, during the early development, platinum and other noble metals have been considered [7]. LSM presents good electrochemical activity for oxygen reduction, a thermal expansion close to that of YSZ, and good stability when the cell operates at about 1000°C. For ITSOFCs, however, the ionic conductivity of LSM is inadequate [3]. Furthermore, chromia vapors generated from the (inexpensive) interconnects of ITSOFC seriously interact with the cathode, thus reducing the oxygen diffusion and limiting the electrode reaction sites [55]. The ionic conductivity issue can be overcome, through mixing LSM with YSZ [56–57].

Alternative materials are also being considered. $La_{1-x}Sr_xFeO_3$ (LSF) and $La_{1-x}Sr_xCo_yFeO_{3-\delta}$ (LSCF), for example, showed acceptable ionic conductivity. Some problems related to the compatibility with YSZ, however can arise [58–59], thus the use of these materials with alternative electrolyte materials is being investigated [60].

1.3.4 Interconnect

There are two main types of materials employed for the interconnect manufacture: ceramic, which are suitable for high temperature usage and metallic alloys for intermediate temperature. Ceramic interconnects are primarily doped lanthanum and yttrium chromites [3], whose electric conductivity typically increases with temperature, thus they are not suitable for ITSOFC. The presence of dopants ensures chemical stability, and the required thermo-mechanical characteristics. The two main drawbacks of the ceramic interconnects are the high costs and the relative rigidity that makes them fragile [3]. When the operating temperature is lower, the use of metallic interconnects becomes feasible. Compared to ceramic materials, metallic alloys are relatively inexpensive and easy to process. In particular, ferritic stainless steels have attracted considerable interest, because the thermal expansion is close to that of the zirconia electrolyte [61]. At the SOFC operating temperature, however, corrosion can be a challenge for metal interconnects. To overcome this problem, chromia is usually used to combat the corrosion phenomenon. Although corrosion is drastically reduced, Cr evaporates and causes electrode poisoning. This phenomenon can be mitigated by coating the interconnect with a perovskite, such as lanthanum strontium-doped-manganite or cobaltite [3].

1.4 Geometrical Designs

1.4.1 Flat Planar

In this configuration, the electrodes, the electrolyte and the current collectors are present as a flat planar geometry, and are arranged in the 'typical' sandwich configuration (Figure 1.1). This configuration is quite simple, relatively easy and inexpensive to construct, and the associated power density can be relatively high. Figure 1.3 depicts a cross section of a flat planar configuration.

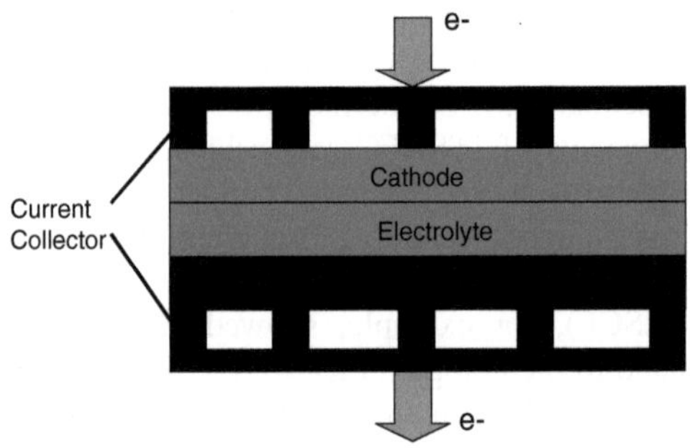

Figure 1.3. Cross section of a flat planar SOFC

Despite the other fuel cell technologies, some problems are associated with flat planar SOFCs. First of all, as mentioned above, the different thermal expansion of the components can lead to cracking problems, thus the maximum active surface is usually limited to some hundred cm^2 and the scale-up to multi MW system is possible only through a modular approach [62]. Another disadvantage is the high contact resistance between electrolyte and electrodes, due to the geometrical configuration and the all solid state components. The non-perfect manufacturing of the current collectors and the electrodes, in fact, leads to non-uniform contact between the two adjacent parts. Although an external pressure can increase the contact area, cracking problems limit the maximum external load. Another issue related to this configuration is ensuring the required sealing. Due to the geometrical configuration, it is a challenge to find a suitable material that presents the chemical and physical proprieties required [3]. For this reason, in 2004, the Solid State Energy Conservation Alliance (SECA) established that the necessity of solving the sealing issues has an extreme priority, in order to develop a ready-to-market product [63].

1.4.2 Monolithic Design

The monolithic SOFC is the most evolved of flat planar configuration (Figure 1.4). It consists of corrugated thin cell components. The result is that the volumetric power density is very high. On the other hand, the structure is more complex to be realized, as well as the manifold system, compared to the flat planar.

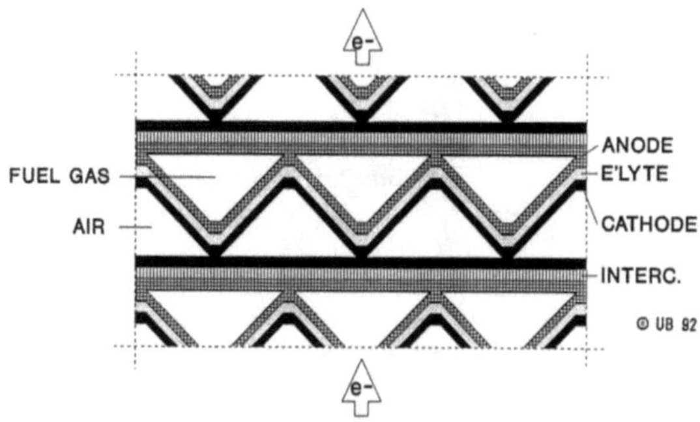

Figure 1.4. Schematic cross section of a monolithic SOFC [64]

1.4.3 Tubular Design

The tubular design currently represents the SOFC design that has reached the most advanced development. In particular, Siemens-Westinghouse (SW) has installed SOFC systems all around the world and demonstrated a 220 kW SOFC-gas turbine hybrid system, at the Fuel Cell National Research Center, University of California at Irvine. Tubular configuration allows the cell to overcome most of the current drawbacks. In particular sealing is required only at the tube ends, thus reducing the sealing problems. Secondly, the cylindrical geometry allows a reduction of the shear stress, thus enabling a cell with better mechanical properties. For this reason, SW is the only SOFC company that realized a scale-up of the systems at the hundred kW level [62].

Figure 1.5 depicts the SW tubular design. The cell is cathode-supported, and thin electrolyte and anode layers cover the cathode, with the exception of a small strip, where a current collector is placed [26]. The cell is closed at one end, thus air is provided through an injector tube. The cathode interconnection, illustrated in Figure 1.5 is then put in contact with the anode of the next cell, thus connecting the cells in an electrical series configuration.

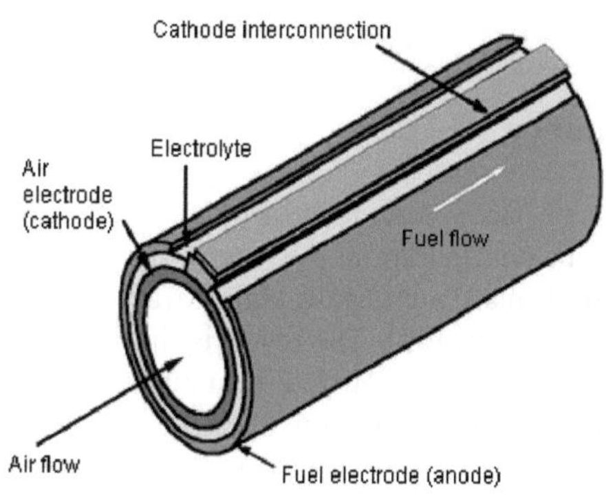

Figure 1.5. Siemens-Westinghouse cell design

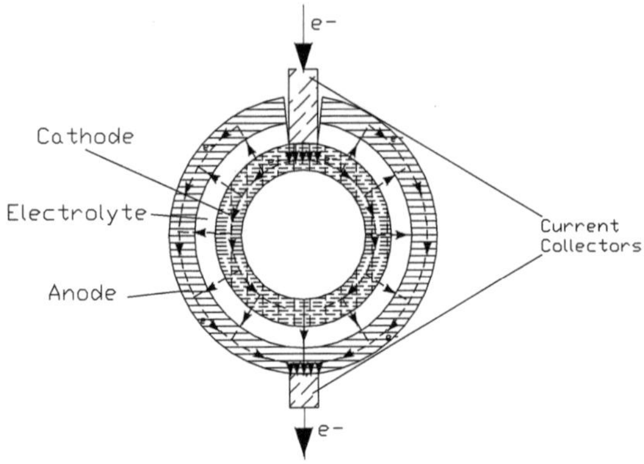

Figure 1.6. Current path in a Siemens-Westinghouse SOFC

Due to the current collectors disposition, the current path is around the tube, as illustrated in Figure 1.6, thus the related performance is lowered, compared to the planar configuration.

In order to overcome this problem, SW has designed a new configuration, called High Power Density (HPD) [27].

Other drawbacks of the tubular design are the low volumetric power density, and the high manufacturing costs.

1.5 Design of an SOFC Stack. The Case of Micro-tubular Solid Oxide Fuel Cell

The term micro-tubular is usually used to indicate tubular fuel cells with a diameter smaller than 1 cm.

In the present section, a 100 W stack design and construction are presented. The stack is composed of micro-tubular, anode supported SOFCs. Problems related to the performance variation of a single cell, when embedded in a stack, are also identified.

1.5.1 Stack Configuration

While planar SOFCs are stacked to form a pile of cells, tubular stacks must be assembled in a different configuration. An easy way to arrange single cells is to align them, in order to form a Planar Multi-cells Array (PMA) and then to stack the PMA as if it were a planar cell [65]. The result is that stacks of different sizes can simply be fabricated by assembling different numbers of PMAs. In Figure 1.7, a PMA is depicted. As explained in section 1.5.2, the anode represents the internal layer of the tube, while the external surface is the cathode. As a consequence of this configuration, oxidant (air) and fuel can be easily managed and the external leakage is mostly limited to air mass loss.

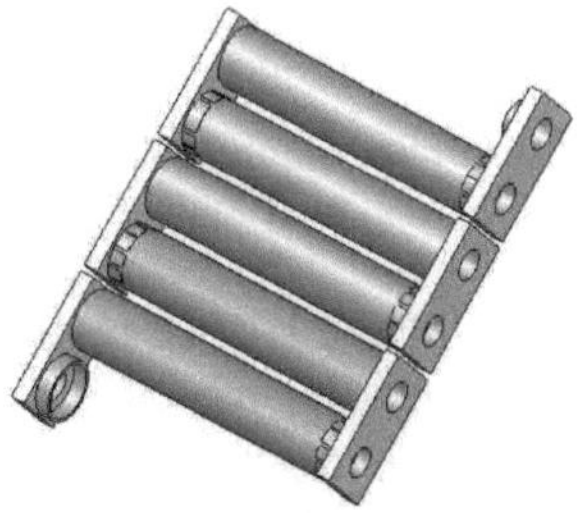

Figure 1.7. Planar multi-cells array (PMA)

In Figure 1.8, a cross-section of a PMA is schematically represented. As can be observed, possible leakages are likely to occur from the environment surrounding the tubes (limited by a box) and the external environment. In the configuration of Figure 1.8, however, air surrounds the stacked tubes, while fuel flows internally along the tubes, thus the fuel leakage can occur only at the tube extremity. A good brazing between the current collector and the tube, however, is needed to avoid fuel leakage. In Section 1.5.3, the tests conducted, and the procedure developed, for brazing the cells to the current collectors are presented.

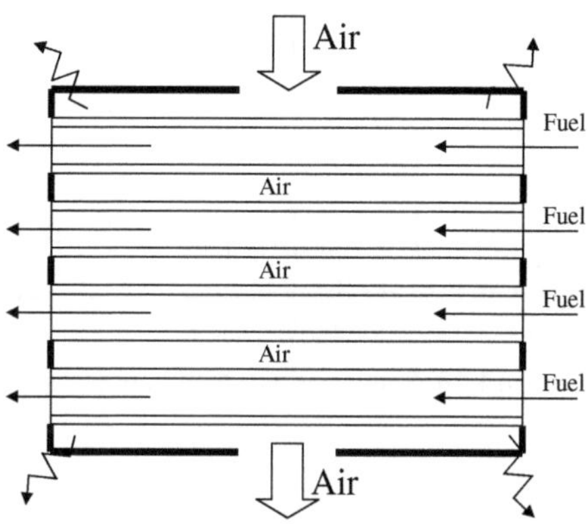

Figure 1.8. Schematic cross section of a planar multi-cells array

Figure 1.9 depicts the current collector. As can be observed, the two cylinders of the current collector are designed so that one is in contact with the inner part of the cell (*i.e.* the anode) and the other with the outer (*i.e.* the cathode), thus every contiguous cell of the PMA is connected in series. Every PMA is then connected in series or parallel, according to the desired current and voltage characteristics.

Figure 1.9. Current collector

1.5.2 Single Cells Construction

For the stack construction, an anode supported fuel cell has been selected. The supporting anode tubes are made of nickel and yttria stabilized zirconia (Ni-YSZ), coated with a thin YSZ electrolyte and a thin coat of lanthanum strontium manganite/cobaltite (LSM) cathode [8]. Figure 1.10 represents the single cell fabrication process. All the information, relative to tube fabrication procedures, are extensively reported in [8]. After the construction process is complete, tubes are cut to a length of 110 mm. Figure 1.11 is a picture of the single fuel cells [8].

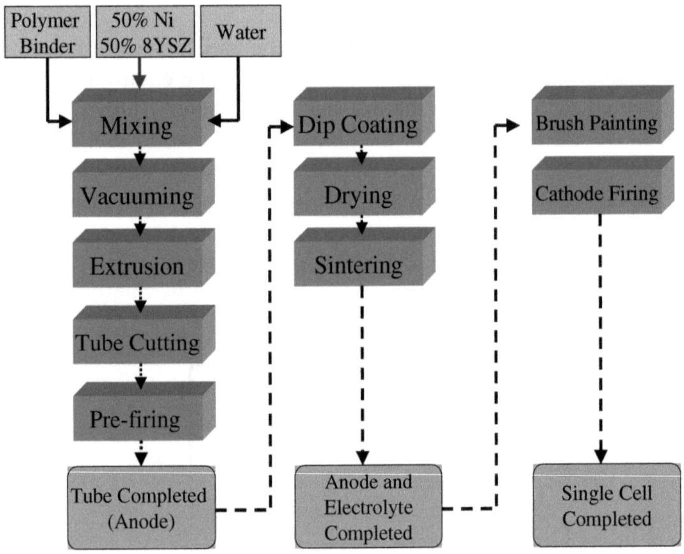

Figure 1.10. Single cell production process

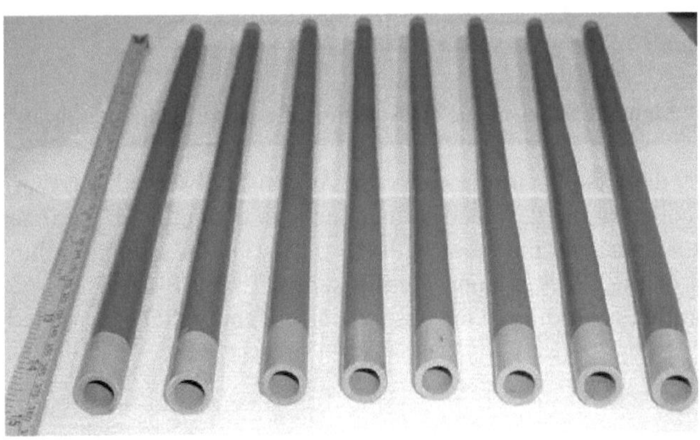

Figure 1.11. Picture of the single cells [8]

1.5.3 Joining Current Collectors and Single Cells

Current collectors are joined to the single cells using a brazing technique. An important issue for the joint integrity is the possibility of internal stress, due to the different thermal expansions of the fuel cell components, the brazing material and the current collector. For this reason, materials selection for stacking the cells is crucial. The thermal coefficient of expansion of the tube is estimated to be $12 \cdot 10^{-6}$ K^{-1} [66].

The material selected for the current collector is nickel 200 (Ni 99.5%, Fe 0.15%, CU 0.05%), whose coefficient of thermal expansion is $14 \cdot 10^{-6}$ K^{-1} [67]. The selection of the brazing material is dictated by the need of a compatible coefficient of thermal expansion and a melting point that is lower than that of the tube and the current collector. Pure silver presents a melting point of 961.78°C and a coefficient of thermal expansion of $18.9 \cdot 10^{-6}$ [68], thus it is an ideal candidate as a brazing material.

Silver braze metal in wire form (0.254 mm and 0.762 mm diameter) is placed in the gap between the tube outer diameter (OD) and the cap inner diameter (ID) at the base of the tube respectively. The interface between the tube OD and the cap ID is designed with enough clearance so that the anode tube and silver wire fits tightly into the nickel end cap. This gap (0.254mm) makes it possible for the molten silver to flow around and fill the joint volume without overflowing. Lap depth required for brazing is calculated using the following relation [69]:

$$X = \frac{W \cdot (D - W) \cdot T}{C \cdot L \cdot D} \tag{1.9}$$

where W, D, T are the wall thickness, outer diameter and tensile strength of anode tube respectively. C is the joint integrity factor with a value of 0.8. L is the shear strength of silver braze alloy and X is the lap depth. The brazing process takes place in a furnace, in specific environmental conditions, *i.e.* in the presence of a slightly reducing/inert atmosphere (98% Ar, 2% H_2) and under the temperature profile of Figure 1.12.

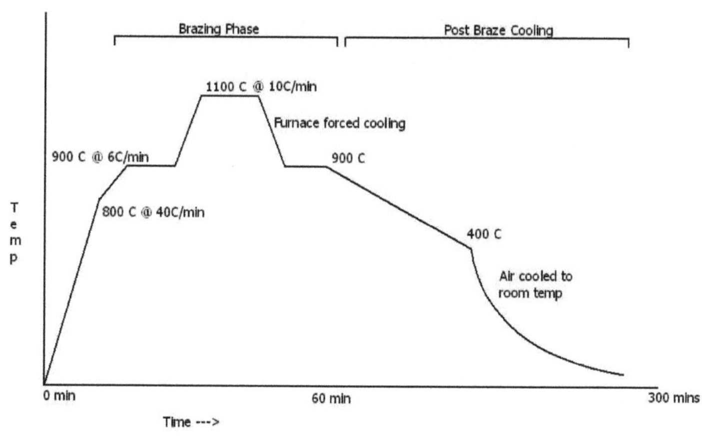

Figure 1.12. Brazing temperature profile

These conditions are set to avoide the oxidation of nickel and silver, thereby enhancing the mechanical and the electrical performance of the joint. As seen from Figure 1.12, the temperature is ramped up to 800°C at a rate of 40°C/min, and again ramped to 900°C at 6°C/min (to avoid excessive overshoot temperature) and is held at that temperature for approximately 15 minutes. This provides enough time for the assembly to come to thermal equilibrium. The temperature is then ramped up at 10°C/min to 1100°C (the brazing temperature), and allowed to soak for 6 minutes.

The assembly is then cooled at 10°C/min to 900°C and allowed to equilibrate for 15 minutes. It is then cooled at 3°C/min to room temperature. Before assembling the stack, sample joints were realized to perform mechanical, leakage and conductivity tests. Microscopic analysis is also conducted to check the uniformity of the joint. The results of these tests allowed the brazing procedure to be optimized, as described in [69].

Table 1.2 briefly shows the optimization history. The sample results are reported in the table in chronological order, thus showing the improvements obtained due to each test feedback. The microscopic analysis of the improved joint (sample ID 3 in Table 1.2.) is reported in Figure 1.13 [69].

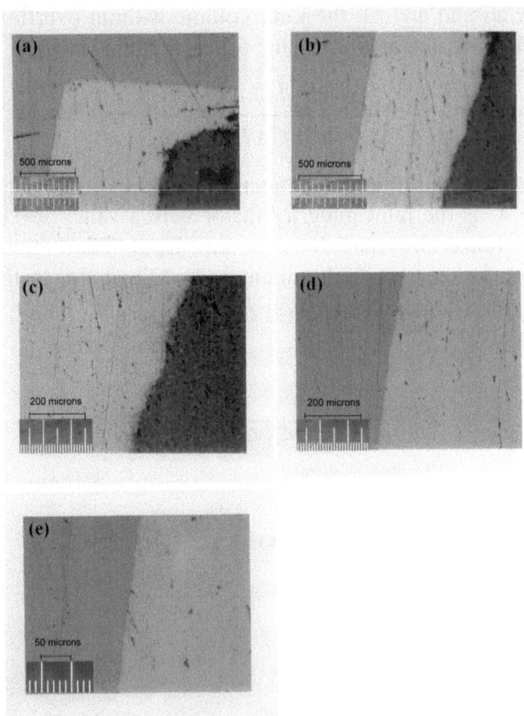

Figure 1.13. Optical microscopy of the bond layer; (a) Ni-Ag-YSZ joint corner (b) Ni-Ag-YSZ joint plane (c) Ag-YSZ interface (d) Ni-Ag interface (e) Ni-Ag interface at higher magnification [69]

Table 1.2. Tests results on joint samples

Sample ID	Leakage	Resistance (milli-ohms)	Torque (kNm)	Micro-structure
1	*Yes*	*noticeable*	*0.003*	*Non-uniform*
2	*No*	*4–5*	*0.0158*	*Uniform*
3	*No*	*4–5*	*0.0160*	*Uniform*

As can be seen, there is very good wetting of both the nickel metal tube and the SOFC anode tube. Silver is observed to diffuse into the ceramic surface as shown in Figure 1.13a and b and more distinctively in Figure 1.13c. The open porosity of the anode tube helped to facilitate diffusion of silver. From Figure 1.13c, it can be seen that the silver has diffused about 50 microns into the ceramic surface.

1.5.4 Stack Design and Expected Performance

Before assembling the stack, tests on the single cells are performed. Although the SOFC can operate with a variety of fuels, hydrogen is considered in the present study. The tests are conducted in an electric furnace, *i.e.* under iso-thermal conditions. A constant inlet flowrate is provided to the cell, while the current density is varied. The first tests are conducted at 850°C, then repeated at 800°C and 750°C. Finally repeated again at 850°C. Figure 1.14 shows the voltage variation during the tests and Figure 1.15 the relative power density.

As is clearly visible, a change in temperature leads to a remarkable performance variation. This behavior is not surprising, because it is well known that conductivity is connected to the operating temperature. Figure 1.14 and Figure 1.15 also show that temperature cycling does not significantly influence the cell performance. From the results of the tests, the operating temperature, chosen for the stack is 850°C. The fuel utilization relative to the tests of Figure 1.14 and Figure 1.15 is always below 20%, thus additional tests have been performed, at different flowrates. Flowrates equal to 25 ml/min, 50 ml/min, 75 ml/min, and 100 ml/min,

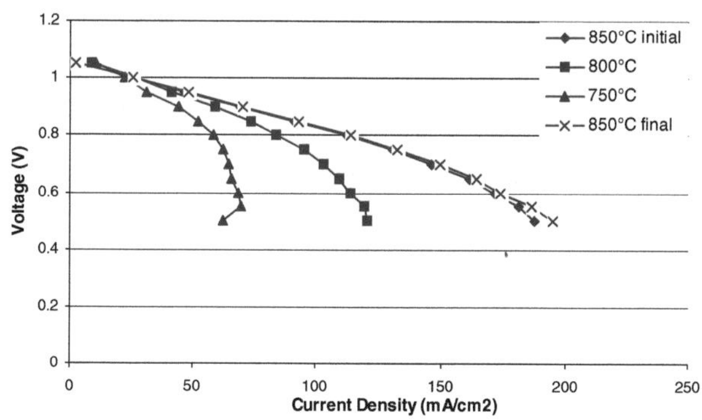

Figure 1.14. Current density-voltage tests results at different temperature and constant inlet flowrate

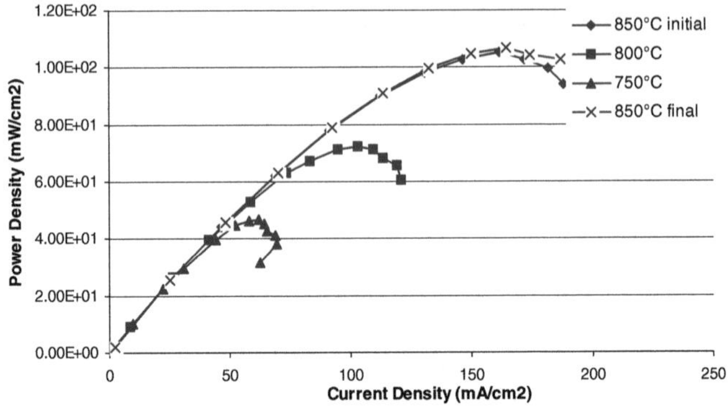

Figure 1.15. Power density-current density at different temperature and constant inlet flowrate

are provided to the cell and the current density is, for each of them, varied. The fuel utilization is computed as:

$$u_f = \frac{I/(2F)}{n_{H_2,inlet}} \tag{1.10}$$

where I is the electric current provided by the cell (express in ampere), F is the Faraday Constant, and $n_{H2,inlet}$ is the hydrogen molar flowrate provided to the cell. Figure 1.16 represents the result for the voltage, and the relative fuel utilization variation. In Figure 1.17, the relative power densities are depicted.

As fully explained in [70], once the performance of the SOFC is known, the choice of the optimal active surface value (*i.e.*, in the present study, the number of single cells to be stacked) must be determined on an economical basis. If the stack, in fact, operates at high current density a reduction of the investment cost is achieved. However, increasing the current density leads to an efficiency reduction, *i.e.* an increase in operating cost. The trade-off between operating and investment costs determines the optimal size of the active surface. At the present time, however, SOFCs are still in an

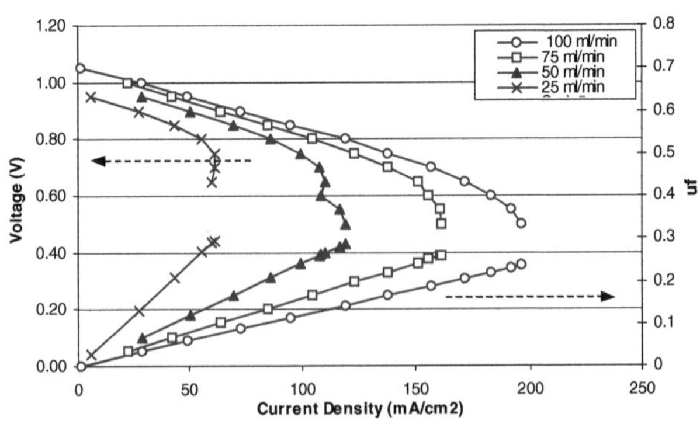

Figure 1.16. Voltage variation and relative fuel utilization, for different inlet flowrates

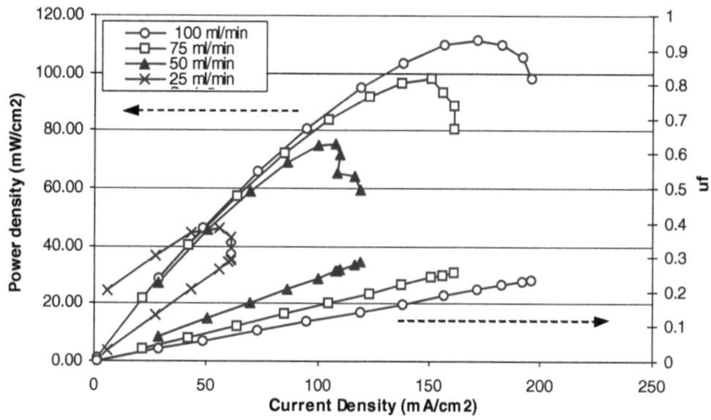

Figure 1.17. Power density variation and relative fuel utilization, for different inlet flowrates

experimental phase, and the construction cost is a long way from that expected on the market. For this reason, the number of single cells to be stacked for realizing the stack is chosen on the basis of the maximum performance achieved of the single cells, at reasonable values for the current density and fuel utilization. The design point chosen is characterized by a current density of 119.3 mA·cm^{-2}, a voltage of 0.5 V and a power density of 59.5 mW·cm^{2}. The main characteristics of the stack are reported in Table 1.3. It should be noted that, once in operation, the stack will be tested under different conditions, and the optimal combination of operating parameters that guarantees good performance and stable conditions will be assessed. For this reason, the value of 100 W should not be considered nominal, but a reference condition.

Table 1.3. Main characteristics of the stack at reference condition

Power	*100 W*
Fuel Utilization	*~30%*
Current Density	*120 mA·cm^{-2}*
Power Density	*59.64 mW·cm^{-2}*
Single Cell Diameter	*1.32 cm*
Single Cell Length	*11 cm*
Electric Current	*5.44 A*
Number of single cells	*40*
Stack Voltage	*20 V*

Although the methodology developed for the single cells fabrication and assembly showed promising results and allows the stack to be fabricated, there are still some issues to be considered when passing from single cells to a complete stack. First, the single cells have been tested in an iso-thermal environment, and, although even the stack can operate under those conditions (*i.e.* inside a furnace), when a complete system is assembled, an iso-thermal condition no longer exists anymore. Secondly, due to the configuration of the current collectors, in-plane ohmic losses can be quite high, thus reducing the overall performance. For a better understanding of this phenomenon, a 2D model of the TSOFC has been implemented and solved using the commercial software FEMLAB®. The model has been implemented in

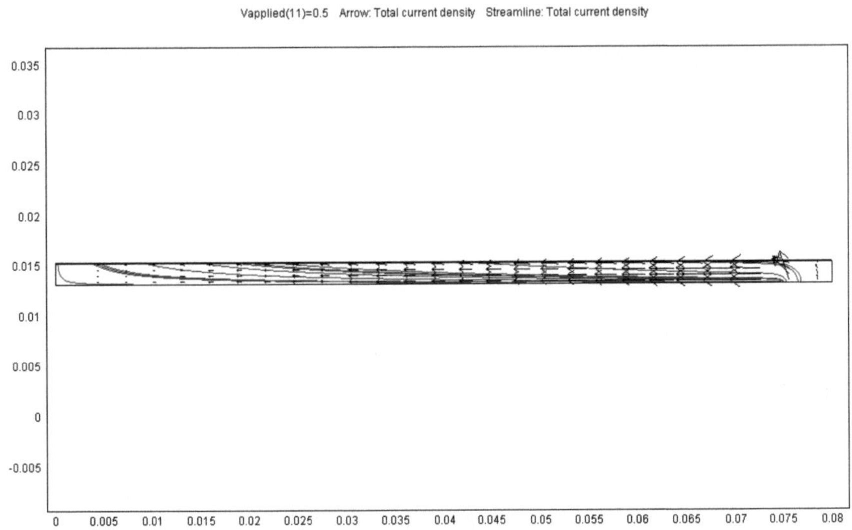

Figure 1.18. Current path and velocity field in a single cell

a 2D environment, thus a cross section of the tube along the axes is considered. First results [65] show that the current mostly runs along the cylinder surface, rather than perpendicularly from one electrode to the other, thus causing a relevant performance reduction. Figure 1.18 shows the simulated current path. Due to the relevant dimension of the anode, compared to the electrolyte and the cathode, only the anode is visible.

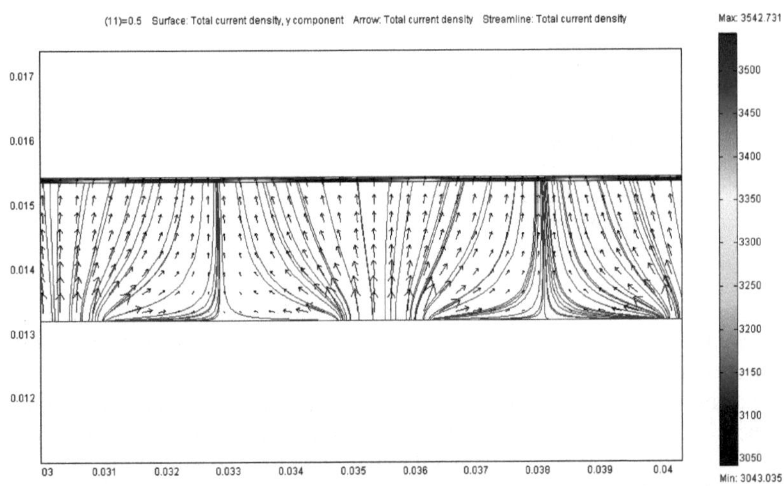

Figure 1.19. Current path with the improved current collectors configuration

A substantial improvement can be made if the current collectors are made with a nickel mesh inside the tube (anode), and a wire wrapped around outside (cathode). In this case, in fact, the current is not only collected at the two ends of the cell, but, similar to the planar cells, close to the reaction zone, thus the current flows fairly straight from one electrode to the other. The simulation result for the current distribution in this case is reported in Figure 1.19 (zoomed on a small part of the tube). Due to the straightness of the current path, the resulting in-plane resistance is drastically reduced.

1.6 Applications

The high operating temperature of SOFCs, and, consequently, the relative long start-up time, limit its applications to stationary and auxiliary power units for transportation.

In Table 1.4, the most important developers, together with the application they are mainly focused on, are reported.

The main issues that currently affect the commercialization of SOFC are the short life-time and the high costs.

In 1999 in the United States of America, the Solid State Energy Conversion Alliance (SECA) was initiated, as an alliance between government, industry and the scientific community. The main goal is to make SOFC commercially available, through achieving a durability ≥ 40,000 hours, and a cost ≤ 400 $/kW. Similar targets are the main driver for the development directions of all the international SOFC developers.

Table 1.4. Main manufacturers and technology status [3]

Manufacturer	Country	Achieved	Year	Technology
Acumentrics Corp	*USA*	*2 kW*	*2002*	*Microtubular*
Adelan	*UK*	*200 W*	*1997*	*Microtubular*
Ceramic Fuel Cells LtD	*Australia*	*5 kW* *25 kW*	*1998* *2000*	*Planar*
Delphi/Battelle	*USA*	*5 kW*	*2001*	*Planar*
Fuel Cell Technologies	*Canada/USA*	*5 kW* *2 kW*	*2002*	*Flat Tubular/Tubular*
General Electric (formerly Honeywell and Allied Signal)	*USA*	*0.7 kW* *1 kW*	*1999* *2001*	*Planar*
Fuel Cell Energy (formerly Global Thermoelectric)	*USA*	*1 kW*	*2000*	*Planar*
MHI/Chubu Electric	*Japan*	*4 kW* *15 kW*	*1997* *2001*	*Planar*
Rolls-Royce	*UK*	*1 kW*	*2000*	*Planar*
Siemens-Westinghouse	*USA*	*25 kW* *110 kW* *220 kW*	*1995* *1998* *2000*	*Tubular*
SOFCo (McDermott Technologies and Cummins Power Generation)	*USA*	*0.7 kW*	*2000*	*Planar*
Sulzer Hexis	*Switzerland*	*1 kW*	*2002*	*Planar*
Tokyo Gas	*Japan*	*1.7 kW*	*1998*	*Planar*
TOTO/Kyushu Electric Power/Nippon Steel	*Japan*	*2.5 kW*	*2000*	*Tubular*

1.6.1 Residential Application

Residential buildings typically show a peak power of a few kW (6–9 for the United States and 2–5 for Europe), thus the use of SOFC as a power unit is attractive in the mid-future, without the need of a big system scale-up. Due to the high operating temperature, the heat content of the outlet gas can be easily recovered for hot water, heating needs and for the air-conditioning system. The low emissions, the low noise level, the fuel flexibility and the high efficiency are additional advantages that make the SOFC an ideal candidate for this application. An SOFC system can operate in a variety of configurations, according to specific economical and technical needs. With respect to the national electric grid, for example, a system can operate independently (this can be the case of remote buildings) or in parallel. An analysis of different configurations under full and partial load requirements was carried out by Bove and Sammes [70]. Currently, most of the SOFC developers are seeking to develop systems for residential applications, due to the compatibility of the state of the art system size and the residential power requirements. Some companies have started testing their products in real cases. Sulzer-Hexis, for example, is currently testing its HXS 1000 PREMIERE units in the employees' homes, thus getting a relevant amount of data of 'real applications', before launching its product to the market.

1.6.2 Power Plant and Grid Support

Due to the high energy conversion efficiency, the SOFC will be a competitor to the current energy conversion systems that feed the electric grid. When coupled with a bottoming cycle, as for example a gas turbine, the electrical efficiency can be up to 75%. The environmental benefits associated with their operation, and the possibility of using a variety of renewable sources of energies, are additional elements that drove some developers to consider the SOFC a value opportunity for power generation.

Figure 1.20 depicts a schematic representation of a pressurized SOFC-GT power plant [71]. In this solution, air is compressed by the compressor of the gas turbine system, heated, using a traditional gas turbine recuperator, and sent to the fuel cell cathode. The unoxidized fuel leaving the anode is mixed with the cathode outlet gas and combusted (additional natural gas can also be introduced into the combustor). The resulting hot gas is expanded in the gas turbine.

As mentioned above, however, only the Siemens-Westinghouse technology has been scaled-up to 200 kW. The SECA alliance target is to start focusing on hybrid SOFC-GT systems from 2006, and implementing the so called 'hybrid-based vision 21 power plants' by 2015.

Another SOFC configuration that attracted particular attention in recent years is the so called Integrated Gasifier Fuel Cell system (IGFC), that is characterized by very high conversion, compared to traditional gasifiers, and ultra-low emissions. In this configuration, coal or biomass is converted into a hydrogen rich gas mixture via an ultra-clean process (gasification) and then converted into electricity through a fuel cell. The high efficiency is due to the heat recovery from the fuel cell to the gasification section.

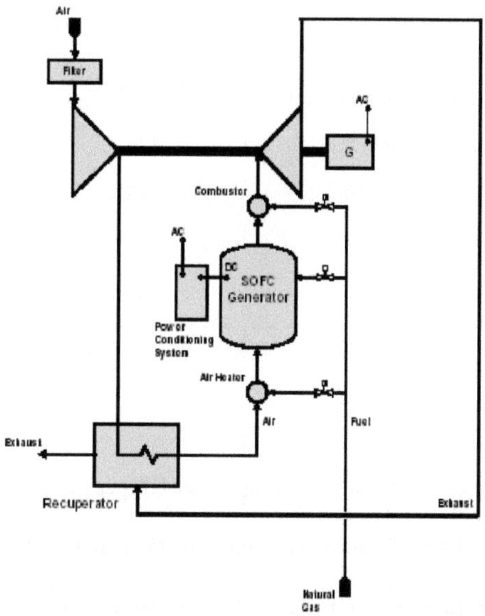

Figure 1.20. Schematic representation of a hybrid solid oxide fuel cell-gas turbine power plant [71]

1.6.3 Auxiliary Power Unit

The term Auxiliary Power Unit (APU) is referred to an energy conversion system that is employed for the power loads of heavy-duty trucks, or any other vehicle whose electrical power requirements are noticeble. Due to the potentially high volume, this application is very attractive. Due to the high power requirements of some types of vehicles, too much power is subtracted from the main engine. Moreover, when the vehicle is not in motion, the poor performance associated with the Internal Combustion Engine (ICE) at reduced load leads to high fuel consumption and high pollution. The use of fuel cells can consederably reduce these problems. The low emissions are also a considerably advantage. Table 1.5 shows the main energetic and environmental advantages of SOFC based APU, compared to the use of the main engine for heavy-duty long haul truck [72].

Although most of the SOFC systems developed by the companies of Table 1.4 are suitable for APU application, only a few companies are focusing their products on APU development. Delphi, in cooperation with Battelle, for example, is developing a 5 kW system, powered with gasoline, diesel and natural gas, to be used as an APU for heavy trucks. BMW and Renault are seeking to apply the Delphi technology in their vehicles.

Table 1.5. Benefits of SOFC based APU versus Idling for Heavy-duty trucks [72]

	Baseline (main engine idle)	SOFC APU	Savings versus idle
Fuel Use (litre/year)	*6327*	*932*	*5394 (85%)*
CO_2 (kg/yr)	*17200*	*2520*	*14700 (85%)*
NO_x (kg/yr)	*274*	*~0*	*274 (>99%)*
CO (kg/yr)	*97*	*~0*	*97 (>99%)*

1.7 References

1. Nernst W. Uber die elektrolytische Leitung fester Körper bei sehr hohen Temperaturen. Z. Elektrochem. 1989;6(2):41–43. (1899).
2. Baur E, Preis H. Uber Brennsto.-Ketten mit Festleitern. Z. Elektrochem., 1937; 43(9):727–732.
3. US Department of Energy (DOE). Fuel Cell Handbook seventh edition. November 2004.
4. Yakabe H, Ogiwara T, Hishinuma M and Yasuda I. 3-D model calculation for planar SOFC. J. Power Sources, 2001;102:144–154.
5. Meusinger J, Riensche E and Stimming U. Reforming of natural gas in solid oxide fuel cell systems. J. Power Sources, 1998;71:315–320.
6. Bove R and Sammes N M. Thermodynamic analysis of SOFC systems using different fuel processors. Proceedings of the 2nd Interational Conference of Fuel Cell Engineering and Technology, 14–16 June 2004, Rochester, NY, pp. 461–466.
7. Singhal S C and Kendall K. High-temperature solid oxide fuel cells: fundamentals, design and applications. Elsevier Science, London, 2004.
8. Du Y and Sammes N M. Fabrication and proprieties of anode-supported tubular solid oxide fuel cells. J. Power Sources 2004;136:66–71.
9. Lee H, Jung H Y, Roh T, Kim H, Choi S H C, Kim W S, Kim J and Lee J. Recent R&D activity at KIST to develop SOFC stack using 10x10 cm^2 anode-supported cells. Proceedings of the sixth European Solid Oxide Fuel Cell Forum, 28 June- 2 July 2004, Lucerne, Switzerland, Vol. 1, pp. 54–59.
10. Yakabe H, Baba Y, Sakurai T and Yoshitaka Y. Evaluation of the residual stress for anode supported SOFCs. J. Power Sources, 2004;135(1–2):9–16.
11. Patel P, Maru H C, Borglum B, Stokes R A, Petri R, Remick R J, Sishtla C, Krist K, Armstrong T and Virkar A. Thermally integrated power systems (TIPS), high power density SOFC generator. Proceedings of the Fuel Cell Seminar 2004, November 1–5 2004, San Antonio, TX, pp. 132–135.
12. Minh N. SECA solid oxide fuel cell program-General Electric SECA industry team. Proceedings of the SECA Annual Workshop and Core Technology Program peer reviewed abstracts, May 11–13, 2004, Boston, MA, p. 2.
13. Yashiro K, Takeda K, Taura T, Otake T, Kaimai A, Nigara Y, Kawada T and Mizusaki J. In situ observation of deposited carbon on anode for solid oxide fuel cells. Proceedings of the eight International Symposium on Solid Oxide Fuel Cells (SOFC VIII), The Electrochemical Society, PV 2003–07, pp.714–719.
14. Tsang S C, Claridge J B and Green M L H. Recent advances in the conversion of methane to synthesis gas. Catalysis Today, 1995;23:3–15.

15. Matsuzaki Y and Yasuda I. The poisoning effect of sulfur-containing impurity gas on a SOFC anode: part I. Dependence on temperature, time and impurity concentration. Solid State Ionics, 2000;132:261–269.
16. Tikekar N M, Armstrong T J and Virkar A V. Reduction and re-oxidation kinetics of nickel-based solid oxide fuel cell anodes. Proceedings of the Eight International Symposium on Solid Oxide Fuel Cells (SOFC VIII), The Electrochemical Society, PV 2003–07, pp.670–679.
17. Pudmich G, Boukamp B A, Gonzalez-Cuenca M, Jungen W, Zipprich W and Tietz F. Chromite/titanate based perovskites for application as anodes in solid oxide fuel cells. Solid State Ionics, 2000;135:433–438.
18. Holtappels P, Bradley J, Irvine J T S, Kaiser A and Mogensen M. Electrochemical characterization of ceramic SOFC anodes. J. Electrochem. Soc., 2001;148(8):A923–A929.
19. Uchida H, Suzuki S and Watanabe M. A high performance electrode for medium-temperature SOFC: mixed conducting ceria based anode with highly dispersed Ni electrocatalysts. Proceedings of the Eight International Symposium on Solid Oxide Fuel Cells (SOFC VIII), The Electrochemical Society, PV 2003–07, pp.728–736.
20. Rösch B, Tu H, Stömer A O, Müller A C and Stimming U. Electrochemical behaviour of Ni-$Ce_{0.9}Gd_{0.1}O_{2-\delta}$ SOFC anodes in methane. Proceedings of the eight International Symposium on Solid Oxide Fuel Cells (SOFC VIII), The Electrochemical Society, PV 2003–07, pp.737–744.
21. Aguilar L, Zha S, Cheng Z, Winnick J and Meilin L. A solid oxide fuel cell operating on hydrogen sulfide (H_2S) and sulfur-containing fuels. J. Power Sources, 2004;135:17–24.
22. Primdahl S, Hansen J R, Grahl-Madsen L and Larsen P H. Sr-Doped $LaCrO_3$ anode for solid oxide fuel cells. J. Electrochem. Soc., 2001;148(1):A74–A81.
23. Trofimenko N, Kuznecov M, Vashook V and Otschik P. Electrochemical proprieties of different anodes for solid oxide fuel cells. Proceedings of the sixth European Solid Oxide Fuel Cell Forum, 28 June- 2 July 2004, Lucerne, Switzerland, Vol. 3, pp. 1494–1501.
24. Fouquet D, Timmermann H, Hennings U, Ivres-Tiffée E and Reimert R. Analysis of internal reforming in SOFC anodes operated with methane. Proceedings of the sixth European Solid Oxide Fuel Cell Forum, 28 June- 2 July 2004, Lucerne, Switzerland, Vol. 3, pp. 1468–1475.
25. Gordon R S. Ionic conducting ceramic electrolytes: a century of progress. Proceedings of the Eight International Symposium on Solid Oxide Fuel Cells (SOFC VIII), The Electrochemical Society, PV 2003–07, pp.141–152.
26. Minh N Q and Takahashi T. Science and technology of ceramic fuel cells. Elsevier Science, 1995.
27. Vora S. Small-scale low cost SOFC power systems. Proceedings of the SECA Annual Workshop and Core Technology Program peer reviewed abstracts, May 11–13, 2004, Boston, MA, p. 3.
28. Agnew G and Spangler A. Reducing fuel cell system cost without lowering operating temperature. Proceedings of the 2nd Interational Conference of Fuel Cell Engineering and Technology, 14–16 June 2004, Rochester, NY, keynote CD.
29. Nakanishi A, Hattori M and Sakaki Y. Development of MOLB type SOFC. Proceedings of the Eight International Symposium on Solid Oxide Fuel Cells (SOFC VIII), The Electrochemical Society, PV 2003–07, pp.53–59.
30. Dinsdale J, Conduit A and Foger K. Development of a residential combined heat and power (CHP) system at Ceramic Fuel Cells Limited. Proceedings of Fuel Cell Seminar, November 1–5, 2004, San Antonio, TX, pp. 101–104.

31. Inagaki T, Nishiwaki F, Kanou J, Miki Y, Hosoi K, Miyazawa T and Komada N. Intermediate temperature SOFC modules and systems. Proceedings of Fuel Cell Seminar, November 1–5, 2004, San Antonio, TX, pp. 113–116.
32. Steinberger-Wilckens R, de Haart B, Buchkremer H-P, Nabielek H, Quadakkers J, Reisgen U, Steinbrech R, Tietz F and Vinke I. Recent results of solid oxide fuel cell development at Forshungszentrum Juelich. Proceedings of Fuel Cell Seminar, November 1–5, 2004, San Antonio, TX, pp. 120–123.
33. Christiansen N, Hansen J B, Kristen S, Holm-Larsen H, Linderoth S, Hendriksen P V, Larsen P H, and Mogensen M. SOFC development program at Haldor-Topsøe/Risø National Laboratory-progress presentation. Proceedings of the Fuel Cell Seminar 2004, November 1–5 2004, San Antonio, TX, pp. 124–127.
34. Sun J, Mao Z, Lin Q, Wang Cand Zhu B. Low temperature (300 to 600°C) SOFCs R&D in China. Proceedings of the sixth European Solid Oxide Fuel Cell Forum, 28 June- 2 July 2004, Lucerne, Switzerland, Vol. 1, pp. 30–39.
35. Myatiev A. The state of the art of SOFC elaboration at Moscow State Institute of Steel and Alloy (MISA). Proceedings of the sixth European Solid Oxide Fuel Cell Forum, 28 June- 2 July 2004, Lucerne, Switzerland, Vol. 1, pp. 40–47.
36. Shaffer S. Solid state energy conversion alliance- Delphi SECA industry team. Annual Workshop and Core Technology Program peer reviewed abstracts, May 11–13, 2004, Boston, MA, p. 1.
37. Kneidel K E, Norrick D, Vesely C, DeBellis C, Kantak M and Palmer B K. Development of SOFC power systems using multi-layer ceramic interconnects. Proceedings of Fuel Cell Seminar, November 1–5, 2004, San Antonio, TX, pp. 109–112.
38. Brandon N P, Blake A, Corcoran D, Cumming D, Duckett A, El-Koury K, Haigh D, Kidd C, Leah R, Lewis G, Matthews C, Maynard N, Oishi N, McColm T, Trezona R, Selcuk A, Schmidt M, and Verdugo L. Development of metal supported solid oxide fuel cells for operation at 500–600°C. J. Fuel Cell Science and Technology, 2004;1(1):61–65.
39. Sarat S, Sammes N M, Yamamoto O. Study of the electrical proprieties of bismuth oxide-doped Scandia-stabilized zirconia. Proceedings of the sixth European Solid Oxide Fuel Cell Forum, 28 June- 2 July 2004, Lucerne, Switzerland, Vol. 3, pp. 1152–1162.
40. Yamamoto O, Arati Y, Takeda Y, Imanishi N, Mizutani Y, Kawai M and Nakamura Y. Electrical conductivity of stabilized zirconia with ytterbia and Scandia. Solid State Ionics, 1995;79:137–142.
41. Mizutani Y, Tamura M and Yamamoto M. Development of high-performance electrolyte in SOFC. Solid State Ionics, 1994;72:271–275.
42. Mori K, Miyamoto H, Takenobu K, Kishizaba H and Sakaki Y. Characteristics and power generation test results of heavy rare earth stabilized zirconia. Proceedings of the sixth European Solid Oxide Fuel Cell Forum, 28 June- 2 July 2004, Lucerne, Switzerland, Vol. 3, pp. 1208–1213.
43. Müller A C, Weber A and Ivers-Tiffée E. Degradation of zirconia electrolytes. Proceedings of the sixth European Solid Oxide Fuel Cell Forum, 28 June- 2 July 2004, Lucerne, Switzerland, Vol. 3, pp. 1231–1238.
44. Godickemeier M and Gaucker L J. Engineering of solid oxide fuel cells with ceria-based electrolytes. J. Electrochem. Soc., 1998;145(2):414–420.
45. Ishihara T, Matsuda H and Takita Y. Doped $LaGaO_3$ perovskite type oxide as a new oxide ionic conductor. J. American Chem. Soc., 1994;116:3801.
46. Goodenough J B and Huang K. Lanthanum gallate as a new SOFC electrolyte. Proceedings of the Fuel Cells '97 review meeting.

47. Iwahara H, Esaka T, Uchida H and Maeda N. Proton conduction in sintered oxides and its applications in steam electrolysis for hydrogen production. Solid State Ionics, 1981;3–4:359–363.
48. Iwahara H. Technological challenges in the application of proton conducting ceramics. Solid State Ionics, 1995;77:289–298.
49. Bonano N, Ellis B and Mahmood M N. Construction and operation of fuel cells based on the solid electrolyte $BaCeO_3$:Gd. Solid State Ionics, 1991;44:305–311.
50. Antoine O, Hatchwell C, Mather G, and McEvoy J. Structure and conductivity of a Yb-doped $SrCeO_3$-$BaZrO_3$ solid solution. Proceedings of the Eight International Symposium on Solid Oxide Fuel Cells (SOFC VIII), The Electrochemical Society, PV 2003–07, pp.379–387.
51. Coors W G, Sidwell R and Anderson F. Characterization of electrical efficiency in protonic ceramic fuel cells. Proceedings of the sixth European Solid Oxide Fuel Cell Forum, 28 June- 2 July 2004, Lucerne, Switzerland, Vol. 1, pp. 117–124.
52. Hassan D, Janes S and Clasen R. Proton-conducting ceramics as electrode/electrolyte materials for SOFC's-part 1: preparation, mechanical and thermal proprieties of sintered bodies. J. Eur. Ceram. Soc., 2003;23:221–228.
53. Shimada T, Wen C, Taniguchi N, Otomo J and Takahashi H. The high temperature proton conductor $BaZr_{0.4}Ce_{0.4}In_{0.2}In_{3-\alpha}$. J. Power Sources, 2004;131:289–292.
54. Smirnova A, Prakash P, Phillips R and Sammes N M. Electrolyte proton-conductive materials for protonic ceramic fuel cells (PFCFs). Proceedings of the sixth European Solid Oxide Fuel Cell Forum, 28 June- 2 July 2004, Lucerne, Switzerland, Vol. 3, pp. 1029–1039.
55. Taniguchi S, Kadowaki M, Kawamura H, Yasuo T, Akiyama Y, Miyake Y and Saitoh T. Degradation phenomena in the cathode of a solid oxide fuel cell with an alloy separator. J. Power Sources, 1995;55(1):73–79.
56. Juhl M, Primdahl S, Manon C and Mogensen M. Performance/structure correlation for composite SOFC cathodes. J. Power Sources, 1996;61(1–2):173–181.
57. Virkar A V, Chen J, Tanner C W and Kim J W. The role of electrode microstructure on activation and concentration polarizations in solid oxide fuel cells. Solid State Ionics, 2000;131(1–2):189–198.
58. Figueiredo F M, Labrincha J A, Frade J R and Marques F M B. Reactions between a zirconia-based electrolyte and $LaCoO_3$-based electrode materials. Solid State Ionics, 1997;101–103(1):343–349.
59. Tu H Y, Takeda Y, Imanishi N and Yamamoto O. $Ln_{0.4}Sr_{0.6}Co_{0.8}Fe_{0.2}O_{3-\delta}$ (Ln=La, Pr, Nd, Sm, Gd) for the electrode in solid oxide fuel cells. Solid State Ionics, 1999;117(3–4):277–281.
60. Esquiro A, Bonanos N, Brandon N, Kilner J and Mogensen M. Electrochemical characterization of a $La_{0.6}Sr_{0.4}Co_{0.2}Fe_{0.8}O_{3-\delta}$ cathode for IT-SOFCs. Proceedings of the Eight International Symposium on Solid Oxide Fuel Cells (SOFC VIII), The Electrochemical Society, PV 2003–07, pp. 580–590.
61. Elangovan S, Hartvigsen J, Lashway R, Balagopal S and Bay I. Metal interconnect development: design and long-term stability. Proceedings of the eight International Symposium on Solid Oxide Fuel Cells (SOFC VIII), The Electrochemical Society, PV 2003–07, pp.851–854.
62. Thissen J H and Rastler D M. Scale-up of SOFC stacks: approaches and impact on performance and cost. Proceedings of the Fuel Cell Seminar 2004, November 1–5 2004, San Antonio, TX, pp. 69–72.
63. Williams M C, Utz B R and Moore K M. DOE FE Distributed Generation Program. Journal of Fuel Cell Science and Technology, 2004;1(1):18–20.
64. Bossel U G. Facts & Figures, an International Energy Agency SOFC Task Report, Baden, October 1992.

65. Bove R, Sammes N M and Lunghi P. Design of a tubular SOFC stack aided by numerical simulations. Proceedings of the Fuel Cell Seminar 2004, November 1–5 2004, San Antonio, TX.
66. Durov A V, Kostjuk B D, Shevchenko A V and Naidich Y V. Joining of zirconia to metal. Material Science and Engineering, 2000;A290:86–189.
67. Hanson W B, Ironside K I and Fernie S A. Active metal brazing of Zirconia. Acta Mater, 2000;48:4673–4676.
68. Lide D R. CRC Handbook of Chemistry and Physics, 1996, 77th Edition. CRC Press, N.Y.
69. Basak A, England R and Sammes N M. Determination of the Mechanical integrity of ceramic-to-metal braze joints in SOFC interconnect applications". Proceedings of the sixth European Solid Oxide Fuel Cell Forum, 28 June- 2 July 2004, Lucerne, Switzerland, Vol. 2, pp. 950–959.
70. Bove R and Sammes N M. Behavior of an SOFC system for small stationary applications under full and partial load operation. Proceedings of the sixth European Solid Oxide Fuel Cell Forum, 28 June- 2 July 2004, Lucerne, Switzerland, Vol. 1, pp. 401–411.
71. Lundberg W L, Israelson G A, Moritz R R, Veyo S E, Holmes R A, Zafred P R, King J E and Kothmann R E. Pressurized solid oxide fuel cell/gas turbine power system. Final report prepared for the US Department of Energy, under contract number DE-AC26–98FT40355, February 2000.
72. Lasher S, Isherwood K, Sriramulu S and Brodrick C-J. Evaluation of the potential for fuel cell APUs. Proceedings of the Fuel Cell Seminar 2004, November 1–5 2004, San Antonio, TX, pp. 57–60.

2

PEM Fuel Cells

Frano Barbir

2.1 Introduction

PEM fuel cells use a proton conductive polymer membrane as electrolyte. PEM stands for Polymer Electrolyte Membrane or Proton Exchange Membrane. Sometimes they are also called polymer membrane fuel cells, or just membrane fuel cells. In the early days (1960s) they were known as Solid Polymer Electrolyte (SPE) fuel cells. This technology has drawn the most attention because of its simplicity, viability, quick start-up, and it has been demonstrated in almost any conceivable application, from powering a cell phone to a locomotive.

At the heart of a PEM fuel cell is a polymer membrane that has some unique capabilities. It is impermeable to gases but it conducts protons (hence Proton Exchange Membrane name). The membrane, which acts as the electrolyte, is squeezed between the two porous, electrically conductive electrodes. These electrodes are typically made out of carbon cloth or carbon fiber paper. At the interface between the porous electrode and the polymer membrane there is a layer with catalyst particles, typically platinum supported on carbon. A schematic diagram of cell configuration and basic operating principles is shown in Figure 2.1 [1,2].

Electrochemical reactions occur at the surface of the catalyst at the interface between the electrolyte and the membrane. Hydrogen, which is fed on one side of the membrane, splits into its primary constituents – protons and electrons. Each hydrogen atom consists of one electron and one proton. Protons travel through the membrane, while the electrons travel through electrically conductive electrodes, through current collectors, and through the outside circuit where they perform useful work and return to the other side of the membrane. At the catalyst sites between the membrane and the other electrode they meet with the protons that went through the membrane and oxygen that is fed on that side of the membrane. Water is created in the electrochemical reaction, and then pushed out of the cell with an excess flow of oxygen. The net result of these simultaneous reactions is current of electrons through an external circuit – direct electrical current.

The hydrogen side is negative and is called the anode, while the oxygen side of the fuel cell is positive and is called the cathode. The electrochemical reactions in

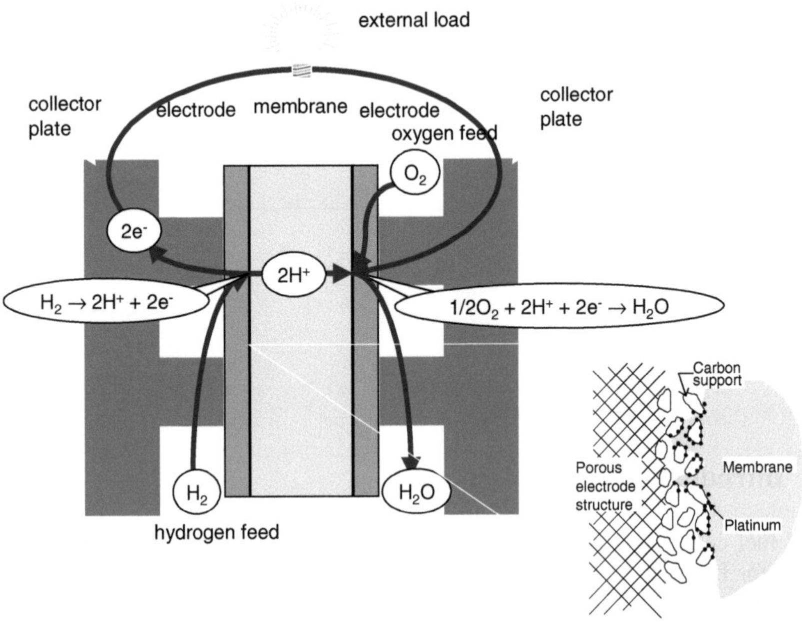

Figure 2.1. The basic principle of operation of a PEM fuel cell [1,2]

fuel cell happen simultaneously on both sides of the membrane – the anode and the cathode. The basic fuel cell reactions are:

At the anode: $$H_2 \rightarrow 2H^+ + 2e^- \tag{2.1}$$

At the cathode: $$\tfrac{1}{2}O_2 + 2H^+ + 2e^- \rightarrow H_2O \tag{2.2}$$

Overall: $$H_2 + \tfrac{1}{2}O_2 \rightarrow H_2O \tag{2.3}$$

The maximum amount of electrical energy generated in a fuel cell corresponds to Gibbs free energy, ΔG, of the above reaction:

$$W_{el} = -\Delta G \tag{2.4}$$

The theoretical potential of fuel cell, E, is then:

$$E = \frac{-\Delta G}{nF} \tag{2.5}$$

Where n is the number of electrons involved in the above reaction, 2, and F is the Faraday's constant (96,485 Coulombs/electron-mol). Since ΔG, n and F are all known, the theoretical hydrogen/oxygen fuel cell potential can also be calculated:

$$E = \frac{-\Delta G}{nF} = \frac{237{,}340}{2 \cdot 96{,}485}\frac{\text{J mol}^{-1}}{\text{As mol}^{-1}} = 1.23\ \text{Volts} \tag{2.6}$$

At 25°C and atmospheric pressure, the theoretical hydrogen/oxygen fuel cell potential is 1.23 Volts.

Assuming that all of the Gibbs free energy can be converted into electrical energy, the maximum possible (theoretical) efficiency of a fuel cell is a ratio between the Gibbs free energy and hydrogen higher heating value, ΔH:

$$\eta = \Delta G/\Delta H = 237.34/286.02 = 83\% \tag{2.7}$$

The theoretical (reversible) cell potential is a function of operating temperature and pressure:

$$E_{T,P} = -\left(\frac{\Delta H}{nF} - \frac{T\Delta S}{nF}\right) + \frac{RT}{nF} ln\left[\frac{a_{H2} a_{O_2}^{0.5}}{a_{H_2O}}\right] \tag{2.8}$$

where, a stands for activity or the ratio between the partial pressures of reactants (H_2 and O_2) or product (H_2O) and atmospheric pressure (for liquid water product $a_{H2O} = 1$).

Table 2.1 shows the theoretical fuel cell potential at different temperatures and pressures.

Table 2.1. Theoretical cell potential at different temperatures and pressures [2]

T(K)	Atm.	200 kPa	300 kPa
298.15	*1.230*	*1.243*	*1.251*
333.15	*1.200*	*1.215*	*1.223*
353.15	*1.184*	*1.200*	*1.209*

Actual cell potentials are always smaller than the theoretical ones due to irreversible losses. Voltage losses in an operational fuel cell are caused by several factors such as:

- kinetics of the electrochemical reactions (activation polarization),
- internal electrical and ionic resistance,
- difficulties in getting the reactants to reaction sites (mass transport limitations),
- internal (stray) currents,
- crossover of reactants.

Figure 2.2 shows typical proportion of these losses and the resulting polarization curve [2]. A polarization curve is the most important characteristic of a fuel cell and its performance. It depends on numerous factors such as catalyst loading, membrane thickness and state of hydration, catalyst layer structure, flow field design, operating conditions (temperature, pressure, humidity, flowrates and concentration of the reactant gases), and uniformity of local conditions over the entire active area. Typically, a fuel cell operating at atmospheric pressure should generate more than 0.6 A/cm^2 at 0.6 V, and more than 1 A/cm^2 at 0.6 V when operated pressurized (300 kPa or higher). The typical operating temperature is between 60°C and 80°C, although small fuel cells for portable power are often designed to operate at lower temperatures, and larger automotive fuel cell should preferably be operated at higher temperatures.

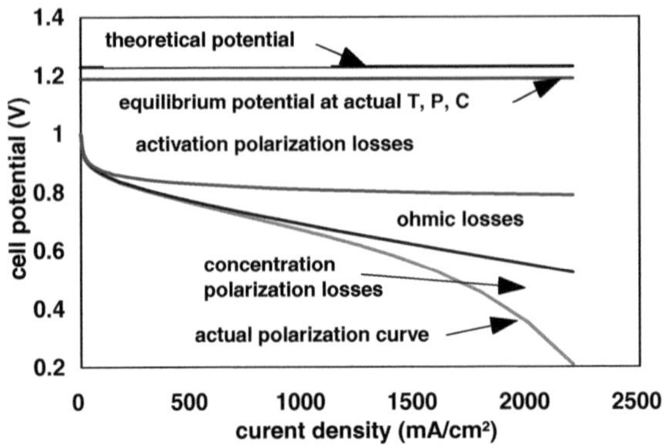

Figure 2.2. Various voltage losses and resulting polarization curve of an operating fuel cell [2]

The fuel cell efficiency, defined as a ratio between the electricity produced and hydrogen consumed is directly proportional to its potential:

$$\eta = \frac{V}{1.482} \tag{2.9}$$

where 1.482 is the thermoneutral potential corresponding to hydrogen's higher heating value. Sometimes, the efficiency is expressed in terms of the lower heating value (LHV):

$$\eta_{LHV} = \frac{V}{1.254} \tag{2.10}$$

In addition, if some hydrogen is lost (i_{loss}) either due to hydrogen diffusion through the membrane, or due to combining with oxygen that diffused through the membrane or due to internal currents, hydrogen consumption will be higher than that corresponding to generated current, and consequently, the fuel cell efficiency would be somewhat lower than given by equation (2.8) [2]:

$$\eta = \frac{V}{1.482} \frac{i}{(i + i_{loss})} \tag{2.11}$$

If hydrogen is supplied to the cell in excess of that required for the reaction stoichiometry, this excess will leave the fuel cell unused. In case of pure hydrogen this excess may be recirculated back into the stack so it does not change the fuel cell efficiency (not accounting for the power needed for hydrogen recirculation pump), but if hydrogen is not pure (such as in reformate gas feed) unused hydrogen leaves the fuel cell and does not participate in the electrochemical reaction. The fuel cell efficiency is then:

$$\eta = \frac{V}{1.482} \eta_{fu} \tag{2.12}$$

where η_{fu} is fuel utilization, which is equal to $1/S_{H2}$, where S_{H2} is the hydrogen stoichiometric ratio, *i.e.*, the ratio between the amount of hydrogen actually supplied to the fuel cell and that consumed in the electrochemical reaction.

2.2 PEM Fuel Cell Components and Their Properties

2.2.1 Membrane

A fuel cell membrane must exhibit relatively high proton conductivity, must present an adequate barrier to mixing of fuel and reactant gases, and must be chemically and mechanically stable in the fuel cell environment [3]. Typically, the membranes for PEM fuel cells are made of perfluorocarbon-sulfonic acid ionomer (PSA). This is essentially a copolymer of tetrafluorethylene (TFE) and various perfluorosulfonate monomers. The best known is Nafion® made by Dupont, which uses perfluoro sulfonylfluoride ethyl-propyl-vinyl ether (PSEPVE). Similar materials have been developed and sold either as a commercial or development product by other manufacturers such as Asahi Glass (Flemion®), Asahi Chemical (Aciplex®), Chlorine Engineers ("C" membrane), and Dow Chemical. W.L. Gore and Associates have developed a composite membrane (GoreSelect®) comprising of a Teflon-like component providing mechanical strength and dimensional stability and a perfluorosulfonic acid component providing protonic conductivity.

Nafion® membranes come extruded in different sizes and thicknesses. They are marked with a letter N followed by a 3 or 4 digit number. The first two digits represent equivalent weight/100, and the last digit or two is the membrane thickness in mills (1 mill = 1/1000 inch = 0.0254 mm). Nafion® is available in several thicknesses, namely 2, 3.5, 5, 7 and 10 mills (50, 89, 125, 178, 250 μm, respectively). For example, Nafion® 117 has an equivalent weight of 1100 and it is 7 mills (0.183 mm) thick.

The main properties of the fuel cell membrane are protonic conductivity, water transport, gas permeation and physical properties such as strength and dimensional stability. All of these properties are directly related to the membrane water content. The water content in a membrane is usually expressed as grams of water per gram of polymer dry weight or the a number of water molecules per sulfonic acid groups present in the polymer, $\lambda = N(H_2O)/N(SO_3H)$. The maximum amount of water in the membrane strongly depends on the membrane (pre)treatment and the state of water used to equilibrate the membrane. A Nafion membrane equilibrated with liquid water (*i.e.*, boiled in water) takes roughly up to 22 water molecules per sulfonate group, while the maximum water uptake from the vapor phase, corresponding to 100% relative humidity in the surrounding gas, is about 14 water molecules per sulfonate group [3].

The protonic conductivity of PFSA membranes is a strong function of water content and temperature. For a fully hydrated membrane ($\lambda = 22$) the protonic conductivity is about 0.1 S cm^{-1} at room temperature, and at $\lambda = 14$ (membrane equilibrated with water vapor) it is about 0.06 S/cm [4]. Protonic conductivity dramatically increases with temperature and at 80 °C reaches 0.18 S cm^{-1} for a membrane immersed in water. Based on these measurements, Springer *et al.* [5] correlated the ionic conductivity (in S cm^{-1}) to water content and temperature with the following expression:

$$\kappa = (0.005139\lambda - 0.00326) exp\left[1268\left(\frac{1}{303} - \frac{1}{T}\right)\right] \quad (2.13)$$

Since water plays a critical role in the membrane's primary function, that is proton conduction, maintaining high water content in the membrane is of critical importance. Several mechanisms affect water transport across a polymer membrane and their combination determines the local water content. These mechanisms include the following:

- Water generation on the cathode side at a rate proportional to current generation
- Electroosmotic drag from anode to the cathode as the protons on their way drag one or more water molecules
- Diffusion due to water concentration gradient across the membrane
- Hydraulic permeation due to pressure difference (if any) between the anode and cathode

For a thin membrane water backdiffusion may be sufficient to counteract the anode drying effect due to the electroosmotic drag. However, for a thicker membrane drying may occur on the anode side, particularly at higher current densities.

In principle, the membrane should be impermeable to reactant species, in order to prevent their mixing before they have a chance to participate in the electrochemical reaction. However, due to the membrane's essentially porous structure, its water content and the solubility of hydrogen and oxygen in water, small quantities of gas do permeate through the membrane. Permeability through wet Nafion is another order of magnitude higher, and as expected, hydrogen has one order of magnitude higher permeability than oxygen. Permeability is a linear function of pressure difference and an exponential function of temperature. Typical value for hydrogen permeability at 25°C is about 50×10^{-10} $cm^{3} \cdot cm \cdot s^{-1} \cdot cm^{-2} \cdot cmHg^{-1}$ [2]. For a 50 µm thick membrane and 300 kPa the loss of hydrogen corresponds to 1.6 mA/cm^2.

2.2.2 Electrode

A fuel cell electrode is essentially a thin catalyst layer pressed between the ionomer membrane and a porous, electrically conductive substrate. It is the layer where the electrochemical reactions take place. More precisely, the electrochemical reactions take place on the catalyst surface. Since there are three kinds of species that participate in the electrochemical reactions, namely gases, electrons and protons, the reactions can take place on a portion of the catalyst surface where all three species have access to. The reaction zone may be enlarged by either "roughening" the surface of the membrane, and/or by reducing the catalyst particle size, and/or by incorporating ionomer in the catalyst layer.

The most common catalyst in PEM fuel cells for both oxygen reduction and hydrogen oxidation reactions is platinum. In the early days of PEMFC development large amounts of Pt catalyst were used (up to 28 mg cm^{-2}). In the late 1990s with the use of supported catalyst structures this was reduced to 0.3–0.4 mg cm^{-2}. It is the catalyst surface area that matters, not the weight, so it is important to have small platinum particles (4 nm or smaller) with large surface area finely dispersed on the surface of the catalyst support, typically carbon powders (cca 40 nm) with high mesoporous area (>75 $m^2\ g^{-1}$). A typical support material is Vulcan XC72R by Cabot, but other carbons such as Black Pearls BP 2000, Ketjen Black Intl. or Chevron Shawinigan have been used [6]. In order to minimize the cell potential losses due to the rate of proton transport and reactant gas permeation in the depth of the electrocatalyst layer, this layer should be made reasonably thin. At the same time, the metal active surface area should be maximized, for which the Pt particles should be as small as possible. For the first reason, higher Pt/C ratios should be selected (>40% by wt.), however smaller Pt particles and consequently larger metal areas are achieved with lower loading.

In general, higher Pt loading results in voltage gain [7], assuming equal utilization and reasonable thickness of the catalyst layer. However, the key in improving the PEM fuel cell performance is not in increasing the Pt loading, but rather in increasing Pt utilization in the catalyst layer.

The catalyst surface active area may be greatly increased if ionomer is included in the catalyst layer either by painting it with solubilized PFSA in a mixture of alcohols and water or preferably by premixing catalyst and ionomer in the process of forming the catalyst layer. The optimum amount of ionomer in the catalyst layer seems to be around 30% by weight [8].

In principle, there are two ways of preparation of a catalyst layer and its attachment to the ionomer membrane. Such a combination of membrane and catalyst layers is called the membrane electrode assembly or MEA. The first way of preparing an MEA is to deposit the catalyst layer on the porous substrate, so called gas diffusion layer, typically carbon fiber paper or carbon cloth, and then hot-press it to the membrane. The second method of preparing an MEA is application of the catalyst layers directly or indirectly (via a decal process) to the membrane, forming a so-called 3-layer MEA or catalyzed membrane. The porous substrate may be added later, either as an additional step in MEA preparation (in that case a 5-layer MEA is formed) or in a process of stack assembly.

Several methods have been developed for deposition of a catalyst layer on either the porous substrate or the membrane, such as: spreading, spraying, sputtering, painting, screen printing, decaling, electro-deposition, evaporative deposition, and impregnation reduction. There are several manufacturers of MEAs, such as Dupont, 3M, Johnson Matthey, W.L. Gore & Associates and (previously dmc^2 and Degussa). Their manufacturing processes are typically trade secrets.

At present there are no alternative cathode electrocatalysts to platinum. Some platinum alloy electrocatalysts prepared on traditional carbon black supports offer a 25 mV performance gain compared to Pt electrocatalysts. However, only the more stable Pt-based metal alloys, such as PtCr, PtZr or PtTi can be used in PEMFC, due to dissolution of the base metal by the perfluorinated sulfonic acid in the electrocatalyst layer and membrane [6]. The focus of the continued search for the elusive electrocatalyst for oxygen reduction in acid environments should be on development of materials with required stability, and greater activity than Pt.

2.2.3 Gas Diffusion Layer

The required properties of the gas diffusion layer follow from its functions [9]:

- It must be sufficiently porous to allow flow of both reactant gases and product water (note that these fluxes are in opposite direction). Depending on the design of the flow field, diffusion in both through plane and in plane is important.
- It must be both electrically and thermally conductive, again both through plane and in plane. Interfacial or contact resistance is typically more important than bulk conductivity.
- Since the catalyst layer is made of discreet small particles the pores of the gas diffusion layer facing the catalyst layer must not be too big.
- It must be sufficiently rigid to support the "flimsy" MEA. However, it must have some flexibility to maintain good electrical contacts.

These somewhat conflicting requirements are best met by carbon fiber based materials such as carbon-fiber papers and woven carbon fabrics or cloths. These diffusion media are generally made hydrophobic in order to avoid flooding in their bulk. Typically, both cathode and anode gas diffusion media are PTFE-treated. A wide range of PTFE loadings have been used in PEMFC diffusion media (5% to 30%), most typically by dipping the diffusion media into an PTFE solution followed by drying and sintering. In addition, the interface with the adjacent catalyst layer may also be fitted with a coating or a microporous layer to ensure better electrical contacts as well as efficient water transport into and out of the diffusion layer. This layer (or layers) consists of carbon or graphite particles mixed with PTFE binder. The resulting pores are between 0.1 and 0.5 μm, thus much smaller than the pore size of the carbon fiber papers (20–50 μm) [9].

2.2.4 Bipolar Plates

The bipolar collector/separator plates have several functions in a fuel cell stack. Their required properties follow from their functions, namely [10]:

- they connect cells electrically in series – therefore they must be electrically conductive;
- they separate the gases in adjacent cells – therefore they must be impermeable to gases;
- they provide structural support for the stack – therefore they must have adequate strength, yet they must be lightweight;
- they conduct heat from active cells to the cooling cells – therefore they must be thermally conductive;
- they typically house the flow-field channels – therefore they must be conformable.

In addition, they must be corrosion resistant in the fuel cell environment, yet they must not be made out of "exotic" and expensive materials. In order to keep the cost down not only must the material be inexpensive, but also the manufacturing process must be suitable for mass production.

In general, two families of materials have been used for PEM fuel cell bipolar plates, namely graphite-composite and metallic. The bipolar plates are exposed to a very corrosive environment inside a fuel cell (pH 2–3 and temperature 60– 80°C). The typical metals such as aluminum, steel, titanium or nickel would corrode in fuel cell environment, and dissolved metal ions would diffuse into the ionomer membrane, resulting in lowering of the ionic conductivity and reducing the fuel cell life. In addition, a corrosion layer on the surface of a bipolar plate would increase electrical resistance. Because of these issues, metallic plates must be adequately coated with a non-corrosive yet electrically conductive layer, such as graphite, diamond-like carbon, conductive polymer, organic self-assembled polymers, noble metals, metal nitrides, metal carbides, indium doped tin oxide, *etc.* Carbon composite bipolar plates have been made using thermoplastics (polypropylene, polyethylene, or polyvinylidenefluoride) or thermoset resins (phenolic, epoxies and vinyl esters) with fillers (such as carbon/graphite powder, carbon black or coke-graphite) and with or without fiber reinforcements. These materials are typically chemically stable in fuel cell environments, although some thermosets may leach and consequently deteriorate. Depending on the rheological properties of these materials they are suitable for compression molding, transfer molding, or injection molding.

One of the most important properties of the fuel cell bi-polar plates is their electrical conductivity. Typical bulk electrical conductivity of graphite-composite bipolar plates is between 50 and 200 S cm^{-1}. Pure graphite has a conductivity of 680 S cm^{-1}, and metallic plates have typically several orders of magnitude higher electrical conductivity. One should distinguish between the bulk and total conductivity/resistivity, the latter including bulk and interfacial contact components. In an actual fuel cell stack contact (interfacial) resistance is more important than bulk resistance. For example, a 3 mm thick molded graphite/composite plate with bulk resistivity as high as 8 mΩcm would result in 2.4 mV voltage loss at 1 A/cm^2, while resistance resulting from the interfacial contacts, such as between the bipolar plate and the gas diffusion layer, may be several times higher [9–11].

2.3 Stack Design Principles

A fuel cell stack consists of a multitude of single cells stacked up so that the cathode of one cell is electrically connected to the anode of the adjacent cell. In this way exactly the same current passes through each of the cells. Note that the electrical circuit is closed with both electron current passing through solid parts of the stack (including the external circuit) and ionic current passing through the electrolyte (ionomer), with the electrochemical reactions at their interfaces (catalyst layers).

The bipolar configuration is the best for larger fuel cells since the current is conducted through relatively thin conductive plates, thus it travels a very short distance through a large area (Figure 2.3). This causes minimum electroresistive losses, even with a relatively poor electrical conductor such as graphite (or graphite polymer mixtures). However, for small cells it is possible to connect the edge of one electrode to the opposing electrode of the adjacent cell by some kind of connector. This is applicable only to very small active area cells because current is conducted in the plane of very thin electrodes, thus traveling relatively long distance through a very small cross-sectional area.

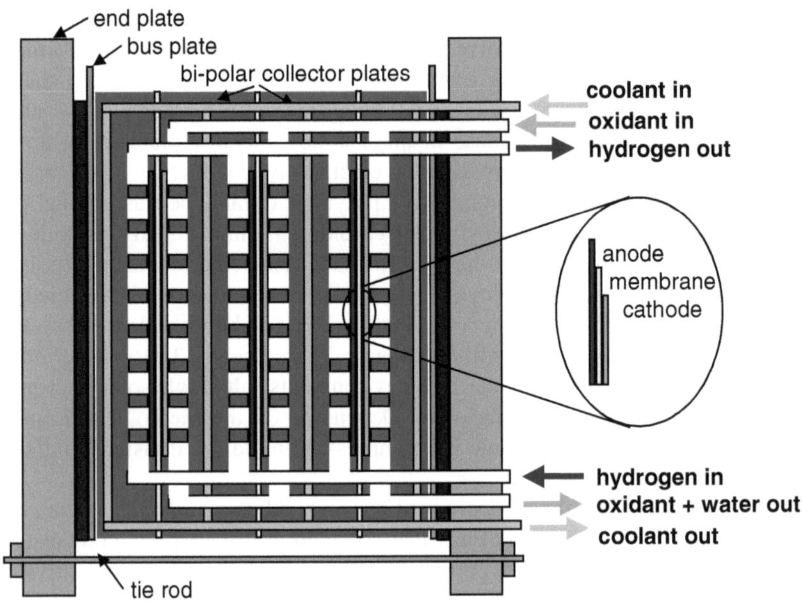

Figure 2.3. Fuel cell stack schematic [2,12]

The key aspects of a fuel cell stack design are the following [2]:

- Uniform distribution of reactants to each cell
- Uniform distribution of reactants inside each cell
- Maintenance of required temperature in each cell
- Minimum resistive losses (choice of materials, configuration, uniform contact pressure)
- No leak of reactant gases (internal between the cells or external)
- Mechanical sturdiness (internal pressure including thermal expansion, external forces during handling and operation, including shocks and vibrations)

Since the fuel cell performance is sensitive to flowrate of the reactants, it is absolutely necessary that each cell in a stack receives approximately the same amount of reactant gases. Uneven flow distribution would result in uneven performance of each cell. Uniformity is accomplished by feeding each cell in the stack in parallel through a manifold that can be either external or internal. External manifolds can be made much bigger to ensure uniformity, they result in a simpler stack design, but they can only be used in a cross-flow configuration and are in general difficult to seal. Internal manifolds are more often used in PEM fuel cell design not only because of better sealing but also because they offer more versatility in gas flow configuration.

The flow pattern through the stack can be either a "U" shape, where the inlet and outlet are at the same side of the stack and the flows in the inlet and outlet manifolds are in opposite direction, or a "Z" shape where the inlets and outlets are on opposite sides of the stack and the flows in the inlet and outlet manifolds are parallel to each other (Figure 2.4) [13]. If properly sized both should result in uniform flow distribution to individual cells. Stacks with more than a hundred cells have been successfully built.

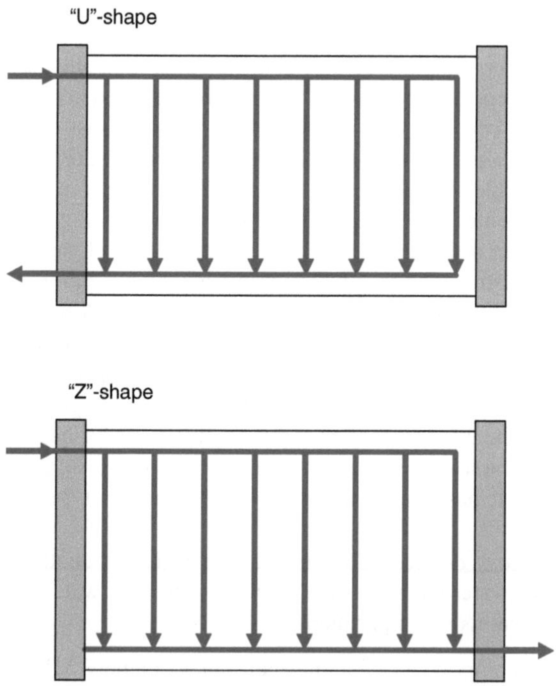

Figure 2.4. Stack flow configurations [13]

Once the reactant gases enter the individual cell they must be distributed over the entire active area. This is typically accomplished through a flow field, which may be in a form of channels covering the entire area in some pattern or porous structures. The following are the key flow field design variables [2]:

- flow field shape
- flow field orientation
- configuration of channels
- channels shape, dimensions and spacing
- pressure drop through the flow field.

One of the most common configurations of the flow field channels is in a serpentine fashion either as a single channel (for smaller active area) or as multiple channels that wind up through the entire active area connecting the inlet to the outlet manifolds. In another configuration, the inlet and outlet channels are not connected but interwined in an interdigitated fashion, thus forcing the flow through the gas diffusion layer. In both configurations, there are several modifications tried in the fuel cell stacks. Typically, the flow field configuration is a well-kept secret by the stack manufacturers. CFD modeling is a great tool for design of the fuel cell flow field and has been increasingly used [14–18].

In order to maintain the desired temperature inside the cells, the heat generated as a byproduct of the electrochemical reactions must be taken away from the cells and from the stack. Different heat management schemes may be applied such as [2]:

- *Cooling with a coolant flowing between the cells.* Coolant may be de-ionized water, antifreeze coolant or air. Cooling may be arranged between each cell, between the pair of cells (in such a configuration one cell has the cathode and the other cell has the anode next to the cooling arrangement), or between a group of cells (this is feasible only for low power densities since it results in higher temperatures in the center cells). Equal distribution of coolant may be accomplished by the manifolding arrangement similar to that of reactant gases. If air is used as a coolant equal distribution may be accomplished by a plenum.
- *Cooling with coolant at the edge of the active area (with or without fins).* The heat is conducted through the bipolar plate and then transferred to the cooling fluid, typically air. In order to achieve a relatively uniform temperature distribution within the active area, the bipolar plate must be very good thermal conductor. In addition, the edge surface may not be sufficient for heat transfer and fins may need to be employed. This method results in a much simpler fuel cell stack, fewer parts, but it has heat transfer limitations, and is typically used for low power outputs.
- *Cooling with phase change.* Coolant may be water or another phase schange medium. Use of water simplifies the stack design since water is already used in both anode and cathode compartments.

The individual components of a fuel cell stack, namely MEAs, gas diffusion layers and bipolar plates must be somehow held together with sufficient contact pressure to (i) prevent leaking of the reactants between the layers, and (ii) to minimize the contact resistance between those layers. This is typically accomplished by sandwiching the stacked components between the two end plates connected with several tie-rods around the perimeter or in some cases through the middle. Other compression and fastening mechanisms may be employed too, such as snap-in shrouds or straps [2].

2.4 System Design

A fuel cell stack is obviously the heart of a fuel cell system, however, without the supporting equipment the stack itself would not be very useful. The fuel cell system typically involves the following subsystems:

- oxidant supply (oxygen or air),
- fuel supply (hydrogen or hydrogen-rich gas),

- heat management,
- water management,
- power conditioning, and
- instrumentation and controls.

The simplest way to supply hydrogen from a tank to a fuel cell is in the dead-end mode. Such a system would only require a preset pressure regulator to reduce the pressure from the stack to the fuel cell operating pressure. The long term operation in a dead-end mode may be possible only with extremely pure gases, both hydrogen and oxygen. Any impurities present in hydrogen will eventually accumulate in the fuel cell anode. This also includes water vapor that may remain (in case when the back-diffusion is higher than the electroosmotic drag, which may be the case with very thin membranes and when operating at low current densities. In addition, nitrogen may diffuse from the air side until an equilibrium concentration is established. In order to eliminate this accumulation of inerts and impurities, purging of the hydrogen compartment may be required. This may be programmed either as a function of cell voltage or as a function of time.

If purging of hydrogen is not possible or preferred due to safety, mass balance or system efficiency reasons, excess hydrogen may be flowed through the stack ($S>1$) and unused hydrogen returned to the inlet, either by a passive (ejector) or an active (pump or compressor) device. In either case, it is preferred to separate and collect any liquid water that may be present at the anode outlet. The amount of liquid water to be collected depends on operating conditions and membrane properties. In thinner membranes, backdiffusion may be higher than the electroosmotic drag, and some of the product water may exit the stack at the anode side.

Hydrogen typically must be humidified up to 100% relative humidity prior to entering the fuel cell stack, in order to avoid drying of the membrane due to electoosmotic drag. In this case a humidifier/heat exchanger is needed at the stack inlet. Hydrogen may be humidified by water injection, and simultaneously or subsequently heated to facilitate evaporation of water, or by membrane humidification.

In hydrogen-air systems, air is supplied by a fan or a blower (for low pressure systems) or by an air compressor for pressurized systems. In the former case the exhaust from the fuel cell is open directly into the environment, while in a pressurized system, pressure is maintained by a preset pressure regulator at the fuel cell exhaust. In either case, a fan, or a blower or a compressor is run by an electric motor that requires electrical power, and thus represents power loss or parasitic load, which may have a significant effect on the system efficiency. Air typically has to be humidified before entering the fuel cell stack. Various humidification schemes may be employed such as:

- bubbling of gas through water
- direct water or steam injection
- exchange of water (and heat) through a water permeable medium
- exchange of water (and heat) on an adsorbent surface (enthalpy wheel).

At the stack outlet, typically there is some liquid water that may be easily separated from the exhaust air, in a simple, off-the-shelf gas/liquid separator. Water collected at the exhaust may be stored and reused, either for cooling and/or for humidification.

Water and heat are the byproducts of the fuel cell operation, and the supporting system must include the means for their removal. Both water and heat from the fuel cell stack may be at least partially re-used, for example for humidification of the reactant gases. Water and heat handling may be integrated into a single subsystem if deionized water is used as a stack coolant. In that case water removes the heat from the stack and the same water and heat are used to humidify the reactant gases. The remaining heat has to be discarded to the surroundings through a heat exchanger, for hydrogen-air systems that is typically a radiator. The amount of heat to be discarded must be calculated from the stack and humidifier energy balances. The size of the radiator heat exchanger depends on the temperature difference between the coolant and the ambient air. For that reason, it is preferred to operate the fuel cell system at a higher temperature, for systems where the size of the components is critical. However, the operating pressure and water balance must be taken into account when deciding on the operating temperature. It should be noted that smaller stacks may be air cooled. In that case a fan replaces the coolant pump. Reuse of waste heat collected by air as coolant is not practical, but it may be convenient to blow the warm exhaust over the metal hydride tanks.

For very small power outputs it is possible to design and operate a fuel cell with a passive air supply, relying only on natural convection due to concentration gradients. Such a fuel cell typically has either the front of the cathode directly exposed to the atmosphere, therefore without the bipolar plates, or in bi-polar configuration the cathode flow field is sideways opened to the atmosphere. In either case an oxygen concentration gradient is formed between the open atmosphere and the catalyst layer where oxygen is being consumed in the electrochemical reaction. The performance of such fuel cells is typically not limited only by the oxygen diffusion rate, but also by water and heat removal, both dependent on the temperature gradient. The maximum

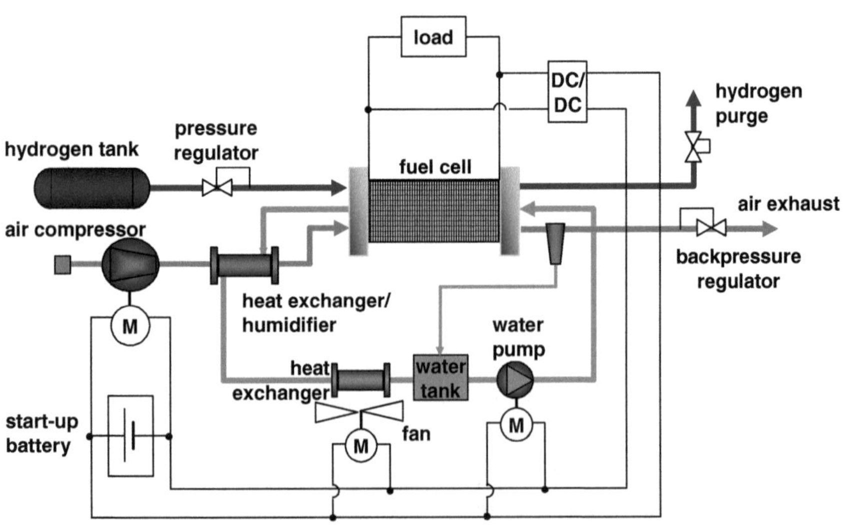

Figure 2.5. An example of a complete hydrogen-air fuel cell system [19]

current density that can be achieved with free-convection fuel cells is typically limited to about 0.1–0.15 A cm^2.

The system configuration greatly depends on the application. In some cases a very simple system, consisting of nothing more than a bottle of hydrogen and a fuel cell may be sufficient. In other cases the system needs most of the components and subsystems discussed above. Figure 2.5 shows a schematic diagram of an actual hydrogen-air fuel cell system employed in a fuel cell utility vehicle [19].

In order to circumvent the issue of hydrogen availability and infrastructure, a fuel cell may be integrated with a fuel processor allowing hydrogen generation from hydrocarbon fuels, such as natural gas, gasoline or methanol. Hydrogen can be generated from hydrocarbon fuels by several processes, such as:

- Steam reformation,
- Partial oxidation,
- Autothermal reformation, which is essentially a combination of steam reformation and partial oxidation.

In addition, several other processes must be employed in order to produce hydrogen pure enough to be used in PEM fuel cells, such as:

- Desulfurization – to remove sulfur compounds present in fuel,
- Shift reaction – to reduce the content of CO in the gas produced by the fuel processor,
- Gas cleanup, involving preferential oxidation, methanation and/or membrane separation – to further minimize the CO content in the reformate gas.

The reforming process needs water and air and thermal management. Figure 2.6 shows such a fuel processor/fuel cell integrated system [2]. It should be noted that

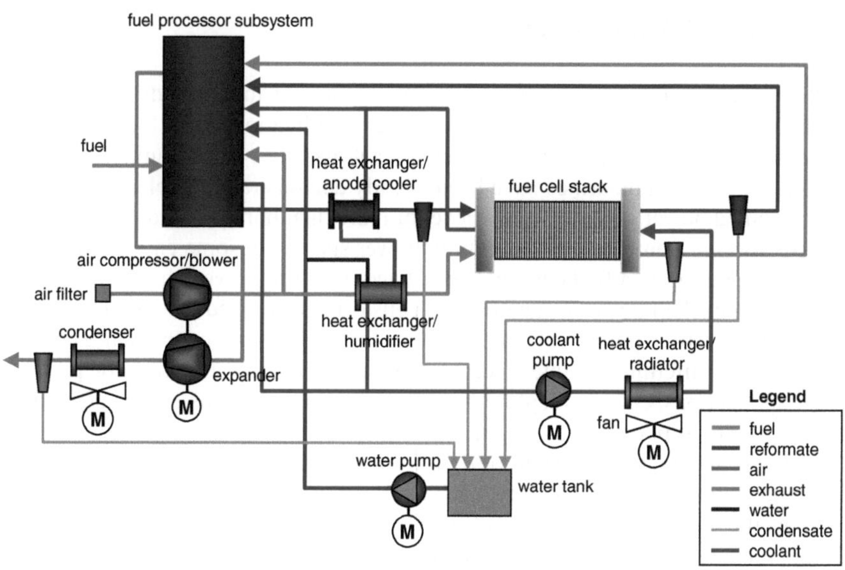

Figure 2.6. Schematic diagram of a fuel cell system integrated with a fuel processor [2]

the fuel processor in Figure 2.6 is shown as a box, while actually it consists of a series of reactors and heat exchangers necessary for the reforming process. There are several issues that must be addressed in the design of such an integrated system, such as CO content in the produced reformate gas, transient behavior, water balance, reforming efficiency, parasitic power needed for the reforming process, controls, *etc.*

The efficiency of the fuel cell system is lower than the fuel cell efficiency given by equation (2.11); it must also include the efficiency of the reforming process, including the preferential oxidation efficiency, the efficiency of power conditioning and the parasitic power requirements (for both fuel cell and fuel processor subsystems). The overall system efficiency is then:

$$\eta_{sys} = \eta_{ref}\eta_{PROX}\eta_{fuel}\eta_{FC}\eta_{PC} \tag{2.14}$$

where the efficiency of power conditioning, η_{PC}, also includes the parasitic power losses, through a coefficient ξ, which is defined as a ratio between power needed to run the parasitic load and fuel cell gross power:

$$\eta_{PC} = \eta_{DC}\left(1 - \frac{\xi}{\eta_{DCaux}}\right) \tag{2.15}$$

where η_{DC} and η_{DCaux} are the effciency of DC/AC or DC/DC power conversions for both the main load and parasitic power, respectively.

The achievable efficiency of such systems is about 40%. The systems using hydrogen as fuel may have an efficiency around 50% and those that use pure oxygen as oxidant may have an efficiency above 50%.

2.5 Fuel Cell Applications

Fuel cells can generate power from a fraction of a watt to hundreds of kilowatts. Because of this, they may be used in almost every application where local electricity generation is needed. Applications such as automobiles, buses, utility vehicles, scooters, bicycles, submarines have been already demonstrated. Fuel cells are ideal for distributed power generation, at a level of individual homes, buildings or a community, offering tremendous flexibility in power supply. In some cases both power and heat produced by a fuel cell may be utilized, resulting in very high overall efficiency. As a backup power generator, fuel cells offer several advantages over either internal combustion engine generators (noise, fuel, reliability, maintenance) or batteries (weight, lifetime, maintenance). Small fuel cells are attractive for portable power applications, either as replacement for batteries (in various electronic devices and gadgets) or as portable power generators.

Fuel cell and fuel cell system design are not necessarily the same for each of these applications. On the contrary, each application, besides power output, has its own specific requirements, such as efficiency, water balance, heat utilization, quick startup, long dormancy, size, weight, fuel supply, *etc.*

2.5.1 Automotive Applications

All major car manufacturers have demonstrated prototype fuel cell vehicles and announced plans for production and commercialization. The major drivers for development of automotive fuel cell technology are their efficiency, low or zero emissions, and fuel that could be produced from indigenous sources rather than being imported. The main obstacles for fuel cell commercialization in automobiles are the cost of fuel cells and the cost and availability of hydrogen.

The fuel cell may be connected to the propulsion motor in several ways [2], namely:

1) Fuel cell is sized to provide all the power needed to run the vehicle. A battery may be present but only for startup (such as a 12V battery). This configuration is typically possible only with direct hydrogen fuel cell systems. A system with a fuel processor would not have as good dynamic response. Also, a small battery would not be sufficient to start up a system with a fuel processor.
2) Fuel cell is sized to provide only the base load, but the peak power for acceleration of the vehicle is provided by the batteries or similar peaking devices (such as ultracapacitors). This may be considered as a parallel hybrid configuration since the fuel cell and the battery operate in parallel – the fuel cell provides cruising power, and the battery provides peak power (such as for acceleration). The presence of a battery in the system results in much faster response to load changes. The vehicle can be started without preheating of the fuel cell system, particularly the fuel processor, and operated as a purely battery-electric vehicle until the fuel cell system becomes operational. A battery allows for recapturing of the braking energy, resulting in a more efficient system. The disadvantages of having the battery are extra cost, weight and volume.
3) Fuel cell is sized only to re-charge the batteries. The batteries provide all the power needed to run the vehicle. This may be considered as serial hybrid configuration (fuel cell charges the battery and battery drives the electric motor). The same advantages and disadvantages of having a battery apply as for the parallel hybrid configuration. The fuel cell nominal power output depends on how fast the batteries would have to be recharged. A smaller battery would have to be recharged faster and would result in a larger fuel cell.
4) Fuel cell serves only as an auxiliary power unit, *i.e.* another engine is used for propulsion, but the fuel cell is used to run the entire or a part of the vehicle electrical system [20]. This may be particularly attractive for trucks, since it would allow operation of an air-conditioning or refrigeration unit while the vehicle is not moving without the need to run the main engine.

In general, a fuel cell propulsion engine is more efficient that a comparable internal combustion engine. However, the efficiency of fuel cells vs. internal combustion engine should not be compared at their most favorable operating point. These two technologies are intrinsically different and have very different efficiency-power characteristics. While an internal combustion engine has its maximum efficiency at or near its maximum power [21], a fuel cell system has its maximum efficiency at partial load [22] (Figure 2.7). Because of this, the efficiency of a hydrogen fueled fuel cell propulsion system in a typical driving schedule, where an automobile engine operates most of the time at partial load, can be about twice that of an internal combustion engine [23–25]. The hydrogen fuel cell system efficiency in a driving schedule can be in the upper 40s and above 50%. The efficiency of a fuel cell propulsion system with an onboard fuel processor is lower than the efficiency of a hydrogen fuel cell system, but still higher than the efficiency of an internal combustion engine. The fuel cell efficiency advantage diminishes if both a fuel cell or an internal combustion engine are used in a hybrid configuration.

One of the biggest problems related to hydrogen use in passenger vehicles is its on-board storage. Hydrogen, can be stored as compressed gas, as a cryogenic liquid or in metal hydrides. Tanks for compressed gaseous hydrogen are bulky, even if hydrogen is compressed to 450 bar. It takes about 40–50 liters of space to store 1 kg of hydrogen. The amount of fuel to be stored onboard depends on the vehicle fuel efficiency and required range.

Automotive fuel cells must survive and operate in extreme weather conditions (-40 to +40°C). This requirement has a tremendous effect on system design. Survival and start-up in extremely cold climates requires specific engineering solutions, such as use of antifreeze coolant and water management. Water cannot be completely eliminated from the system, because water is essential for the membrane ionic conductivity.

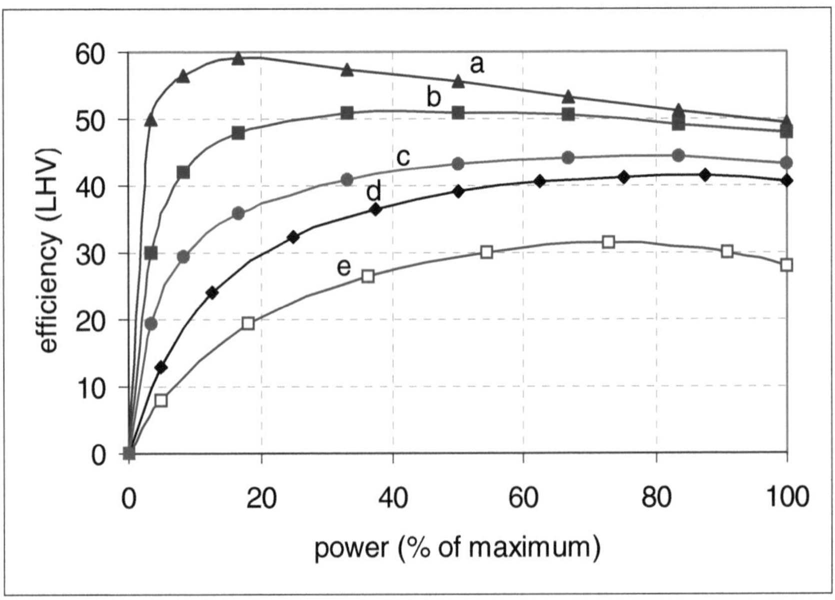

Figure 2.7. Comparison of the efficiency of fuel cells and internal combustion engines [2]
a) fuel cell system operating at low pressure and low temperature
b) fuel cell system operating at high pressure and high temperature
c) fuel cell system with an on-board fuel processor
d) compression ignition internal combustion engine (diesel)
e) spark ignition internal combustion engine (gasoline) (compiled from [21] and [22])

The fuel cell heat rejecting equipment (radiator and/or condenser) must be sized for heat rejection in extremely hot weather (typically from 32 to 40°C [21]). Although a fuel cell system is more efficient than an internal combustion engine it has similar or larger cooling loads. More importantly, because of the fuel cell's low operating temperature (60 to 80°C) the heat rejection equipment is typically much larger than that for a comparable internal combustion engine [21,26].

Water balance requirements result in additional cooling loads [21,26]. Although water is produced in a fuel cell, water is needed for humidification of reactant gases and for fuel processing (in the case of an on-board fuel processor) and it has to be reclaimed from the exhaust gases.

Buses for city and regional transport are considered to be the most likely types of vehicles for an early market introduction of fuel cell technology. Buses require more power than passenger automobiles, typically about 250 kW or more. They operate in a more demanding operating regime with frequent starts and stops. Nevertheless, the average fuel economy of a bus fuel cell system is roughly 15% better than that of a diesel engine [27]. Buses are almost always operated in a fleet, and refueled in a central facility. This makes refueling with hydrogen much easier. In addition, storing larger quantities of hydrogen on board (typically above 20 kg) is less of a problem. Fuel cell buses typically store hydrogen in composite compressed gas cylinders at 250 to 300 bar, located on the roof. Because hydrogen is much lighter than air, the roof location is considered to be very safe.

Utility vehicles, such as fork-lifts, material handling industrial vehicles, airport ground support tow vehicles, lawn maintenance vehicles, golf carts and airport people movers, may be another early adapter of the fuel cell technology. This application is not as demanding as the passenger vehicles or the buses. The competing technology are typically the batteries, most often the lead acid batteries, that require frequent and lengthy charging and pose significant maintenance problems. Early demonstrations of fuel cell powered utility vehicles have shown that such vehicles offer lower operating cost, reduced maintenance, lower down-time, and extended range [28].

Scooters and bicycles may be a significant market for fuel cell technologies, particularly in developing countries. Despite the stringent requirements relating to weight, size and low cost, fuel cells have been successfully demonstrated in various scooters and bicycles [28]. The power requirement is considerably less than for the automobiles – up to 3 kW for scooters and up to 1 kW for bicycles. Although the range may be smaller than for automobiles too, the volume of hydrogen storage is one of the critical issues. Some prototypes were demonstrated with direct methanol fuel cell systems. The fuel cells for scooters and bicycles are almost always air-cooled. The refueling issue of these vehicles in mass markets is equally as complex as the issue of automobile refueling. However, because of significantly smaller quantities of hydrogen to be stored onboard, additional options are possible, such as distribution of metal hydride tanks or home refueling devices (probably electrolyzers).

2.5.2 Stationary Power Applications

Although development and demonstrations of fuel cells in automobiles usually draws more attention, applications for stationary power generation offer even greater market opportunity. The drivers for both market sectors are similar – higher efficiency and lower emissions. The system design for both applications is also similar in principle. The main differences are in the choice of fuel, power conditioning, and heat rejection [29]. There are also some differences in requirements for automotive and stationary fuel cell systems. For example, size and weight requirements are very important in automotive applications, but not so significant in stationary applications. The acceptable noise level is lower for stationary applications, especially if the unit is to be installed indoors. The fuel cell itself of course does not generate any noise, noise may be coming from air and fluid handling devices. Automobile systems are expected to have a very short start-up time (fraction of a minute), while the startup of a stationary system is not time limited, unless operated as a backup or emergency power generator. Both automotive and stationary systems are expected to survive and operate in extreme ambient conditions, although some stationary units may be designed for indoor installation only. And finally, the automotive systems for passenger vehicles are expected to have a lifetime of 3000 to 5000 operational hours, systems for buses and trucks somewhat longer, but the stationary fuel cell power systems are expected to operate for 40000 to 800000 hours (five to ten years).

Stationary fuel cell power systems will enable the concept of distributed generation, allowing the utility companies to increase their installed capacity following the increase in demand more closely, rather than anticipating the demand in huge increments by adding gigantic power plants. Presently, obtaining the permissions and building a conventional power plant have become very difficult tasks. Fuel cells, on the other hand, do not need special permitting and may be installed virtually everywhere – inside the residential areas, even inside the residential dwellings. To the end users the fuel cells offer reliability, energy independence, "green" power, and, ultimately, lower cost of energy.

Stationary fuel cells may be used in different applications, namely:

- As the only power source, thus competing with or replacing the grid, or providing electricity in the areas not covered by the grid.
- As a supplemental power source working in parallel with the grid covering either the base load or the peak load.
- For both power and heat cogeneration in either power load following or heat load following mode.
- In combined systems with intermittent renewable energy sources (such as photovoltaics or wind turbines) generating power in periods when these energy sources cannot meet the demand.
- As a backup or emergency power generator providing power when the grid (or any other primary power source) is down.

Accordingly, the fuel cell system, and particularly its power conditioning and interconnect module, may be designed as [2]:

- Grid parallel – allowing power from the grid to the consumer when needed, but not allowing power from the fuel cell back to the grid. The fuel cell system may be sized to provide most of the consumer's energy needs but the grid is used to cover the short-term demand peaks. Such a system essentially does not need batteries (except for start-up when the grid is down), and does not need interconnect standards.
- Grid interconnected – allowing power flow in both directions, namely power from the grid to the consumer when needed, and power from the fuel cell back to the grid. Such a system may be designed as load following or as constant power, since excess fuel cell power can be exported to the grid. Of course, this design option requires interconnect standards.
- Standalone – providing power without grid. The system must be capable of load following. Very often a sizeable battery bank is used to enable load following.
- Backup or emergency generator – the system must be capable of quick startup and is also often combined with the batteries or other peaking device. Batteries are typically superior for low power/low duration backup power, but a fuel cell system becomes competitive for higher power (several kW) and longer duration (over 30 minutes). A backup power system may be equipped with an electrolyzer-hydrogen generator and hydrogen storage [30]. In this case the unit generates its own fuel during periods when electricity from the grid is available.

Commercialization of stationary fuel cells greatly depends on their economics, which besides the selling price includes their annual (not maximum) efficiency, capacity factor (depending on application), lifetime, maintenance, but most of all on a ratio between the prices of electricity and natural gas. In order for a stationary fuel cell to be feasible, the ratio between the prices of electricity and natural gas must be larger than the reciprocal of its annual efficiency (*i.e.* > 3 and preferably > 4). The prices of electricity and natural gas vary with time and from region to region. The economics of stationary fuel cells may be improved if the waste heat is utilized in a cogeneration manner. Some residential fuel cells are being developed to operate in a heat load following mode.

2.5.3 Portable Power Applications

A portable power system is a small grid-independent electric power unit ranging from a few watts to roughly one kilowatt, which serves mainly the purpose of convenience rather than being a primarily a result of environmental or energy-saving considerations [31]. These devices may be divided into two main categories:

1) battery replacements, typically well under 100 W
2) portable power generators, up to 1 kW

The key feature of small fuel cells to be used as battery replacements is the running time without recharging. Obviously, by definition, the size and weight are also important. Power units with either significantly higher power densities or larger energy storage capacities than those of existing secondary batteries may find applications in portable computers, communication and transmission devices, power tools, remote meteorological or other observation systems, and in military gadgets. Besides the size of the fuel cell itself, the critical issue is the fuel and its storage. Hydrogen, although being a preferred fuel for PEM fuel cells is rarely used because of the bulkiness or weight of its storage, even in the small quantities required by those small devices. Hydrogen may be stored in room temperature metal hydride storage tanks. Some chemical hydrides offer higher energy density, however, they must be equipped with suitable reactors where hydrogen is released in a controlled chemical reaction [32]. Most of portable fuel cells use methanol as fuel, or more precisely aqueous methanol solutions, either directly (so called direct methanol fuel cells) or via microreformers.

The military market is particularly attractive as it often may be a sympathetic early adopter of new technologies, willing to accept high prices and limited performance if other, application-specific, requirements can be met (such as low noise, low thermal signature, long duration both in operation and dormant, size and weight, safety) [33]. Some examples of early military fuel cell products or prototypes include battery chargers, soldier power, telecommunications, navigation systems, computers, various tools, exoskeletons, auxiliary power unit for vehicles, unmanned aerial vehicles, small autonomous robot vehicles, unattended sensors and munitions, and ocean sensors and transponders [33].

Development of small fuel cells for portable power applications have resulted in a myriad of stack configurations. Some stacks are miniaturized replicas of the larger automotive or stationary power fuel cells with the same components, MEAs, gas diffusion layers, bipolar plates and end plates. Some use a planar configuration where the cells are connected with conductive strips. Recently, microfluidic cells manufactured on silicon based chips have emerged [34, 35]. The fuel cell systems for these applications are extremely simplified. The simplicity of the system is more important than the cell/stack size. The power density generated is often below 0.1 W/cm^2. These cells/stacks do not need active cooling, those using hydrogen mainly operate in a dead-end mode, and air is often supplied passively.

2.6 Summary

PEM fuel cells are the most attractive of all the fuel cell types for many applications due to their simplicity, quick startup, load following capabilities, efficiency, modularity and versatility. An overwhelming majority of fuel cell related patents and publications refer to PEM fuel cells. Only for large (>250 kW) stationary power applications, high temperature fuel cells have significant advantage.

Although the PEM fuel cells have been demonstrated in numerous applications, the key barriers to their widespread commercial use are:

- Nonexistance of hydrogen fuel infrastructure, and difficulties related to hydrogen storage,
- Relatively high cost of early fuel cell prototypes and products,
- Insufficient lifetime for some applications (particularly stationary power generation).

Commercialization of fuel cells, particularly for transportation and stationary electricity generation markets, must be accompanied by commercialization of hydrogen energy technologies, *i.e.* technologies for hydrogen production, distribution and storage. In other words, hydrogen must become a readily available commodity (not as a technical gas but as an energy carrier) before fuel cells can be fully commercialized. Fuel cells runing on natural gas may also be a viable option but only if their efficiency and economics are better that that of conventional technologies for electricity generation. On the other hand, fuel cells have many unique properties, such as high energy efficiency, no emissions, no noise, modularity, and potentially low cost, which may make them attractive in many applications even with limited hydrogen supply. Fuel cells may very well become a driver for development of hydrogen energy technologies. Fuel cells will change the way energy is converted to useful power and are likely one of those powerful technologies that could create the next revolution – the energy revolution, similar to the steam engine that created the industrial revolution, the Ford Model T that created automobile revolution or the computer that created the information revolution.

2.7 References

1. F. Barbir, Fuel Cell Tutorial, presented at Future Car Challenge Workshop, Dearborn, MI, October 25–26, 1997
2. F. Barbir, *PEM Fuel Cells: Theory and Practice*, Elsevier Academic Press, New York, 2005
3. S.Gottesfeld and T.A. Zawodzinski, Polymer Electrolyte Fuel Cells, in R.C. Alkire, H. Gerischer, D.M. Kolb, and C.W. Tobias (Eds.) *Advances in Electrochemical Science and Engineering*, Volume 5, Wiley-VCH, New York, 1997
4. T.A. Zawodzinski, Jr., T.E. Springer, J. Davey, R. Jestel, C. Lopez, J. Valerio, and S. Gottesfeld, A Comparative Study of Water Uptake By and Transport Through Ionomeric Fuel Cell Membranes, Journal of the Electrochemical Society, Vol. 140, 1993, p. 1981–1985
5. T.E. Springer, T.A. Zawodzinski, and S. Gottesfeld, Polymer Electrolyte Fuel Cell Model, Journal of the Electrochemical Society, Vol. 138, No. 8, pp. 2334–42, 1991.
6. T.R. Ralph and M.P. Hogarth, Catalysis for Low Temperature Fuel Cells, Part I: The Cathode Challenges, Platinum Metals Review, Vol. 46, No. 1, pp.3–14 , 2002
7. H.A. Gasteiger, W. Gu, R. Makharia and M.F. Mathias, Catalyst utilization and mass transfer limitations in the polymer electrolyte fuel cells, Electrochemical Society Meeting, Orlando, September, 2003
8. T.A. Zawodzinski, Jr., M. Eikerling, L. Pratt, R.Antonio, R. Tommy, M. Hickner, J. McGrath, Membranes for Operation Above 100°C, in *Proc. 2002 National Laboratory R&D Meeting DOE Fuel Cells for Transportation Program* (Golden, CO, May 9, 2002).[M.F. Mathias, J. Roth, J. Fleming and W. Lehnert, Diffusion media materials and characterization, in W. Vielstich, A. Lamm, and H.A. Gastegier (Eds.) *Handbook of Fuel Cells, Fundamentals, Technology and Applications*, Vol. 3 Fuel Cell Technology and Applications, pp. 517–537, John Wiley & Sons, Ltd., New York, 2003.
10. F. Barbir, J. Braun, and J.Neutzler, Properties of Molded Graphite Bi-Polar Plates for PEM Fuel Cells, *International Journal on New Materials for Electrochemical Systems*, No. 2, pp. 197–200, 1999

11. V. Mishra, F. Yang and R. Pitchumani, Electrical contact resistance between gas diffusion layers and bi-polar plates in a PEM fuel cell, Proc. 2nd Int. Conf. Fuel Cell Science, Engineering and Technology. Rochester, NY, 2004
12. T.N. Veziroglu and F. Barbir, *Hydrogen Energy Technologies*, UNIDO - Emerging Technologies Series, United Nations Industrial Development Organisation, Vienna, Austria, 1998
13. J.-H. Koh, K. Seo, C. G. Lee, Y. -S. Yoo, and H. C. Lim, Pressure and Flow Distribution in Internal Gas Manifolds of a Fuel Cell Stack, *Journal of Power Sources,* Vol. 115, 2003, pp. 54–65.
14. V. Gurau, F. Barbir, and H. Liu, Two-Dimensional Model for the Entire PEM Fuel Cell Sandwich, in *Proton Conduction Membrane Fuel Cells II*, S. Gottesfeld and T.F. Fuller (eds.), Proc. Vol. 98–27, pp. 479–503, The Electrochemical Society, Pennington, NJ, 1999.
15. H. Naseri-Neshat, S. Shimpalee, S. Dutta, W.K. Lee and J.W. Van Zee, Predicting the effect of gas-flow channel spacing on current density in PEM fuel cells, Advanced Energy Systems Vol. 39, pp. 337–350, ASME, 1999
16. S.Um, C.-Y. Wang and K. S. Chen, 2000, "Computational fluid dynamics modeling of proton exchange membrane fuel cells," *J. Electrochem. Soc.,* Vol. 147, pp. 4485–4493, 2000.
17. L. You and H. Liu, A two-phase flow and transport model for the cathode of PEM fuel cells, *Int. J. Heat and Mass Transfer*, 45, pp. 2277–2287, 2002
18. S. Shimpalee, S. Greenway, D. Spuckler and J. W. Van Zee, Predicting water and current distributions in a commercial-size PEMFC, *J. Power Sources*, Vol. 135, pp. 79–87, 2004
19. F. Barbir, M. Nadal, and M. Fuchs, Fuel Cell Powered Utility Vehicles, in Buchi, F. (editor), *Proc. of the Portable Fuel Cell Conference* (Lucerne, Switzerland, June 1999), pp. 113–126.
20. J. Tachtler, T. Dietsch, and G. Goetz, Fuel Cell Auxiliary Power Unit – Innovation for the Electric Supply of Passenger Cars, SAE Paper No. 2000-01-0374, in *Fuel Cell Power for Transportation 2000* (SAE SP-1505) (SAE, Warrendale, PA, 2000), pp. 109–117.
21. D.A. Masten and A. D. Bosco, System Design for Vehicle Applications: GM/Opel, in W. Vielstich, A. Lamm, and H. Gasteiger (editors), Handbook of Fuel Cell Technology – Fundamentals, Technology and Applications, Vol. 4 (J. Wiley, New York, 2003), pp. 714–724.
22. R. Stone, Competing Technologies for Transportation, in G. Hoogers (editor), Fuel Cell Technology Handbook (CRC Press, Boca Raton, FL, 2003).
23. C.E. Thomas, B. D. James, F. D. Lomax, Jr., and I. F. Kuhn, Jr., Fuel Options for the Fuel Cell Vehicle: Hydrogen, Methanol or Gasoline? International Journal of Hydrogen Energy, Vol. 25, No. 6, 2000, pp. 551–568.
24. Well-to-Wheel Energy Use and Greenhouse Gas Emissions of Advanced Fuel/Vehicle Systems, North American Analysis. Report by General Motors in cooperation with Argonne National Laboratory, BP Amoco, ExxonMobil and Shell, 2001.
25. Weiss, M. A., J. B. Heywood, E. M. Drake, A. Schafer, and F. F. AuYeung, *On the Road in 2020. A Life-Cycle Analysis of New Automobile Technologies* (Massachusetts Institute of Technology, Boston, 2000).
26. Fronk, M. H., D. L. Wetter, D. A. Masten, and A. Bosco, PEM Fuel Cell System Solutions for Transportation, SAE Paper No. 2000-01-0373, in *Fuel Cell Power for Transportation 2000* (SAE SP-1505) (SAE, Warrendale, PA, 2000), pp. 101–108.
27. Hoogers, G., Automotive Applications, in G. Hoogers (editor), *Fuel Cell Technology Handbook* (CRC Press, Boca Raton, FL, 2003).

28. M. Croper, Fuel Cell Market Survey: Niche Transport, *Fuel Cell Today,* Article 823, June 2004: http://www.fuelcelltoday.com/FuelCellToday/FCTFiles/FCTArticleFiles/Article_823_NicheTransport0604.pdf (accessed February 2005)
29. F. Barbir, System Design for Stationary Power Generation, in W. Vielstich, A. Lamm, and H. Gasteiger (editors), *Handbook of Fuel Cell Technology – Fundamentals, Technology and Applications,* Vol. 4 (J. Wiley, New York, 2003), pp. 683–692.
30. F. Barbir, T. Maloney, T. Molter, and F. Tombaugh, Fuel Cell Stack and System Development: Matching Market to Technology Status, in *Proc. 2002 Fuel Cell Seminar* (Palm Springs, CA, November 18–21, 2002), pp. 948–951.
31. G. Hoogers, Portable Applications, in G. Hoogers (editor), *Fuel Cell Technology Handbook* (CRC Press, Boca Raton, FL, 2003).
32. A. Heinzel and C. Hebling, Portable PEM Systems, in W. Vielstich, A. Lamm, and H. Gasteiger (editors), *Handbook of Fuel Cell Technology – Fundamentals, Technology and Applications,* Vol. 4 (J. Wiley, New York, 2003), pp. 1142–1151.
33. S. Geiger and D. Jollie, Fuel Cell Market Survey: Military Applications, *Fuel Cell Today,* Article 756, April 2004: http://www.fuelcelltoday.com/FuelCellToday/FCTFiles/FCTArticleFiles/Article_756_MilitarySurvey0404.pdf (accessed February 2005)
34. K. Shah, W. C. Shin, and R. S. Besser, A PDMS Micro Proton Exchange Membrane Fuel Cell by Conventional and Non-Conventional Microfabrication Techniques, *Sensors and Actuators B: Chemical,* Vol. 97, No. 2–3, 2004, pp. 157–167.
35. Y. Yamazaki, Application of MEMS Technology to Micro Fuel Cells, *Electrochimica Acta,* Vol. 50, No. 2–3, 2004, pp. 659–662.

3

Durability and Accelerated Characterization of Fuel Cells

Ken Reifsnider and Xinyu Huang

3.1 Introduction

Durability, in the present context, is a concept that relates how fuel cells and fuel cell systems are made and operated to how long they last. The science and methodology of this concept are the subject of this chapter. The approach we will discuss has its roots in the history of technical development of such concepts for composite material systems in general [1]. It is appropriate as a starting point for fuel cells, since fuel cells are functional composite material systems.

The definition of the concept for fuel cells and fuel cell systems is illustrated in Figure 3.1. The foundation of the concept is a "performance metric." The performance metric is defined as a function or functional that relates operating conditions (inputs, or "extensive variables") to cell, stack, or system properties and characteristics, to predict (calculate) an output that defines performance. The output, for a fuel cell, may be voltage, current, power, power density, or any other single or multiple output that defines the performance. One may write an expression such as

$$\text{Performance} = \text{fn}\left(\frac{\text{OperatingConditions}}{\text{FuelCellCharacteristics}}\right) \tag{3.1}$$

As Figure 3.1 shows, if we ask what is the (maximum) performance available from our fuel cell (or system), we can expect a level defined by how the fuel cell is made. If we have mature and robust representations (balance equations and constitutive equations) of the multiphysics processes that define that performance, it can be calculated. We can also expect that the (maximum) available performance will decrease as a function of time or history of use (otherwise, durability is not a concern). The operating performance for a given set of inputs will, in general, be less than the maximum available level; it will be defined by what we need from the fuel cell for a given set of loads. When (and if) the available performance reduces to a level that will no longer provide the needed operating performance, then the fuel cell "fails," for this application, defining the "life" curve in Figure 3.1.

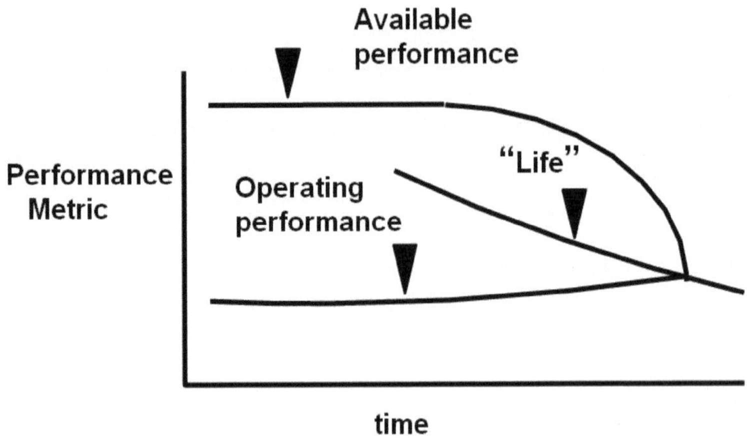

Figure 3.1. Schematic representation of a definition of durability based on a performance metric

Another basic part of the durability concept discussed here is the idea that the rate at which the performance degrades is proportional to the level of operating performance required, *i.e.* if we need higher levels of power, degradation is more rapid, and if we do not require any performance (no power, for example) from the system, then the performance doesn't degrade. It should be noted, however, that the (maximum) available performance will be influenced by cyclic variations in operating conditions, even for OCV conditions. In any case, this element of the concept results in the shape of the "life" curve shown in Figure 3.1, which slopes up on the left (shorter life for higher required operating performance) and down on the right (longer life for lower required operating performance).

A final basic element of this approach is the idea of the definition of "failure," *i.e.* the definition of a "failure mode." As we noted in Figure 3.1, the definition of the performance metric depends on what we regard as "unacceptable" performance, *i.e.* the performance that we decide defines successful operation of the fuel cell, or failure. Of course, simple examples of such a metric are apparent. If we select a strength metric, then an event such as mechanical failure (*e.g.*, crack formation or growth) may define fuel cell failure. In that case, the metric may be a mechanical failure criterion, applied in regions of specific interest. We will look at an example of such a metric in the next section. If we select a performance requirement such as maximum available power, then the metric may be the product of the voltage and current for a predicted V-I curve for a given cell. There are many other examples, some of which will be discussed below. From the standpoint of our concept foundations, we need to realize that the durability approach developed here is dependent on precise and definitive experimental data, for the definition of the degradation mechanisms and failure modes that are to be analyzed and modeled. Moreover, since we wish to use mechanistic modeling to calculate performance metrics based on inputs as well as the details of how a cell (or stack or system) is manufactured, we need to have specific material properties, as a function of operating time and conditions. This is a tall order, and compromises in that requirement define many of the frontiers in this methodology. They also define needed research and development. However, modeling can be done at a variety of levels, so the approach can be executed with levels of success that mature with available material information. Nevertheless, there is a critical need for determining the properties and characteristics (including micro- or nanostructure) of the materials in a fuel cell after long-term use, especially at the end of life. This information has proved to be critical to durability and life prediction for other high temperature systems [2].

The advantages of the present approach are well established [1]. First, if material-specific properties and their changes in fuel cell environments are determined, the performance of fuel cells of all possible designs can be estimated with practical utility, greatly reducing the cost of development and enabling optimization and trade studies without the cost of trial and failure. Secondly, accelerated characterization is enabled since mechanistic models for known failure modes can be used to assess the effect of changes in input (operating) conditions. And thirdly, the probabilistic chance of failure of specific fuel cells operated with a specific history of operating loads and conditions can be estimated if those operating conditions are used as inputs to a simulation of performance based on the models. This "stimulated simulation" results in orders of magnitude improvement in the accuracy of predicted life for a specific fuel cell, which is a direct increase in reliability, compared to statistical estimation of norms and confidence intervals of fleet performance based on sample population behavior.

We will examine the specific nature of this approach, and demonstrate its utility with illustrations for two classes of failure modes, mechanical and electrochemical. Polymer-based fuel cell systems will be considered for the mechanical failure illustration, and ceramic fuel cell systems will be used to illustrate the electrochemical failure philosophy and methodology.

3.2 Strength-based Performance Metrics

To illustrate the principles introduced above, and to develop the associated methodology for those concepts, we will develop two discussions. The first will focus on strength-based performance metrics that are used when mechanical failure modes define durability and life. The second discussion, in the next section, will focus on performance metrics that are appropriate for "failure" defined by degradation of electrochemical performance below acceptable levels. While these failure modes are very different, it should be emphasized that the physical processes and degradation modes that precipitate those failure modes are not necessarily different in any fundamental way. Chemical, electrochemical, thermal, mechanical, and other physical processes act and interact to degrade performance of all types in the general case. "Failure" results when a particular combination of these degradation events causes performance to fall below an acceptable level for one or more required performance definitions. The fact that we will look at only two such definitions here relates only to the limited space available for the discussion, and to the limited data available to support our illustrations.

Strength-based performance metrics are widely discussed in the structural integrity literature. A brief outline of those discussions is presented below, followed by an example application for polymer-based fuel cells.

3.2.1 Failure Functions for Damage Accumulation

For mechanical behavior, the performance metric becomes a mechanical failure criterion, having the general form given below

$$Fa\left(\sigma_{ij} / X_{ij} \right) \tag{3.2}$$

where the stress components in the numerator of the fraction, σ_{ij}, match the direction and sense of the material strength parameters, X_{ij} in the denominator (the indices, ij, are not summed). This form is assumed to be a generalization of relationships such as classical criteria such as Von Mises or Tresca, in which the stress in a given direction is compared to the strength of the material in that direction. However, we will require the following interpretations of this expression:

1) The stress and strength components used in Fa will be "local," in the sense that they represent the state of stress and state of material that control the final material failure event that defines strength, with the definition that is appropriate for a given failure mode.
2) Fa will be defined for a specific failure mode, *i.e.* when we talk of Fa, we will be discussing the controlling failure mode. Each failure mode will have a different Fa, possibly a different form, and certainly a different value for a given set of applied conditions. Determination of Fa for each failure mode will require the solution of a separate and distinct mechanical stress/strain analysis.
3) The local stress state and material state, represented by σ_{ij} and X_{ij}, will be assumed to be functions of time, in some general sense, *i.e.* the local stress state and material state will be assumed to be functions of the history of the applied conditions and the response of the material. This is a substantial departure from the classical concept; in our case, changes in material state (as well as stress state) are considered to follow degradation as it occurs.

Indeed, we make the claim that all physical phenomena that result in degradation that is important for mechanical failure can be represented in terms of changes in stress state (the numerator of equation 3.2) or changes in material state (represented by the strength of the material in principal directions, X_{ij} (the denominator in equation 3.2). Microcracking, for example, changes material stiffness, which alters the stress state (especially in the region of cracking); viscoelastic behavior results in time-dependent stiffness changes resulting in creep or relaxation and time-dependent stress/strain state alterations; physical aging of polymer components results in changes in stiffness with resultant stress/strain state variations. Material state changes are induced by a myriad of processes, including diffusion (resulting in creep rupture, for example), chemical degradation (such as corrosion, demixing, and compound formation), or thermodynamic changes (including sintering, or other morphological changes).

To complete our durability methodology for mechanical failure modes, we also need to construct a method of following the time dependent changes that embraces the rate information defined by the physical phenomena involved in changing the stress state of the material state, in such a way that we can estimate the remaining strength, a measurable quantity that can be used to validate the methodology and to estimate life (*i.e.* when remaining strength drops to the level of applied conditions at an instant of time in the history of loading, failure -and therefore life - is predicted).

For this purpose, we will appeal to the concept of damage accumulation, as developed in the literature [1,2]. The concept is summarized in Figure 3.2. Reading from the top of the figure, a list of possible physical phenomena that cause changes in the failure function, $Fa(\sigma_{ij}/X_{ij})$ that we chose for the failure mode of interest are shown at the top of the figure. They might include degradation due to cyclic application of environments (fatigue), time-dependent strain under constant applied conditions (creep), failure under constant conditions (creep rupture), changes in

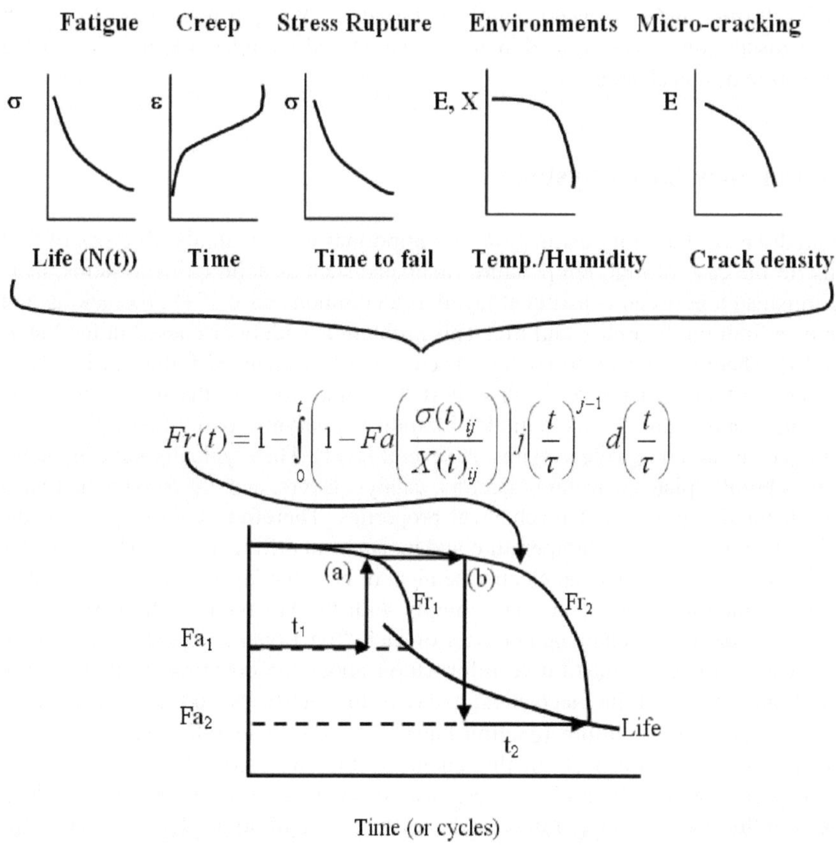

Figure 3.2. Schematic representation of a definition of durability based on a performance metric

stiffness (E) or material strength (X) with environments like temperature and moisture, or changes in stiffness with microdamage like cracking or crazing. All of these changes either alter the stress state, σ_{ij}, or the material state, X_{ij}, in the failure function. The failure function, $Fa(\sigma_{ij}/X_{ij})$, appears in the integral in the middle of Figure 3.2, and has the effect of reducing the calculated remaining strength, Fr, from the initial value (unit value for our normalized calculation). The only other unknown in the integral is τ, which is a characteristic time to failure if all conditions were constant, rather than changing with time; so τ can be measured for constant current, temperature, moisture, *etc.* The integral predicts the remaining strength, Fr, which is plotted on the curve below the integral. When the remaining strength reduces to the level of applied conditions, Fa, then mechanical failure of the fuel cell is predicted. If conditions are not constant (the normal situation when using a fuel cell), then Fa changes with time. One such change is shown in Figure 3.2. We see that we can still calculate life when Fa_1 changes to Fa_2 after a period of time, t_1 since the methodology assumes that the strength reduction due to Fa_1 is point (a) on Fr_1, which is the same as point (b) on Fr_2. So the remaining life must be t_2, and the total life must be $t_1 + t_2$. Note that this method is sensitive to the specific history and sequence of applied conditions. We will make use of this fact in our discussion of accelerated characterization and stimulated simulation in the last section of this chapter.

3.3 Polymer-based Systems

Strength-based durability estimations are important for essentially all types of fuel cells. In the case of high temperature fuel cells, such as solid oxide systems, thermal mismatch between constituent layers is a common cause of microcracking and can be a limit on durability and life. This problem is widely discussed in the literature [3]. There is comparatively less discussion of mechanical failure in low temperature systems. Therefore, in this short discourse, we will discuss some of the challenges and requirements to model durability in polymer-based fuel cells.

Fuel cells are composite systems, at several levels. They typically have elements such as bipolar plates, current collectors, catalyst layers, and electrolytes that have very distinctively different mechanical properties. Therefore, they respond to the fuel cell environments of temperature and moisture in different ways. However, for the fuel cell to function as an electrochemical device, the layers must act and interact as a functional composite, and comply with the behavior of the system as a unit. To model the mechanical behavior of such "functional composite systems," it is necessary to have constitutive information about the constituents under those conditions, and about the mechanical nature of the interfaces, so that we can calculate the inputs to the failure function $Fa(\sigma_{ij}/X_{ij})$ in the integral in Figure 3.2. We will discuss just two aspects of this challenge that are especially important. The first aspect is the highly nonlinear response of polymers, and the manner in which this is influenced by temperature and moisture or hydration. The second is the difficulty one faces when trying to define mechanical strength under such complex enviro-mechanical conditions.

For the purposes of illustration of material nonlinearity, we will concentrate on polymer membranes that are formed from perfluorosulfonate ionomers (PFSI); the data shown will be representative of commercial membrane materials such as Nafion™. Although there is a robust literature that discusses the effect of temperature and moisture on the mechanical response of polymers, PFSI materials are a special case, especially with respect to the effect of moisture. In the presence of water (or other polar solvents) the membrane swells and the sulfonic acid groups ionize, protonating the sorbed solvent molecules that are responsible for conducting the protons [4,5]. The water uptake is usually described in terms of the number of water molecules per acid site, λ. However, water uptake from pure liquid (λ values of the order of 22) is quite different from that from saturated vapor (λ values of the order of 14), a phenomenon known as Schroeder's paradox [6,7].

Gebel suggests that the morphology of the ionomer inverts from a distribution of ionic groups and their counterions in a matrix of perfluorinated hydrophobic phase at low water contents to a network of ionic domains connected with cylinders of water dispersed in the polymer matrix for intermediate hydration (λ values of the order of 14) [8]. However, for higher values of hydration associated with boiling of the membranes (λ values of the order of 24) he found evidence of a structure of polymer rods or rod-like particles. The stiffness and strength of polymer materials is known to be a strong function of such changes in the network structure [9]. Therefore, it is no surprise that the mechanical response of these materials is highly dependent on hydration, properly measured by a direct determination of λ. An example of a uniaxial loading response of such materials is shown in Figure 3.3. The tests were conducted at 25°C. These data represent the behavior of a Nafion™ material in a high hydration state (λ values from 14 ~ 22) and a low

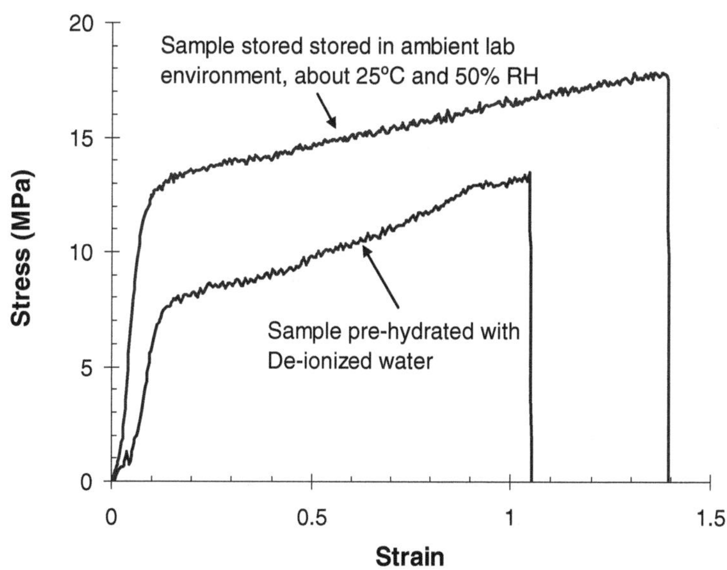

Figure 3.3. Typical stress-strain response of Nafion™ membrane material at high and low hydration levels

hydration state (λ values of the order of 3~4). Several aspects of the observed behavior are important to the present discussion.

First, the elastic stiffness of the membrane material is significantly lower for the hydrated material, as expected. And secondly, the behavior of the membrane material is highly nonlinear, with large strains to failure. In some commercial products, the behavior is also anisotropic, *i.e.*, the mechanical properties in the "machine direction" are different from those in a direction perpendicular to the machine direction associated with the manufacturing process. For nonlinear deformation, as observed in Figure 3.3, Ogihara and Reifsnider [10] introduced a three parameter generalization of the 2D single plasticity model originated by Sun and Chen [11] to construct a constitutive representation of the nonlinear quasi-static behavior of anisotropic materials. A quadratic yield function is assumed in the form

$$2f(\sigma_{ij}) = a_{ij}\sigma_{ij}^2 = k \tag{3.3}$$

where k is a state variable and the stresses are referred to the principal material directions, such as the 'machine direction' and 'transverse direction' identified in some membrane materials. The yield function is taken as the plastic potential function from which the incremental plastic strain can be derived in the usual way. Introducing an effective stress, $\sigma^* = \sqrt{3f}$, one can write

$$dW^p = \sigma_{ij}\, d\varepsilon_{ij}^p = 2f d\lambda = \sigma^* d\varepsilon^{*p} \Rightarrow d\varepsilon^{*p} = \frac{2}{3}\sigma^* d\lambda$$

$$\text{and } d\lambda = \frac{3}{2}\left(\frac{d\varepsilon^{*p}}{d\sigma^*}\right)\left(\frac{d\sigma^*}{\sigma^*}\right) \tag{3.4}$$

where ε^{*p} is the effective plastic strain and W^p is the plastic work per unit volume. For plane stress (in the x_1,x_2 plane) equation (3.1) reduces to four terms. For unidirectional loading, σ_x, in a direction that forms a positive angle θ with the principle material direction, x_1, the stress components in the material system are

$$\sigma_{11} = \sigma_x \cos^2(\vartheta)$$

$$\sigma_{22} = \sigma_x \sin^2(\vartheta)$$

$$\sigma_{12} = -\sigma_x \sin(\theta)\cos(\vartheta) \tag{3.5}$$

whereupon

$$\sigma^* = \sigma_x h(\vartheta) \text{ and } h(\vartheta) = \sqrt{\frac{3}{2}\left(a_{11}\cos^4\theta + a_{22}\sin^4\theta + 2(a_{12} + a_{66})\sin^2\theta\cos^2\theta\right)} \tag{3.6}$$

Ogihara introduced the requirement that $a_{11} = 1$, without loss of generality, resulting in the most general three parameter plasticity model for planar problems and uniaxial loading [10]. Then the plastic potential function reduces to

$$2f = \sigma_x^2\, h^2(\vartheta);\ \text{where } h(\vartheta) = \sqrt{\frac{3}{2}\left(\cos^4\theta + a_{22}\sin^4\theta + 2(a_{12} + a_{66})\sin^2\theta\cos^2\theta\right)} \tag{3.7}$$

Since the effective stress-strain relation should be a material property under monotonic loading, the material parameters in (3.6) must be chosen so that the $\sigma^*(\varepsilon^{*p})$ relations are independent of loading angle. If (3.6) is cast in the following form, it is possible to determine C_1 and C_2 from tensile tests on coupons cut from the membrane in different material directions (for anisotropic materials).

$$h(\theta) = \sqrt{3/2*(\cos(\theta)^4 + C_1 \sin(\theta)^4 + 2*C_2 \sin(\theta)^2 * \cos(\theta)^2)} \tag{3.8}$$

We can determine the parameters of the plastic potential function so that the effective stress, $\sigma \cdot h(\theta)$, vs. effective strain, $\varepsilon^n/h(\theta)$, data form a master curve. Hence, a generalized representation of the fully nonlinear response of the membrane material can be achieved in this manner. This representation can be used directly in mechanical analysis codes such as ANSYS and ABAQUS, which we are implementing in our laboratory in order to calculate the inputs to the remaining strength equation shown in Figure 3.2.

The second issue has to do with the concept of strength. The special microstructure that manifests itself in the mechanical stress-strain response of the membrane materials being considered also affects the strength of those materials in a variety of ways. If one accepts the concept of phase inversion and the consequential formation of a connected network of polymer rods suggested by Gebel [9], and regards that as a molecular composite, then the literature suggests that the polymer microstructure is an active participant in the development of deformation and failure modes that define strength [12].

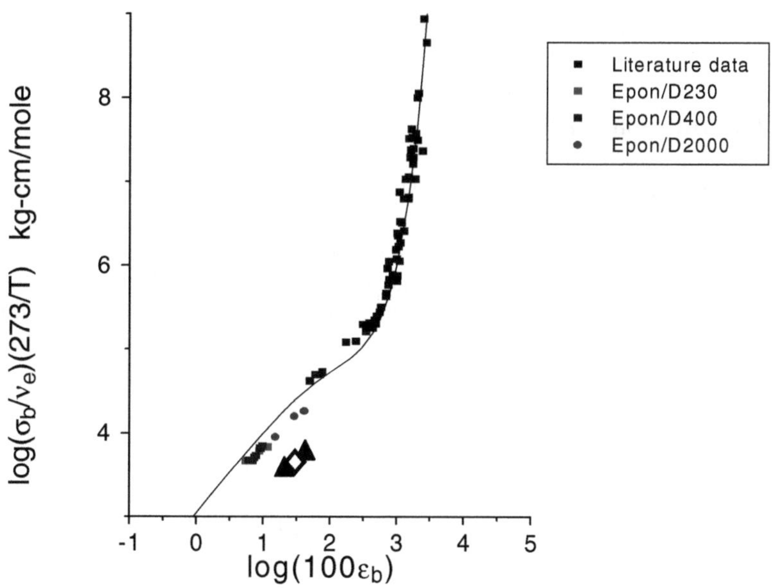

Figure 3.4. Ultimate tensile strength properties from literature data and Epon 828/Jeffamine network systems from reference [12] fitted to a modified Martin, Roth, Stiehler equation [9]

Figure 3.4 shows a well-known representation of the relationship between strain to break and stress to break for a wide range of polymer materials, as noted by Nielsen and Landel [12]. Inomers are sometimes referred to as "reversible crosslinkers." The data at the bottom of the curve in Figure 3.4 were obtained from polymers that we modified to alter crosslinking in a controlled way. As the figure shows, if the break strength is normalized by absolute temperature and the cross link density, and plotted against the strain to break, a very wide range of polymers and elastomers fall on a common curve. Further, we have reported in earlier work that all of the data on that curve, including the modified polymers that we created, can be represented by a modified MRS equation [9]. A version of that representation is quoted below.

$$\left(\frac{\sigma b}{\upsilon_e}\frac{273}{T}\right) = 3RT \cdot \left(\frac{\varepsilon_b}{\lambda_b}\right) \cdot \exp\left[A \cdot \left(\lambda_b - \frac{1}{\lambda_b}\right)\right] \tag{3.9}$$

In this equation, $\lambda_b = (1+\varepsilon)$ is the extension ratio, and ν_e is the effective cross link density, which reflects the influence of the internal structure. If stress and strain to break in equation (3.9) are grouped on one side of the equation, the material constants on the other side form a location parameter for a given material on the failure curve in Figure 3.4. That is equivalent to the statement of a failure criterion for the polymer material that includes a microstructure parameter and temperature as explicit variables. We do not have sufficient data to determine if such a representation is appropriate for the membrane materials discussed above. However, we offer the postulate that the strength of polymers with microstructure may be expected to be represented by a relationship such as equation (3.9).

3.4 Ceramic-based Systems

3.4.1 Electrochemical Performance Metrics

We have defined durability in terms of a performance metric that can be based on mechanical failure modes or on electrochemical modes, as explained in Section 3.1. We used polymer-based systems, operating at low temperatures and high hydration levels to discuss mechanical failure modes. For electrochemical failure modes, we will illustrate the concepts using solid oxide fuel cell (SOFC) examples.

Electrochemical metrics can be based on any measurable parameter that is a criterion for satisfactory performance of a fuel cell (or fuel cell system), such as voltage, current, power density, energy density, *etc.*, as mentioned earlier.

For the present illustration, we will select maximum energy density as the metric, *i.e.* we define "failure" of the cell or stack by the reduction of the maximum energy density available from the cell below an acceptable limit for the application at hand. Figure 3.5 illustrates the maximum available power definition. If the performance of the system degrades, we can track that degradation, and thereby the performance life of the system, by following changes in that maximum available power. In fact, when that available maximum falls below acceptable limits for our given application, we will call that event the 'life' of the fuel cell or fuel cell system.

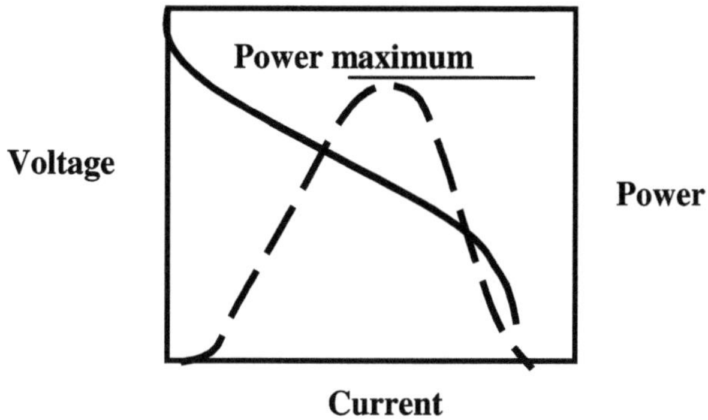

Figure 3.5. Schematic diagram of the maximum power for a given measured or predicted V-I curve

That concept is illustrated in Figure 3.6 below. We will assume that the net effect of degradation modes is the decrease of the maximum available power from the fuel cell. In Figure 3.6 we have normalized that power (with either volume or mass, and the area of the working electrodes) to obtain the maximum specific power density. For SOFCs, damage modes may include a wide variety of physical processes and phenomena. Of course, mechanical failure such as localized microcracking can reduce performance (and we could analyze that process with a mechanical metric, as above). But if the mechanical degradation modes do not cause mechanical failure, but only reduce electrochemical performance, then the appropriate metric is an electrochemical performance measure like the specific maximum power available. Ohmic resistance in all parts of the SOFC, especially in the electrolyte and the interfaces may be significant with time and history. Polarization losses in the cathode and anode may increase due to a variety of processes, including demixing, migration of impurities, changes in porosity, formation of insulating phases, *etc.* [13–15] The reduction in performance due to the collective effect of the action and interaction of these (and possibly other) degradation processes can

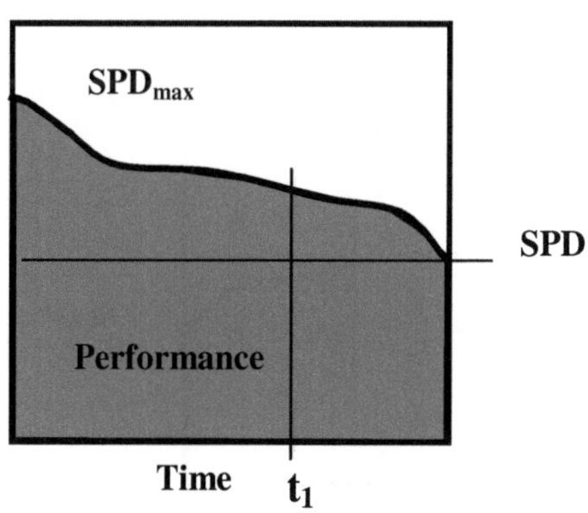

Figure 3.6. Concept of degradation defined by maximum specific power density (SPD_{max}) decrease

be measured by the maximum available power metric, and predicted if we have a thorough knowledge of the changes in the physical constants in the balance equations and constitutive laws that describe the operation of the fuel cell [15]. It is important, however, that the metric be a measurable quantity, so that predictive modeling of the effects can be validated.

Figure 3.7 illustrates an example of such an electrochemical metric, and how it is evaluated. The concepts are analytically similar to the mechanical case, and can be done with the same computational code. Only a few degradation modes are listed in this small space. Each of them alters the value of the kernel of the metric, the current operating condition, SPD(t). If microcracking, for example, increases Ohmic loss, then we may adjust operating conditions (increase our operating point on the V-I curve, for example) to maintain the same SPD(t), but the maximum available power from the cell will be reduced, as suggested by the fact that the integral reduces the initial (normalized) value. Other mechanisms enter the performance function, SPD(t) in similar ways, and produce their respective remaining

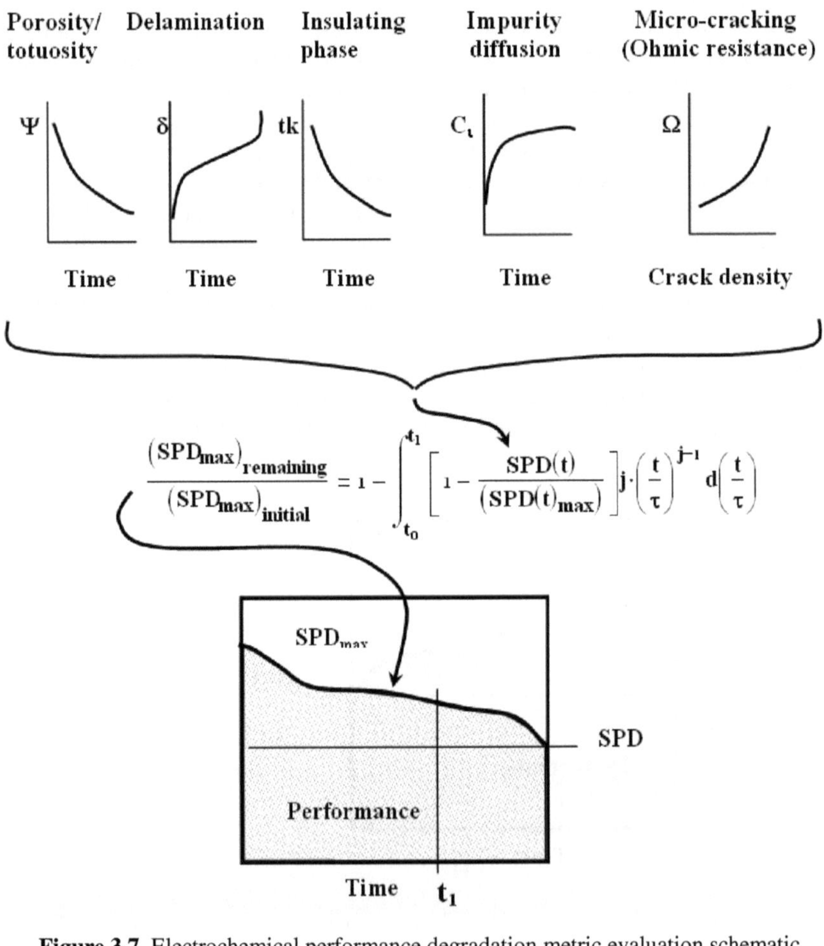

$$\frac{(SPD_{max})_{remaining}}{(SPD_{max})_{initial}} = 1 - \int_{t_0}^{t_1} \left[1 - \frac{SPD(t)}{(SPD(t)_{max})}\right] j \cdot \left(\frac{t}{\tau}\right)^{j-1} d\left(\frac{t}{\tau}\right)$$

Figure 3.7. Electrochemical performance degradation metric evaluation schematic performance reductions. The resulting SPD_{max} reduction is the curve shown earlier in Figure 3.6, and when it reaches the operating condition we need, and just falls below it, life is defined.

3.4.2 Applying the Electrochemical Model

How do we relate all of the degradation mechanisms to the performance function, SPD(t)? If we can predict the V-I curve for a cell or stack, in terms of the material constants that represent the way in which the fuel cell was made, then the changes in those constants (which we can in theory measure) define how the performance will change. Although all analytical representations of the voltage-current curve are not yet completely based on the local physics of the thermal, electrochemical, and mechanical processes that define performance, progress is being made in this venture [15].

Figure 3.8 shows an indication of predicted performance for a specific SOFC cell, which compared well with measured results [15]. The figure also shows the losses associated with various aspects of the internal processes that make the fuel cell work. The system of equations that was solved to determine the nature of those losses as a function of how the fuel cell was made, can also be used to predict the degradation of performance if changes in the material constants in these equations can be measured with time or operating history. Figure 3.9 shows the predicted dependence of current in the fuel cell on the distributed porosity of the electrodes (at a constant voltage).

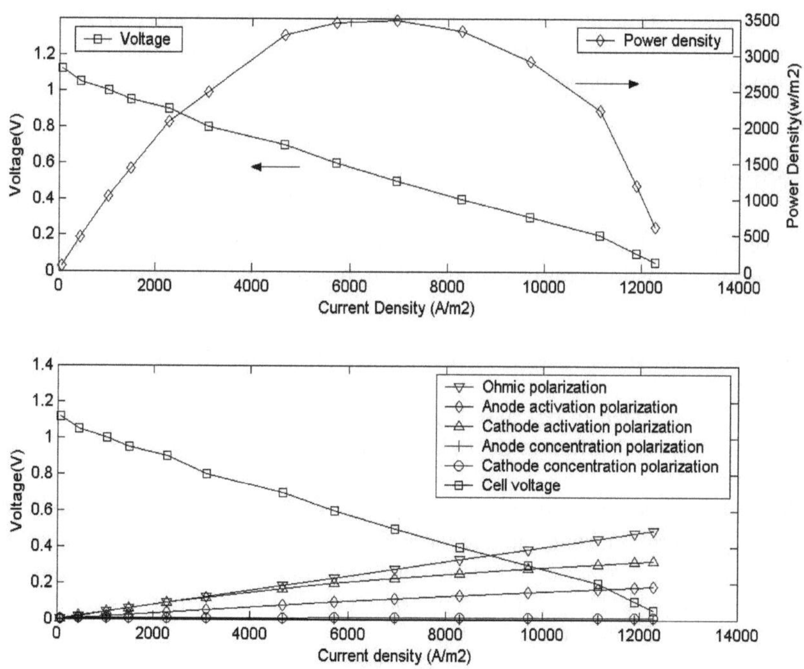

Figure 3.8. Predicted SOFC performance curves showing contributions to losses from several sources, [13]

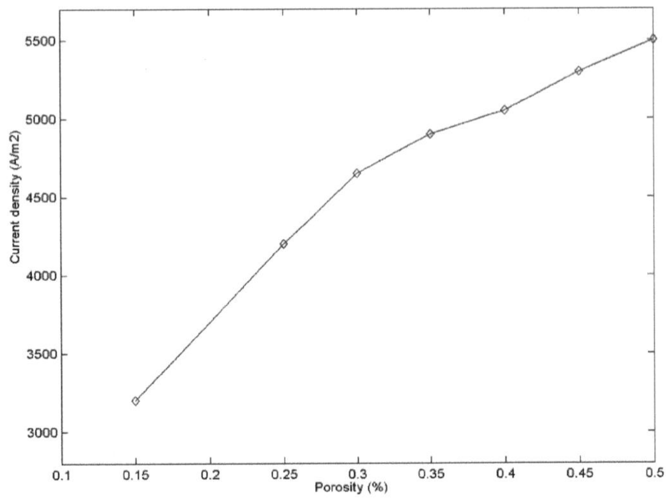

Figure 3.9. Current density vs. spatially uniform distributed porosity of the electrodes at the voltage= 0.7V

We were able to measure porosity changes precisely, in this case, but only by disassembling the fuel cell. Obviously, this is not a real-time health monitoring method at this point. Nevertheless, methods are under development to provide such real-time data so that the inputs to analysis representations can be updated in real time to create a "stimulated simulation" of the operation of the fuel cell or stack, or system. The present approach is a foundation for the prediction of durability of fuel cells and stacks that supports such development.

3.5 Summary

We have discussed an approach to durability for fuel cells and fuel cell systems that is substantially different from classical approaches. Salient features include the following:

- 'Failure' is defined as 'failure to perform' at an acceptable level, not as a discrete event
- 'Failure' is assumed to be caused, generally, by an accumulation of the consequences of several different degradation mechanisms, not by a single degradation event.
- The collective action and interaction of electrochemical, physical, micro- (or nano-) structural, mechanical, and thermal degradation processes are represented by changes in the material or kinetic constants in the balance equations and constitutive equations that predict the performance of a fuel cell or system in terms of how it is made and how it is operated.
- Experimental effort is focused on the determination of the changes in those material or kinetic constants as a function of local conditions.

The thrust of this philosophy is 'models assisted by experiments.' The motivation for this approach is the reduction of cost and acceleration of technology insertion. Fuel cells and fuel cell systems are complex and involve highly integrated multiple disciplines. Trial and error, or test-fail-fix approaches to durability are not feasible for such functional composite systems. Modern systems-based design and manufacturing methods that use the full power of modeling and computation are required if we are to bring the fundamental science and methodology of durability to the applications arena.

At the same time, the role of physical observations is even more basic and important. The discussion above is founded on the idea of a 'performance metric,' a measurable quantity like power, energy, voltage, strength, *etc.* This is quite different from equating durability to 'how long did it last.' Having a measurable performance metric means that it is possible to follow the degradation process, to validate models of performance, at all times.

And physical observations are even more basic to the determination of the degradation mechanisms and rates. Indeed, the present approach requires more science, and less phenomenology. If equations of the mechanics, electrochemistry, and multiphysics are to be properly set and solved to estimate performance based on how the fuel cells and systems are manufactured and used, they must be formulated on the basis of a strong foundation of understanding of the physical processes that make a fuel cell work, at the local and basic level. These understandings must be obtained from careful laboratory work to establish the science of the subject.

Nevertheless, the present approach shows promise as a proven methodology for system-based design based on science foundations.

3.6 References

1. Reifsnider, K.L. and Case, S.W., *Damage Tolerance and Durability of Material Systems,* Wiley, New York, 2003
2. *Materials Research to Meet 21st-Century Defense Needs,* National Materials Advisory Board, Division on Engoineering and Physical Sciences, The National Academies Press, 2003
3. Huang, X. and Reifsnider, K.L. "Thermal Stress and Long-Term Behavior of Layered Thin Film Composites: A Foundation for Solid Oxide Fuel Cell Technology," 16th Annual Conference, American Society for Composites, Blacksburg, Virginia, Sept. 9–12, 2001, CD ROM.
4. Grot, W., *Encyclopedia of Polymer Science and Engineering,"* Vol. 6, 2nd Ed., John Wiley, New York (1989)
5. Thampan, T., Malhotra, S., Tang, H., and Datta, R., J. Electrochem. Soc., 147, 3242 (2000)
6. Schroeder, P.V., Z. Phys. Chem., 45, 75 (1903)
7. Choi, P. and Datta, R., "Sorption in Proton-Exchange Membranes," J. Electrochem. S., 150, no. 12, 601–607 (2003)
8. Gebel, G., "Structural Evolution of Water Swollen Perfluorosulfonated Ionomers from Dry Membrane to Solution," Polymer 41, 5829–5838 (2000)
9. Shan, L., Verghese, K.N.E., Robertson, C. G., and Reifsnider, K.L., "Effect of Network Structure of Epoxy DGEBA-Poly(oxypropylene) diamines on Tensile Behavior," J. of Polymer Science: Part B: Polymer Physics, Vol. 37, 2815–2819 (1999)

10. Ogihara S., and Reifsnider K. L.. Characterization of Nonlinear Behavior in Woven Composite Laminates. Applied Composite Materials 9: 249–263 (2002.)
11. Sun, C.T. and J.L.Chen. "A Simple Flow Rule for Characterizing Nonlinear Behavior of Fiber Composites,"*J. Composite Materials*, Vol. 23, 1009–1020 (1989)
12. Nielsen, L.E., and Landel, R.F., *Mechanical Properties of Polymers and Composites*, Marcel Dekker, New York, 1994
13. Ju G., Reifsnider K., Huang X. and Du Y., *"Time Dependent Properties and Performance of a Tubular Solid Oxide Fuel Cell,"* J. of Fuel Cell Science and Technology, Vol. 1, 2004, pp. 1–9
14. An K., Halverson H., Reifsnider K., Case S.and McCord M., "Comparison of Methodologies for Determination of Fracture Strength of 8 mol % Yttria-Stabilized Zirconia Electrolyte Materials," J. Fuel Cell Science and Technology, May 2005, Vol. 2, 99–103
15. Reifsnider K., Huang X., Ju G., Feshler M.and An K., "Mechaics of Composite Materials in Fuel Cell Systems," Mechanics of Composite Materials, Vol. 41, No. 1, 2005, 3–16
16. Halverson, H. (1996), "Improving Fatigue Life Predictions: Theory and Experiment on Unidirectional and Cross-ply Polymer Matrix Composites," M.S. Thesis, Department of Engineering Science and Mechanics, College of Engineering, Virginia Polytechnic Institute and State University

4

Transport and Electrochemical Phenomena

F. Yang and R. Pitchumani

4.1 Introduction

Fuel cell technology, a core component in a hydrogen-based energy economy, has the advantages of high energy conversion efficiency, low pollution, and no dependency on depleting fossil resources. Significant progress has been made in the state-of-the-art in materials, design, and fabrication, however, the widespread commercialization of most fuel cells is still limited by issues such as high cost and low durability [1–4]. Mathematical models are effective tools in understanding and optimization of various transport and electrochemical processes, leading to cost reduction and improved performance and durability. The purpose of this chapter is to summarize the current status of fundamental models for fuel cells that have been actively studied in the past decade.

A typical fuel cell can be schematically represented by the layered structure in Figure 4.1, where an electrolyte is placed between an anode and a cathode backing layer, or gas diffusion layer (GDL). A thin catalyst layer exists between the anode (or cathode) GDL and the electrolyte, referred to as the anode (or cathode) catalyst layer. The anode-electrolyte-cathode assembly is clamped between two bipolar plates, which house the flow channels for fuel and oxidant feed. Based on the charge carriers in the electrolytes, *i.e.*, proton (H^+), hydroxide (OH^-), carbonate (CO_3^{2-}), and oxide (O^{2-}), fuel cells may be classified into four categories: acid fuel cell, alkaline fuel cell (AFC), molten carbonate fuel cell (MCFC), and solid oxide fuel cell (SOFC), respectively. The acid fuel cells may be further divided into proton exchange membrane fuel cell (PEMFC), direct methanol fuel cell (DMFC), and phosphoric acid fuel cell (PAFC).

The transport and electrochemical processes of a PEM fuel cell involve the following: The hydrogen fuel is fed through the anode flow channel, and is distributed to the catalyst layer via the anode gas diffusion layer. Hydrogen molecules are oxidized in the anode catalyst layer to produce protons and electrons, which, in turn, are transported to the cathode through the membrane and an external circuit, respectively. In the cathode catalyst layer, the oxidant molecules (transported from the cathode flow channel and gas diffusion layer) combine with the protons and the electrons to produce water. The processes in other types of fuel cells may be similarly discussed, and

Table 4.1 presents a summary of the various half-cell reactions [3]. This Chapter covers the modeling studies on three types of fuel cells, namely, the proton exchange membrane fuel cell, solid oxide fuel cell, and the direct methanol fuel cell.

During the operation of fuel cells, coupled transport and electrochemical processes occur simultaneously on multiple length scales, ranging from microscopic-level in the cell components (cell-level) to the stack-level, and finally to the system-level. Modeling approaches at the microscopic/cell levels form the focus of this chapter, which is organized as follows. Sections 4.2–4.4 review the prominent studies on PEMFCs, DMFCs, and SOFCs, respectively. Section 4.5 discusses the use of computational models in addressing application considerations of optimization and process design under uncertainty. The Chapter concludes with a summary of topics for future work pertaining to the three types of fuel cells.

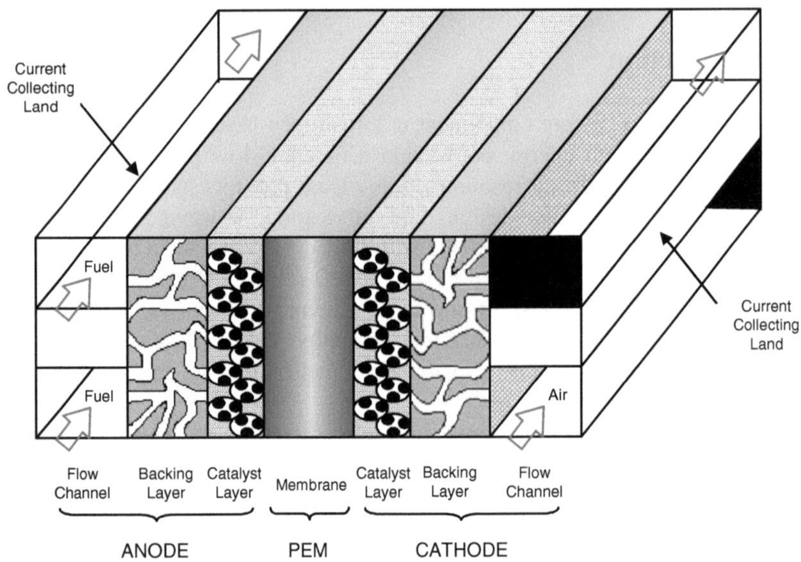

Figure 4.1. Three-dimensional schematic illustration of a PEM fuel cell (redrawn from [5])

4.2 Modeling of Proton Exchange Membrane Fuel Cells

In a PEMFC, the electrolyte is a perfluorosulfonic acid membrane with higher proton conductivity than any conventionally strong acid (*e.g.*, H_2SO_4, HCl, and $HClO_4$). The backbone structure of the membrane polymer is the polytetrafluoroethylene (PTFE), also known as the strongly hydrophobic Teflon. Side chains ending with the sulphonic acid (HSO_3) are tethered to the backbone polymer, forming the Nafion type membrane materials. Since the HSO_3 group is ionically bonded, the side chains tend to cluster within the overall structure through the strong attraction between the H^+ and SO_3^- ions. The sulphonic acid is highly hydrophyllic; consequently, the Nafion materials consist of a generally hydrophobic bulk with hydrophyllic regions (*i.e.* sulphonic sites). With adequate adsorption of water molecules, acid solution forms within the sulphonic sites, and protons may be conducted through the network of hydrated regions. The catalyst layer is a porous mixture of the electrocatalyst and the proton-conducting membrane, where fine particles of platinum are dispersed on carbon supports to maximize the reaction area. Porous carbon cloth or paper are the materials for the electrodes, which may further contain Teflon to prevent water flooding. The bipolar plates in the cell stack are usually made of graphite, while stainless steel or titanium may also be used.

Table 4.1. Electrochemical reactions in the different types of fuel cells [3]

Type	Anode reaction	Cathode reaction	Conducting ion
AFC	$H_2 + 2OH^- \rightarrow 2H_2O + 2e^-$	$1/2O_2 + H_2O + 2e^- \rightarrow 2OH^-$	OH^-
PEMFC	$H_2 \rightarrow 2H^+ + 2e^-$	$1/2O_2 + 2H^+ + 2e^- \rightarrow H_2O$	H^+
DMFC	$CH_3OH + H_2O \rightarrow CO_2 + 6H^+ + 6e^-$	$1/2O_2 + 2H^+ + 2e^- \rightarrow H_2O$	H^+
PAFC	$H_2 \rightarrow 2H^+ + 2e^-$	$1/2O_2 + 2H^+ + 2e^- \rightarrow H_2O$	H^+
MCFC	$H_2 + CO_3^{2-} \rightarrow 2H_2O + CO_2 + 2e^-$	$1/2O_2 + CO_2 + 2e^- \rightarrow CO_3^{2-}$	CO_3^{2-}
SOFC	$H_2 + O^{2-} \rightarrow H_2O + 2e^-$	$1/2O_2 + 2e^- \rightarrow O^{2-}$	O^{2-}

Several technical challenges for the application of PEM fuel cells include (i) managing water and heat in the cells, (ii) finding alternative low-cost materials and manufacturing techniques for the cell components, and (iii) minimizing CO poisoning at the anode. To address the development issues, modeling investigations are powerful and affordable alternative to experimental studies and iterative trial-and-error development. During the last decade, the number of theoretical models on PEM fuel cell has increased significantly with progressively increasing complexity and scope. Owing to easier access to faster computers, full three-dimensional and two phase flow models are becoming common practice. In this section, the major modeling approaches in the literature are summarized, starting with the simplest performance models. Built upon the basic concepts obtained from the performance models, detailed mechanistic models on the single cell and microscopic levels are discussed. Finally, significant findings on the operating mechanisms in fuel cells from the mathematical models are presented, which provide useful information on cell optimization and design under uncertainty.

4.2.1 Performance Models

The performance models characterize the electrical performance of a cell by using a single equation, *i.e.* cell voltage versus current density. The models are useful in determining kinetic parameters and general ohmic resistance from experimental data [6–18]. However, the drawbacks of the performance models include the lack of fundamental description of mechanistic behavior and detailed inner-cell operating conditions, which are critical for accurate prediction of cell performance or optimization. Accounting for various voltage losses, or overpotentials, involved in the operating of a fuel cell, a typical expression of the cell voltage, E_{cell}, may be written in terms of the current density, i, as [1]:

$$E_{cell} = E - (i + i_n)r - A\ln\left(\frac{i + i_n}{i_0}\right) + B\ln\left(1 - \frac{i + i_n}{i_l}\right) \tag{4.1}$$

where E is the reversible open circuit voltage, r is the total Ohmic resistance of the fuel cell components, A represents the slope of a Tafel plot (*i.e.*, a curve of voltage loss versus log of current density), B is a constant in the mass transfer overpotential, and i_n, i_0, and i_l denote the equivalent internal current density, the exchange current density, and the limiting current density, respectively. The physical interpretation of the various terms in Eq. (4.1) is discussed as follows.

From the thermodynamic consideration that the electrical work done by the fuel cell equals the Gibbs free energy released by the electrochemical reactions, the reversible open circuit voltage, E, is given by the Nernst equation (1)

$$E = E^0 + \frac{RT}{2F}\ln\left(\frac{a_{H_2} \cdot a_{O_2}}{a_w}\right) \tag{4.2}$$

where E^0 is the electromotive force (emf) at the standard pressure (*i.e.* 1 bar), R is the universal gas constant, T is the temperature, F is the Faraday constant, and a_{H_2}, a_{O_2}, and a_w are the activities of the hydrogen, oxygen, and water species, respectively. Note that for ideal gases, the activity is defined as $a = P/P^0$, where P is the partial pressure, and P^0 is the standard pressure (*i.e.* 1 bar). The Ohmic resistance r consists of the electrical resistance in the electrodes and flow channels, the ionic resistance in the membrane, and the contact resistance between the cell components.

An ideal polymer membrane is an ion conductor with high conductivity for the protons only, however, parasitic transport of other species is always present in a practical fuel cell. *Fuel crossover* refers to the case in which hydrogen molecules diffuse across the membrane from the anode to the cathode, and react with the oxygen at the cathode catalyst sites without generating useful current. Furthermore, small amount of electrons may be directly conducted from the anode through the membrane to the cathode, known as the *internal current*. Both effects – fuel crossover and internal current – are equivalent in the sense that electrons are wasted inside the cell without going through the external circuit. Consequently, an equivalent internal current, i_n, may be defined to account for the two parasitic transports. Note that i_n is added to the external current density i for the last three terms in Eq. (4.1), since the internal current density contributes to the voltage losses in a similar way as the external current density.

The third term on the right hand side of Eq. (4.1) represents the activation losses, which are caused by the *charge double layer* schematically shown in Figure 4.2. The rates of half-cell electrochemical reactions depend on the density of the charges at the interface between the proton conducting electrolyte and the electron conducting electrode. The accumulation of the electrons and proton ions leads to the generation of an electrical voltage at the electrode/electrolyte interface. To

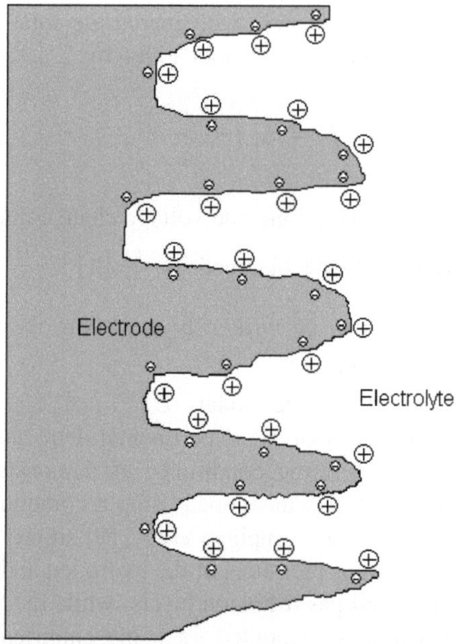

Figure 4.2 The charge double layer near the electrode/electrolyte interface (redrawn from [1])

complete the electrochemical reactions in the first line of Table 4.1, electric work must be done for the charges to overcome the interfacial voltage, leading to the activation overpotentials. Tafel observed in 1905 that the activation overpotentials are proportional to the log of the current density. Due to the fact that the half-cell reactions are reversible in nature, the electrons continuously flow back and forth across the electrode/electrolyte interface. Consequently, an exchange current density, i_0, is present even for the case of an open electrical circuit (*i.e.*, $i = 0$).

Owing to the flow resistances and the consumption of reactants during the cell operation, the concentration (or partial pressure) of the hydrogen and the oxygen may be significantly reduced, resulting in the concentration voltage losses. The last term in Eq. (4.1) denotes the concentration overpotential, and may be *qualitatively* derived based on the following steps. Considering the voltage change, ΔV, caused by the pressure change of the hydrogen from P_1 to P_2, and assuming ideal gas behavior and constant concentration of other species, the Nernst equation, Eq. (4.2), yields:

$$\Delta V = \frac{RT}{2F} \ln\left(\frac{P_2}{P_1}\right)$$

Define a limiting current density i_l at which the hydrogen fuel is completely consumed, *i.e.*, the partial pressure of H_2 in the anode catalyst layer is zero. Let P_1 in the above equation represent the pressure when the total current density (defined as the sum of the external and internal current densities, $i + i_n$) is zero, and assume the pressure decreases *linearly* with increasing total current density; the pressure P_2 at any total current density is then given by:

$$P_2 = P_1\left(1 - \frac{i + i_n}{i_l}\right)$$

Combining the above two equations, the voltage change due to the reduction of hydration partial pressure may be obtained as $\frac{RT}{2F}\ln\left(1 - \frac{i+i_n}{i_l}\right)$, and the general concentration overpotential may be written by replacing the fraction $\frac{RT}{2F}$ with a constant B, as seen in Eq. (4.1).

The parameters in the performance models (*e.g.* r, A, and i_l) are typically determined from a least squares fit of the experimental data, and the fitting may be very good under different operating conditions, as demonstrated in Figure. 4.3. More complicated models are also developed using a combination of mechanistic and empirical techniques [19–24]. Amphlett *et al.* [19] solved the Stefan-Maxwell equation to determine the partial pressures of the hydrogen and oxygen at the interfaces between the catalyst and gas diffusion layers, while the temperature dependence of the activation losses is accounted for using empirical fitting parameters. Pisani *et al.* [20] and Kulikovsky [21] have adopted a semi-empirical approach for the polarization behavior, where the analytical form of the limiting current density

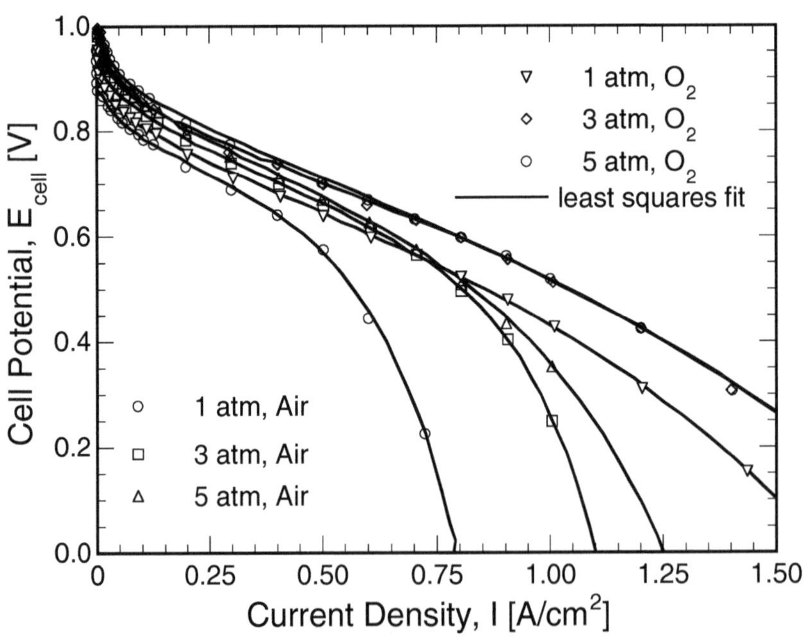

Figure 4.3. Empirical model and experimental comparison of polarization curves for air or oxygen at different gas ressures and at 70°C (redrawn from [10])

is determined by solving the mass diffusion equations. Analytical solutions for the polarization performance are also reported by Gurau *et al.* [22] and Standaert *et al.* [23, 24]. The performance models are useful for gross understanding of the inter-cell processes, while in-depth examination of the underlying phenomena is needed for the purpose of design and optimization. Thus, mechanistic models for the detailed inter-cell operating conditions form the focus of the discussion in the remainder of the section.

4.2.2 Mechanistic Modeling of PEM Fuel Cell

As explained previously, the transport and electrochemical processes occur throughout various layers in Figure 4.1. To simulate the coupled phenomena, five types of relations must be solved simultaneously: (i) the conservation equations, (ii) constitutive relations for various fluxes, (iii) kinetic equations for reactions, (iv) equilibrium relationships, and (v) auxiliary relations such as variable definitions and Faraday's law [25]. The conservation equations are applicable for all the layers in the fuel cell, and are discussed in the Section 4.2.2.1. The other four types of relations are presented specifically for the membrane, gas diffusion layers, and the catalyst layers in Sections 4.2.2.2–4.2.2.4, respectively. Significant findings from the modeling studies are discussed in Section 4.2.2.5.

4.2.2.1 Conservation Equations

A fundamental description of fuel cell operation involves the five conservation principles of mass, momentum, species, electrical charge, and thermal energy. With a unified-domain approach, a set of governing equations valid for all the fuel cell layers may be written in the vector form as [26–29]:

mass:
$$\frac{\partial(\varepsilon\rho)}{\partial t}+\nabla\cdot(\rho\mathbf{u})=S_m \tag{4.3}$$

momentum:
$$\frac{1}{\varepsilon}\left[\frac{\partial(\rho\mathbf{u})}{\partial t}+\frac{1}{\varepsilon}\nabla\cdot(\rho\mathbf{u}u)\right]=-\nabla p+\nabla\cdot\tau+S_u \tag{4.4}$$

species:
$$\frac{\partial(\varepsilon C_k)}{\partial t}+\nabla\cdot(\mathbf{u}C_k)=\nabla\cdot(D_k^{eff}\nabla C_k)+S_k \tag{4.5}$$

charge:
$$\nabla\cdot(\kappa^{eff}\nabla\Phi_e)+j=0 \tag{4.6}$$

$$\nabla\cdot(\sigma^{eff}\nabla\Phi_s)-j=0 \tag{4.7}$$

energy:
$$\frac{\partial[(\rho c_p)_m T]}{\partial t}+\nabla\cdot(\rho c_p\mathbf{u}T)=\nabla\cdot(k^{eff}\nabla T)+S_T \tag{4.8}$$

The charge transport equations, Eqs. (4.6) and (4.7), involve the volumetric transfer current, *j*, as a source term. Depending on the characteristic of the half-cell reactions, the general Butler-Volmer equation, Eq. (4.13), may be simplified for the anode and cathode catalyst layers as [22, 26, 30]:

anode: $$j_a = A_{av} i_{0,a}^{ref} \left(\frac{C_{H_2}}{C_{H_{2,ref}}} \right)^{1/2} \left(\frac{\alpha_a + \alpha_c}{RT} F \eta_a \right) \quad (4.15)$$

cathode: $$j_c = A_{av} i_{0,c}^{ref} \left(\frac{C_{O_2}}{C_{O_{2,ref}}} \right) \exp\left(-\frac{\alpha_c}{RT} F \eta_c \right) \quad (4.16)$$

where the sub/superscripts a, c, and ref, denote the anode, cathode, and reference state, respectively. Since the hydrogen oxidation reaction (HOR) in the anode catalyst layer is quick and the overpotential η_a is typically small, j_a in Eq. (4.15) is approximately proportional to η_a. For a PEM fuel cell operating on reformate feed at the anode, the electrochemical reactions in the anode catalyst layer involve the adsorption process of CO, leading to decreased active area A_{av} at the anode [31, 32]. The oxygen reduction reaction (ORR) is typically slow with high η_c, and the expression for j_c, Eq. (4.16), may be obtained by neglecting the anodic reaction term of Eq. (4.13). Note that the ORR is treated as a first-order reaction based on the experimental work of Bernardi and Verbrugge [33] and Gottesfeld and Zawodzinski [34].

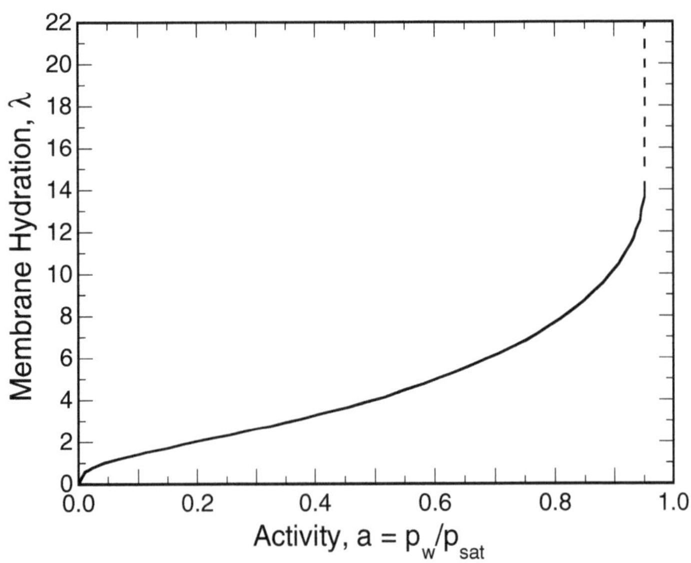

Figure 4.4. Equilibrium water-uptake or isotherm curve at 30°C (redrawn from [25]). The dashed line denotes the effect of Schroeder's paradox: when the membrane is equilibrated in saturated vapor or liquid water (both of which correspond to $a = \mathbf{1}$), values of λ are 14 or 22, respectively

In the energy equation, Eq. (4.8), k^{eff} is the effective thermal conductivity, and the heat capacitance in a porous material, $(\rho c_p)_m$, is a volume-averaged volumetric specific heat capacity over the solid matrix and the fluid in the micropores:

$$(\rho c_p)_m = \varepsilon(\rho c_p) + (1-\varepsilon)(\rho c_p)_s \tag{4.17}$$

where the subscript s refers to the solid material and (ρc_p) pertains to the fluid. The source term S_T in Eq. (4.8) consists of contributions from three mechanisms, *i.e.*, irreversible heat from the electrochemical reaction, reversible or entropic heat, and Joule heating. Detailed discussion on the simulation results of the energy equation and other conservation equations are presented in Section 4.2.2.5.

To solve the conservation equations discussed so far, two more issues must be carefully examined, *i.e.* [1] the material property characterization and [2] computational techniques, which form the focus of the presentation in Sections 4.2.2.1.1 and 4.2.2.1.2, respectively. It must be pointed out that some material properties reported in the literature have a large scatter [27], and more accurate experimental measurements are needed in future work.

4.2.2.1.1 Material Properties

The material properties affecting the transport processes in a PEM fuel cell are characterized by experimental measurements or empirical correlations. In general, four types of properties can be identified: (i) transport properties of the membrane, (ii) kinetic data for electrochemical reactions, (iii) effective parameters for porous materials, and (iv) properties of the reactants and products.

The proton conductivity of the membrane is strongly dependent on the water content (or hydration λ), which is defined as the ratio of the number of water molecules to the sulfonic charge sites. Several studies [35–37] measured the water content of membrane exposed to water vapor at various concentration or immersed in liquid water, and the results are shown as the water uptake curve in Figure 4.4. As physically expected, the membrane hydration increases monotonically with water activity, a, which is defined as the ratio of partial pressure of water to the saturation pressure, $a = \frac{p_w}{p_{sat}}$. When the membrane is equilibrated in saturated vapor or liquid water (both of which correspond to $a = 1$), different values of λ are observed. The discontinuity of the hydration value in the membrane when $a = 1$ is called "Schröeder's paradox" in the literature [25, 26]. An empirical correlation for the water uptake behavior of Nafion 117 is commonly used with an allowance for the activity to exceed unity [35]:

$$\begin{aligned} \lambda &= 0.043 + 17.81a - 39.85a^2 + 36.0a^3, \ 0 < a \le 1 \\ &= 14. + 1.4(a-1), 1 \le a \le 3 \end{aligned} \tag{4.18}$$

Note that $a > 1$ may be interpreted as the supersaturated condition.

Three membrane properties critical to proton and water transport are: the proton conductivity, κ, the water diffusion coefficient, D_w, and the electro-osmotic drag coefficient, n_d. Springer *et al.* [35] presented a correlation for κ in $[\Omega\text{-}cm]^{-1}$ as a function of temperature and hydration for Nafion 117 as:

$$\kappa = \exp\left[1268\left(\frac{1}{303}-\frac{1}{T}\right)\right](0.005139\lambda - 0.00326) \tag{4.19}$$

Since the water content varies spatially, the proton conductivity has a corresponding variation in the membrane. The distribution of water concentration depends on the transport parameters D_w and n_d, which are also extensively studied in the literature [28, 35–37]. Springer *et al.* [35] developed an empirical expression for the water diffusion coefficient in $[cm^2/s]$ as:

$$D_w = 10^{-6}\times\exp\left[2416\left(\frac{1}{303}-\frac{1}{T}\right)\right]\times(2.563-0.33\lambda+0.0264\lambda^2-0.000671\lambda^3) \tag{4.20}$$

In the membrane, protons travel in a complex form, $H^+(H_2O)_n$, and the number of water molecules dragged by a proton, n_d, depends on the water content. Springer *et al.* [35] found that $n_d = 2.5$ for a fully hydrated membrane in equilibrium with liquid water at 30° C or 50° C. By assuming linear variation with water content, the electro-osmotic drag coefficient is written as:

$$n_d = 2.5\frac{\lambda}{22} \tag{4.21}$$

A quadratic expression for n_d was adopted by Dutta *et al* [38, 39], with $n_d = 0.0029\lambda^2 + 0.05\lambda - 3.4\times10^{-19}$.

Important parameters for the electrochemical reactions include the electrochemically active area per volume, A_{av}, in the catalyst layer, the exchange current density, and the charge transfer coefficient. The specific active area is proportional to the platinum loading and inversely proportional to the catalyst layer thickness. Furthermore, A_{av} may significantly change with the fabrication methods of the catalyst layer. The exchange current density for the anode is much higher than that of the cathode, and consequently, the anode activation polarization is negligible as compared to that at the cathode. Parthasarathy *et al.* [7] determined the temperature dependence of the cathode exchange current density, $i_{0,c}$, as:

$$i_{0,c} = 7.33\times10^{-10}\exp\left[8999\left(\frac{1}{313}-\frac{1}{T}\right)\right] \tag{4.22}$$

The exchange current density and charge transfer coefficient for the oxygen reduction reaction are well documented in refs. [15, 34, 40, 41].

The effective thermal conductivity, electrical conductivity and mass diffusion in porous layers may be approximated by a volume-fraction-weighted harmonic averaging method [28]. For example, the effective thermal conductivity, k_{eff}, for the catalyst layer can be written as:

$$1/k_{eff} = \varepsilon_e/k_e + \varepsilon_s/k_s + \varepsilon_w/k_w \tag{4.23}$$

where the conductivity is determined by the volume fractions of the components in the layer, namely, the electrolyte (e), solid (s), and water (w). It must be mentioned that the compression of the porous layer may change the pore sizes and distribution, which, in turn, affect the transport properties, as discussed in refs. [26, 42]. Furthermore, the effects of the compression on properties are critical input information for the study of porous structure-flow interactions [26]. It must be pointed out that more accurate evaluations of effective properties such as k_{eff} are available in the composite literature, where generalized unit cells and statistical analyses are employed [43, 44]. In general, the literature lacks experimental data on the transport properties for deformed porous media. A review of the porous medium properties for the PEM fuel cell system was given by Mathias *et al.* [45].

The binary diffusion coefficients for gas species i and j, D_{ij}, may be determined with the empirical correlation of Fuller, Schettler, and Giddings [46]:

$$D_{ij} = 10^{-3}[(MW_i + MW_j)/MW_iMW_j]^{0.5} \frac{T^{1.75}}{p[(\sum v)_i^{1/3} + (\sum v)_j^{1/3}]^2} \tag{4.24}$$

where MW, T, p, and $(\sum v)$ are the molecular weight, temperature, total pressure, and molecular diffusion volume, respectively. Note that when the gas species are dissolved in the membrane or water phases, diffusivity values may be a few orders of magnitude lower than those in the gaseous phases. The thermal conductivity of an n-component gaseous mixture, k_{mix}, may be computed by the Wassiljewa equation [46]:

$$k_{mix} = \sum_{i=1}^{n} \left(\frac{x_i k_i}{\sum_{j=1}^{n} x_j A_{ij}} \right) \tag{4.25}$$

where x_i and k_i are the mole fraction and thermal conductivity for a component i; the coefficient A_{ij} is a function of temperature determined by a modification of the Sutherland model [46].

4.2.2.1.2 Computational Techniques

All the conservation equations, Eqs. [4.3– 4.8], may be written in a common form as:

$$\frac{\partial(\rho\phi)}{\partial t} + \nabla \cdot (\rho \mathbf{u} \phi) = \nabla \cdot (\Gamma \nabla \phi) + S \tag{4.26}$$

where ϕ is a general dependent variable, Γ is an appropriate diffusion coefficient, and S is a source term that may include effects such as the drag on fluid by porous electrodes and the electro-osmotic drag on water molecules in the membrane. Numerical solution of Eq. (4.26) with advection, diffusion, and source terms is discussed by Patankar [47]. Since the geometry of fuel cells has an aspect ratio (*e.g.* flow channel length divided by cell thickness) on the order of 100, the upwind scheme is effective for the flow-field solution [26]. A particular problem in the solution of governing equations for fuel cells is the strong coupling between the electrical and membrane potentials, Φ_s and Φ_e. For one-dimensional problems, it is found that an efficient iterative algorithm is the Newton's method, which can be used to solve the discretized difference equations *simultaneously* [48, 49]. However, the inversion of the Jacobian matrix involved in Newton's method may be computationally expensive for two- or three-dimensional problems. A set of advanced iterative numerical algorithms, including the generalized minimal residual subroutine (GMRES), was presented by Wu *et al.* [49] to handle large, nonsymmetric Jacobian systems. Furthermore, when using commercial computational fluid dynamics package, the two potential equations must be appropriately linearized [50].

Fully three-dimensional simulations for a fuel cell with a typical bipolar plate geometry shown in Figure 4.5 demand a large mesh [about 2–6 million grid points] [26]. The majority of numerical studies use no more than a few hundred thousand grid points. To address the computational challenge, Meng and Wang [51] adopted a parallel computational technique. The computational domain is divided into a number of subdomains, and each is assigned to one processor with proper load balancing.

4.2.2.2 Membrane Modeling

The discussion presented so far pertains to the simulation of an entire fuel cell, and the modeling studies for individual cell components are covered in the remainder of Section 4.2.2. One of the most important components of a PEM fuel cell is the electrolyte, which serves to effectively separate the anode fuel from the cathode oxidant and to conduct protons at high rates during cell operation. Various types of electrolytes have been examined experimentally, however, the modeling studies focus primarily on the Nafion membrane. Owing to the fact that the governing equations for Nafion are generally valid for other types of membranes (only with different property values), the following discussion is limited to the Nafion.

4.2.2.2.1 Microscopic and Physical Models

Due to the key role of the membrane in a fuel cell, extensive studies have been reported in the literature from microscopic and physical models to macroscopic descriptions [52–94]. Fundamental understanding of the transport processes such as diffusion and conduction in the membrane on a single-pore level has been provided by the microscopic models based on *ab initio* quantum chemistry [52–58], molecular dynamics [59, 60], and nonequilibrium statistical mechanics [61, 62]. Quantum chemistry applies the principles of quantum mechanics to calculate the

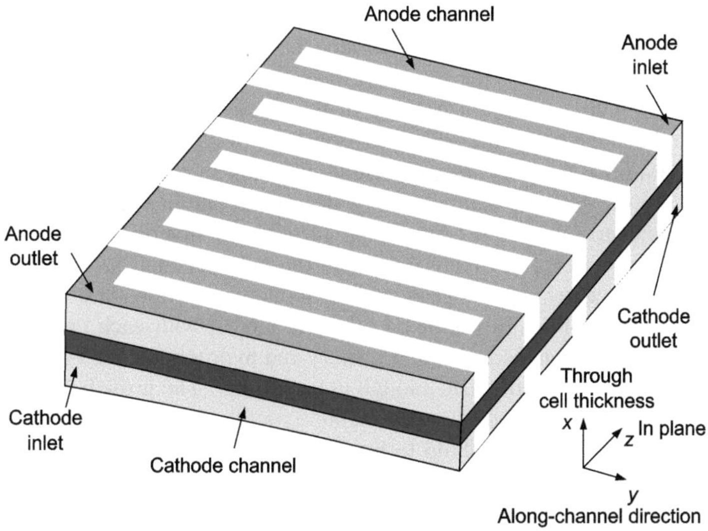

Figure 4.5. Typical configuration of a single PEM fuel cell with multiple flow channels (redrawn from [26])

stationary and transitional states of molecules through explicit treatment of the electronic structure [63, 64]. The modeling provides information on local structure, aggregation of the sulfonic groups through the hydrogen bonds, and the proton dissociation in the fragments of polymers without any assumptions on the bonding of a system. Owing to excessive computational requirements, the system size of the quantum chemistry modeling is typically limited to less than 100 atoms.

Molecular dynamics simulations solve Newton's equation of motion with empirical potentials for a typical system consisting of thousands of particles over time periods of nanoseconds [65–69]. In addition to the determination of the mass and charge transport properties as functions of the temperature, water content, and chemical and physical characteristics of the polymer chains, the calculations also provide new insight into molecular level mechanisms. Statistical mechanical analysis [70–72] has been applied to the diffusion of protons in a single hydrated pore in polymer membranes with several assumptions on the pore geometry, the distribution of the fixed charge groups, and the vehicular mechanism of proton transport through the center of the pore. With membrane-specific morphology input information from small-angle X-ray scattering (SAXS) experiments and electronic structure information from quantum chemistry, the proton self-diffusion coefficient in both Nafion and other types of membranes is correctly computed over a wide range of hydration without any fitting parameters. The readers are referred to a review paper by Kreuer *et al.* [73] for a detailed discussion on the microscopic models.

While the microscopic models provide valuable information on the local mechanisms, the direct application to an overall fuel-cell is limited by the technical difficulties such as high computational costs. Alternatively, macroscopic models are adopted to describe the transport and relevant parameters in a more empirical fashion, and the discussion of which forms the focus of the subsection. Before the presentation of the macro homogeneous models, it is instructive to look at a general physical description of the transport in Nafion by Weber and Newman [74]. Figures 4.6(a)–4.6(d) show a schematic of the physical model, which focuses on the structural change of the membrane as a function of the water content, λ.

In Figure 4.6(a), the gray area represents the bulk membrane material and the black dots are the sulfonic acid sites. The dry membrane in Figure 4.6(a) absorbs water molecules to solvate the acid groups, forming the inverted micelles [denoted by the light gray circles in Figure 4.6(b)] in the polymer matrix. With increasing water uptake, the water clusters grow and form interconnections or collapsed channels [*i.e.* the dotted lines in Figures 4.6(b) and 4.6(c)] with each other. The collapsed channels are considered to be transitory and hydrophobic, and can only exist when the water clusters are close enough to each other. The growth of the cluster-channel network is governed by a percolation-type phenomenon, and conductivity data shows a percolation threshold to be $\lambda = 2$ [75]. Figure 4.6(c) shows a membrane in equilibrium with saturated water vapor, where a complete cluster-channel network is formed. When the membrane is immersed within liquid water, the collapsed channels are expanded and stabilized by the water molecules, resulting in the liquid pore structure in Figure 4.6(d). Since the expanded channels are filled with liquid water, the uptake of the membrane increases without an increase in

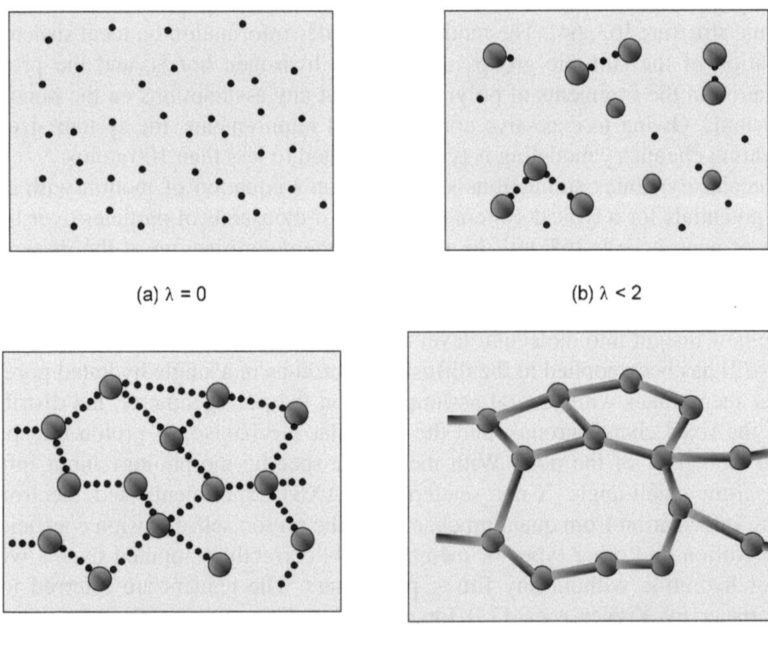

Figure 4.6. Evolution of the membrane structure at four cases of membrane hydration, (a) $\lambda = 0$, (b) $\lambda < 2$, (c) $\lambda = 14$, and (d) $\lambda = 22$. The gray area represents the bulk membrane material, the black dots are the sulfonic acid sites, the light gray circles denote absorbed water clusters, and the dotted lines are collapsed channels (redrawn from [26])

the water activity; consequently, "Schröeder's paradox" is explained by the physical model.

Note that two types of membrane structures are proposed in the physical model, *i.e.* the vapor-equilibrated membrane with collapsed channels and the liquid-equilibrated membrane with expanded channels. The two structures correspond to the two major types of macroscopic models of the membrane, namely, the single phase and the two-phase models discussed in Sections 4.2.2.2.2, and 4.2.2.2.3, respectively. Most macroscopic models consider a membrane system consisting of three species: the membrane polymer, proton, and water. Other types of ions are neglected in the three-species system, which, in turn, may be described by three transport properties [76]. Furthermore, the hydrogen and oxygen crossover does not have significant influence on the water and proton transport, and are also commonly ignored.

4.2.2.2.2 Single-Phase Models

The single-phase (or diffusive) models consider the vapor-equilibrated membrane shown in Figure 4.6(c). As the name indicates, the system is treated as a single, homogeneous phase where the water and proton dissolve in the polymer matrix and transport by diffusion. The membrane matrix is considered to be stationary in the space, and the fluxes of the dissolved species may be obtained by dilute solution theory [48] or concentrated solution theory [77, 78]. Dilute solution theory only considers the interactions between each dissolved species and the solvent (*i.e.*, the polymer matrix), and the analysis shows that the general motion of charge carriers is governed by the Nernst-Planck equation:

$$\mathbf{N}_i = -z_i u_i F c_i \nabla \Phi_e - D_i \nabla c_i + c_i \mathbf{v}_e \tag{4.27}$$

The first term in Eq. (4.27) represents the migration of the charged species i in an electrolyte potential gradient $\nabla \Phi_e$, where z_i is the charge number, and u_i and c_i are the mobility and the concentration, respectively. The diffusive and convective fluxes are governed by the second and the third term, respectively, and the diffusion coefficient D_i is related to the mobility u_i by the Nernst-Einstein equation [48]

$$D_i = RTu_i \tag{4.28}$$

Since the one-phase analysis considers the polymer matrix as stationary solvent, the convective velocity $\mathbf{v}_e = 0$, Eq. (4.27) reduces to Ohm's law (*i.e.*, Eq. (4.10)) for the case of zero proton concentration gradient, and to Fick's law when $z_i = 0$ for the case of water transport. Note that the fluxes obtained from Eq. (4.27) may be used in macroscopic mass conservation analysis.

It is also reported in the literature that the flux of proton across the membrane induces a flow of water in the same direction via the electro-osmotic drag effect. The electro-osmotic flow is a result of the proton-water interaction, and can not be modeled by the dilute solution theory. Consequently, the three-species membrane system mentioned above may be modeled more rigorously by the concentrated solution theory, which accounts for the interactions among all the species. Consider the thermodynamic driving force to be a sum of frictional interactions among different species, the following expressions for the ionic and water fluxes are obtained [77, 78]:

$$i_e = -\frac{-\kappa n_d}{F}\nabla\mu_w - \kappa\nabla\Phi_e \tag{4.29}$$

$$N_w = n_d\frac{i_e}{F} - \alpha_w\nabla\mu_w \tag{4.30}$$

where μ_w denotes the chemical potential of water, and α_w is the transport coefficient of water. Note that the proton-water interaction is taken into account by the two terms containing the electro-osmotic drag coefficients, n_d, in Eqs. (4.29) and (4.30). Similar flux equations to those in Eqs. (4.29) and (4.30) were derived in the literature using different approaches [79, 80]. The single phase models do not consider the water flux driven by the pressure gradient, which may be accounted for by the two-phase (or hydraulic) models.

4.2.2.2.3 Two-Phase Models

The hydraulic models consider the liquid equilibrated membrane shown in Figure 4.6(d), where the system consists of two-phases, namely, the polymer matrix and the liquid water. A major assumption of the two-phase models is the fully hydrated state of the membrane with $\lambda = 22$, corresponding to the complete filling of liquid water in the membrane micro-pores. Consequently, the concentration gradient and the diffusion transport of the water species are zero. The proton species is assumed to be dissolved in water and moves along with the water molecules. One of the first two-phase models was presented by Bernardi and Verbrugge [33, 81], which again adopted the Nernst-Planck equation, Eq. [4.27], to describe the proton flux. However, the convective velocity, $\mathbf{v}_w$, is non-zero and is given by Schlogl's equation for the water species [25, 82]:

$$\mathbf{v}_w = -\left(\frac{K}{\mu}\right)\nabla p_L - \left(\frac{K_\Phi}{\mu}\right)z_f c_f F\nabla\Phi_e \tag{4.31}$$

where K and K_Φ represent the effective hydraulic and electrokinetic permeability, respectively, p_L is the liquid pressure, μ is the liquid viscosity, and z_f and c_f denote the charge and concentration of fixed ionic sites, respectively.

Some hydraulic models [29, 83–87] also assume a constant gas volume fraction in the membrane to allow for the gas crossover mechanism. Since the diffusion coefficients of oxygen and hydrogen in Nafion are known, Fick's law is readily employed to calculate the gas fluxes in well-hydrated membranes. In the two-phase models, the water flux is attributed to the combined effects of a potential gradient and a pressure gradient. The portion of water flux driven by the pressure gradient is primarily due to the permeation of water through the micro-pore network of a fully hydrated membrane. When the membrane is partially hydrated, a water concentration gradient may exist across the membrane thickness, and modifications of the hydraulic models are required.

4.2.2.2.4 Combined Models

As seen in the above presentation, the one-phase models consider the water diffusion without pressure gradient, while the two-phase models describe pressure-driven (convective) flow without diffusive flux. Several combined models try to incorporate both diffusive and convective flows simultaneously, and the following expression for the water flux was obtained from concentrated solution theory [88–93]:

$$N_w = n_d \frac{i_e}{F} - D_w \nabla c_w - c_w \frac{K}{\mu} \nabla p_G \tag{4.32}$$

The adoption of the gas pressure p_G indicates that a gas phase exists in the membrane, which is inconsistent with experimental data [25]. Consequently, Eq. (4.32) is only considered as a fitting model without rigorous theoretical derivation.

More rigorous combination of models was developed by Weber and Newman [75] and Eikerling *et al.* [94] by using a capillary framework. The two transport modes (diffusion and convection) are assumed to occur in parallel, and microscopic effects such as the pore-size distribution and percolation phenomena are taken into account. Overall, the rigorous combination of models is a physically-based description of transport in membrane that agrees with experimental data for various sources and operating conditions.

4.2.2.3 Gas Diffusion Layer Models

As mentioned previously, the porous diffusion media between the catalyst layer and the gas channel are the gas diffusion layers (GDL), which provide structural support, uniform distribution of the reactant gases, and a pathway for electrons and liquid water to or from the catalyst layer. Owing to the high conductivity of carbon in the gas diffusion layer, the conduction of electrons is ignored in most GDL models. However, electronic conduction may become an important factor for the current distribution due to small contact areas with the gas channels [50] or the composition of the diffusion media [95]. Ohm's law accounting for porous media effect may be adopted for the electrical current, $\mathbf{i}_s$, in the GDLs:

$$\mathbf{i}_s = -\sigma^{eff} \nabla \Phi_s = -\frac{\varepsilon_s}{\tau_s} \sigma_0 \nabla \Phi_s = -\varepsilon_s^{1.5} \sigma_0 \nabla \Phi_s \tag{4.33}$$

where σ^{eff} denotes the effective conductivity of the porous GDL, σ_0 is the intrinsic conductivity of the GDL material, and ε_s and τ_s are the volume fraction and tortuosity of the solid conducting phase, respectively. Note that the Bruggeman expression is used to relate τ_s to ε_s as $\tau_s = \varepsilon_s^{-0.5}$ [96–99].

Under the assumptions of one-dimensional steady-state flow, the fluxes of the various reactants in the gas diffusion layers are constant, and are related to the current density by the stoichiometric coefficients [28, 32]. However, the water flux may not be constant due to evaporation or condensation in the porous media, and a rate term for the phase change, r_w, may be written as [28]:

$$r_w = h_m(\rho_w^{sat} - \rho_w) = Sh\frac{D_{vg}^{eff}}{L} f_e(\rho_w^{sat} - \rho_w) \tag{4.34}$$

where h_m is the mass transfer coefficient, ρ_w is the partial density of water vapor, and ρ_w^{sat} is the saturation partial density of water, which may be corrected for the pore effects via the Kelvin equation [100]. The mass transfer coefficient, h_m, is related to the Sherwood number, Sh, via the correlation $h_m = Sh\frac{D_{vg}^{eff}}{L} f_e$, where L is a characteristic length scale, D_{vg}^{eff} is the effective diffusion coefficient of water vapor, and f_e is the specific area of the liquid/vapor interface. Evidently, two-phase flow exists in the gas diffusion layers, and the treatment of the gas- and liquid-phase transport are discussed separately in the Sections 4.2.2.3.1 and 4.2.2.3.2, respectively.

4.2.2.3.1 Gas-Phase Transport in Gas Diffusion Layers

The transport of a multicomponent gas mixture through a porous media is usually described by the Stefan-Maxwell equations [101]:

$$\nabla x_i = \sum_j \frac{x_i \mathbf{N}_j - x_j \mathbf{N}_i}{c_T D_{ij}^{eff}} \tag{4.35}$$

where x_i and $\mathbf{N}_i$ are the mole fraction and molar flux of species i, respectively, c_T is the total concentration or molar density of all the gas species, and D_{ij}^{eff} is the effective binary diffusion coefficient for species i and j, which may be related to the binary diffusion coefficient D_{ij} via the Bruggeman relation

$$D_{ij}^{eff} = \varepsilon_G^{1.5} D_{ij} \tag{4.36}$$

where ε_G denotes the volume fraction of the gas phase, and equals to the bulk porosity of the media when the liquid water is ignored. For the case of two-phase flow, ε_G must be determined from the liquid saturation condition in the gas diffusion layer. Note that the effects of porosity and tortuosity on the effective properties may be modeled differently from the Bruggeman correction, as discussed in Refs. [102, 103]. It must also be pointed out that the Stefan-Maxwell equation reduces to Fick's law for a two component system.

With the decrease in the pore size in the GDL, the gas molecules collide more often with the pore wall than with each other, resulting in Knudsen diffusion from the intensified gas-wall interaction [104]. From an order-of-magnitude analysis, it is noted that the bulk diffusion dominates when the mean-free path of a molecule is less than 1% of the pore radius, while Knudsen diffusion dominates when the mean-free path is more than 10 times the pore radius [104]. To account for the gas-wall interaction, a Knudsen diffusion term is added to the Stefan-Maxwell equation based on a dusty-gas analysis [105]:

$$\nabla x_i = -\frac{\mathbf{N}_i}{c_T D_{K_i}^{eff}} + \sum_j \frac{x_i \mathbf{N}_j - x_j \mathbf{N}_i}{c_T D_{ij}^{eff}} \tag{4.37}$$

where the effective Knudsen diffusion coefficient $D_{K_i}^{eff}$ is proportional to the mean pore radius, r_p, and the mean thermal velocity of the gas molecules [106, 107]:

$$D_{K_i}^{eff} = \frac{2}{3} r_p \sqrt{\frac{8RT}{\pi MW_i}} \tag{4.38}$$

In Eq. (4.38), R, T, and MW_i are the gas constant, temperature, and molecular weight of species i, respectively.

While most models treat the gas transport in the GDL as pure diffusion where the total gas pressure remains constant through the thickness of the porous media, some models (primarily the computational fluid dynamics models) compute the convective velocity, $\mathbf{v}_G$, by adopting Darcy's law for the gas phase [101]:

$$\mathbf{v}_G = -\frac{K_G}{\mu_G} \nabla p_G \tag{4.39}$$

where K_G and μ_G are the permeability and viscosity for the gas mixture, respectively, and most computational fluid models incorporate Eq. (4.39) as a source term into the momentum equation. Alternatively, the dusty-gas models include the convective flow in the Stefan-Maxwell equation as [108–110]:

$$\nabla x_i = -\frac{x_i K_G}{D_{K_i}^{eff} \mu_G} \nabla p_G - \frac{\mathbf{N}_i}{c_T D_{K_i}^{eff}} + \sum_j \frac{x_i \mathbf{N}_j - x_j \mathbf{N}_i}{c_T D_{ij}^{eff}} \tag{4.40}$$

In general, the pressure difference through the gas diffusion layers is small according to the results of most simulations, and the assumption of uniform pressure may be valid for typical operating conditions [33, 111–113]. The results are physically expected since the gas mixture has convective flow in the channel direction (y-direction in Figure 4.1), and can only transport by diffusion in the porous media due to a no-slip condition at the pore walls. However, a small pressure gradient coupled with thermal gradients may significantly affect the water transport. Furthermore, for an interdigitated flow field shown in Figure 4.7, the gas channels are not continuous and the gases are forced by convection through the diffusion media to the next gas channel [114–116].

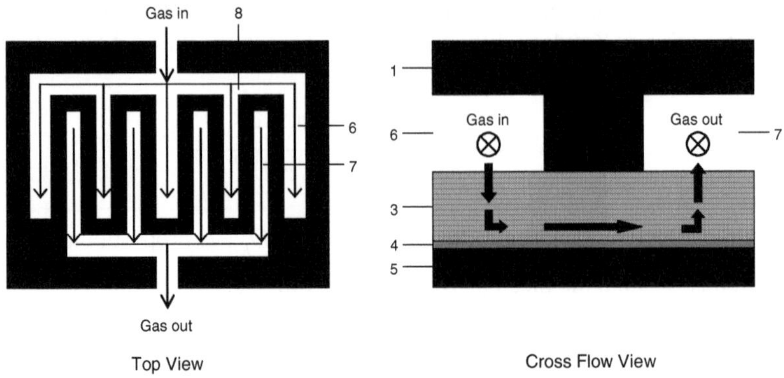

Figure 4.7. Schematic diagram of a PEM fuel cell with interdigitated flow field (redrawn from [116])

4.2.2.3.2 Liquid Water Transport in Gas Diffusion Layers

Sufficient liquid water is desirable for high membrane conductivity, while excessive liquid may block the pores in the gas diffusion layer, preventing the reactants from reaching the reaction sites. Consequently, liquid water transport is a critical issue to cell performance and various approaches were adopted to alternate the problem. In some simplified models, liquid water is treated as a stationary species that occupies a certain volume fraction in the GDL pores [35], [84, 90, 117, 118]. The effective binary diffusivities are thus decreased (see Eq. (4.36)), and the flooding effect of liquid is accounted for to some extent. Note that most of the simplified models use the liquid volume fraction as a fitting parameter [84, 90, 117]. More elaborate models treat the liquid water to be fine droplets that flow with the gas mixture [29, 89, 119]. Evaporation and condensation may take place; however, a separate liquid phase is not considered. Instead, the liquid is assumed to be a component of the gas and exerts negligible influence on the gas flow field. Without resorting to complicated two-phase transport analyses, the models keep track of the liquid water volume fraction at various locations.

The two types of models above essentially simulate single-phase transport, while more accurate treatment of liquid water flow calls for two-phase flow models. To treat the liquid water flux, some simple two-phase models assume isolated gas and liquid pores in the media [28, 33, 83, 85, 87, 115]. This assumption is based on the fact that the GDL is a mixture of hydrophobic Teflon and hydrophyllic carbon solid. Similar to the pressure-driven gas flow (Eq. (4.39)), the flux of liquid water follows Darcy's law as:

$$\mathbf{N}_{w,L} = -\frac{K}{V_w \mu} \nabla p_L \tag{4.41}$$

where the subscript L denotes the liquid phase, and V_w is the molar volume of water. Furthermore, a phase mixture approach is adopted by some models, where all the properties pertain to a gas-liquid mixture [38, 39, 120]. The models also use Eq. (4.41) to perform liquid flux computation, which is omitted by the single-phase models. An oversimplification in the approach is that the liquid flows with the same velocity as the gas, consequently, the interaction between the gas and the liquid is not adequately accounted for.

Gas-liquid two-phase flow in porous media is a well-known problem in a wide range of engineering applications, and rigorous modeling of the phenomenon has been extensively reported in the literature. An exhaustive presentation of the various modeling approaches is beyond the scope of this chapter, and the readers are referred to books and reviews on this subject [100, 121, 122]. In the following presentation, rigorous modeling studies specifically for the diffusion medium in fuel cells are briefly discussed [85, 91, 111, 112, 123–125]. Several basic definitions for two-phase flow, such as the capillary pressure and liquid saturation, are first introduced, followed by the presentation of the constitutive and conservation relations. In these models, the interaction between the gas and the liquid is characterized by a capillary pressure, p_C, defined as [100, 122]:

$$p_C = p_L - p_G = -\frac{2\gamma\cos\theta}{r} \tag{4.42}$$

where γ is the surface tension of water, θ is the contact angle of a water droplet with a pore wall, and r is the pore radius. Depending on the wetting characteristic of the GDL material, the contact angle has a range of $0^o \le \theta < 90^o$ for a hydrophobic material, and $90^o \le \theta \le 180^o$ for a hydrophillic one.

One of important goals of the two-phase models is to predict the distribution of liquid saturation, s, which is defined as the portion of pore volume filled with liquid. Thus, the volume fraction of the gas phase, ε_G, is related to the porosity of the gas diffusion layer, ε_0 as:

$$\varepsilon_G = \varepsilon_0(1-s) \tag{4.43}$$

Equation (4.43) indicates that the increase in saturation results in decreases in gas phase volume fraction and effective diffusion coefficients (see Eq. (4.36)).

To determine the liquid saturation s, several emprical constitutive equations are adopted to relate the capillary pressure, p_C, to the saturation, s [85, 87, 91, 111, 112, 126]. An example correlation for p_C as a function of s is given by Wang and Cheng [126]:

$$p_C = \gamma\cos\theta\left(\frac{\varepsilon_0}{K}\right)^{0.5}[1.417(1-s)-2.120(1-s)^2+1.263(1-s)^3] \tag{4.44}$$

where the surface tension is taken to be 0.0625 Nm^{-1} for the liquid water-air system at 80°C, and K is the effective permeability of the gas diffusion layer. The functional form for the $p_C - s$ relationship is also determined using a bundle-of-capillary model [100, 127].

The capillary pressure in Eq. (4.44) at various locations in the porous media must be known for the determination of the liquid saturation. In typical two-phase flow models, Darcy's law, *i.e.* Eqs. (4.39) and (4.41), is employed to calculate the pressure fields for both liquid and gas phases, which, in turn, are used in Eqs. (4.42) and (4.44) to obtain the liquid saturation distribution. It must be mentioned that some models utilize the capillary pressure as the driving force for the liquid-water flow:

$$\mathbf{N}_{w,L} = -\frac{K}{V_w \mu} \nabla p_L = -\frac{K}{V_w \mu} (\nabla p_G + \nabla p_C) = -\frac{K}{V_w \mu} \nabla p_C \tag{4.45}$$

The last expression in Eq. (4.45) empolys the assumption that the gas pressure is constant within the gas diffusion layer. The effective permeability K in Eqs. (4.44) and (4.45) is commonly related to a relative permeability, K_r, as:

$$K = K_r K_{sat} \tag{4.46}$$

where the permeability at complete saturation, K_{sat}, depends only on the stucture of the porous medium, and may be used as a fitting parameter [111, 125] or estimated from a Carman-Kozeny equation [102].

Various relations for the relative permeability, K_r, as a function of the saturation are reported in the literature. Nguyen and coworkers [111, 125] and Berning and Djilali [128] adopt a linear dependence of K_r on saturation. Most other models represent the relative permeabilities for liquid and gas phases, namely, K_{rl} and K_{rg}, with the following expressions [91, 112, 124].

$$K_{rl} = s^3 \tag{4.47}$$

$$K_{rg} = (1-s)^3 \tag{4.48}$$

An analytical model of Weber *et al.* [127] also reveals that K_r may be proportional to the cube of the saturation. It must be mentioned that different constitutive relationships for capillary pressure, permeability, and saturation may result in orders of magnitude difference in effective permeabilites, and reliable experimental data are needed to determine the validity of each model.

The constitutive relations in Eqs. (4.42–4.48) are commonly incorporated in the multiphase mixture model to simulate two-phase flow in the PEM fuel cells [91, 112, 126]. Similar conservation equations as Eqs. (4.3–4.8) may be written for the two-phase transport phenomena with different definition of variables and properties involved [91, 112]. In the multiphase mixture model, the parameters are usually averaged by using the saturation; for example, the mixture density ρ_m and concentration C_m are given as:

$$\rho_m = \rho_l s + \rho_g (1-s) \tag{4.49}$$

$$\rho_m C_m = \rho_l C_l s + \rho_g C_g (1-s) \tag{4.50}$$

where the subscripts m, l, and g denote the mixture, liquid, and gas, respectively. The conservation equations solve for the mass-average mixture velocity,

$\mathbf{u}_m$, and the individual velocities, $\mathbf{u}_l$ and $\mathbf{u}_g$, of each phase may be obtained from the interfacial drag between the two phases [112].

$$\mathbf{u}_l = \lambda_l \frac{\rho_m}{\rho_L}\mathbf{u}_m + \frac{K\lambda_l(1-\lambda_l)}{\varepsilon_0 \rho_l v_m}\left[\nabla p_C + (\rho_l - \rho_g)\mathbf{g}\right] \tag{4.51}$$

$$\mathbf{u}_g = (1-\lambda_l)\frac{\rho_m}{\rho_g}\mathbf{u}_m - \frac{K\lambda_l(1-\lambda_l)}{\varepsilon_0 \rho_g v_m}\left[\nabla p_C + (\rho_l - \rho_g)\mathbf{g}\right] \tag{4.52}$$

where $\mathbf{g}$ is the constant of gravity, and the relative mobility of the liquid phase, λ_l, the velocity and kinetic viscosity of the mixture, $\mathbf{u}_m$ and v_m, are defined by the following expressions:

$$\lambda_l(s) = \frac{K_{rl}/v_l}{K_{rl}/v_l + K_{rg}/v_g} \tag{4.53}$$

$$\rho_m \mathbf{u}_m = \rho_l \mathbf{u}_l + \rho_g \mathbf{u}_g \tag{4.54}$$

$$v_m = \frac{1}{K_{rl}/v_l + K_{rg}/v_g} \tag{4.55}$$

The advantage of the multiphase mixture model is that the two-phase flow is solved with a set of simplified pseudo-one-phase equations for the mixture. However, the pseudo-one-phase treatment can not account for the effects of pore-size distribution and mixed wettability [25].

4.2.2.4 Catalyst Layer Models

The catalyst layer, with thickness around 10 μm, is a critical component of a fuel cell, since the hydrogen oxidation reaction (HOR) occurs in the anode catalyst layer and the oxygen reduction reaction (ORR) takes place in the cathode catalyst layer. In the catalyst layer, all the phases discussed so far (including carbon supported catalyst solid, membrane, gas, and liquid) exist, and consequently, the analyses for the membrane and the gas diffusion layer may be applied along with the kinetics relations for the electrochemical reactions. The physical processes in a catalyst layer include the electron conduction in the solid phase, the proton transport in the membrane phase, the gas diffusion in the gas, liquid, and membrane phases, and the electrochemical reactions on the active catalyst sites.

Both microscopic and macroscopic models are reported for the catalyst layer in the literature. The microscopic models include the quantum mechanical models, pore-level models, as well as agglomerate-based ones. Electrochemical reaction mechanisms, elementary transfer reactions, and transition states are simulated by the quantum models, which are beyond the scope of this chapter and the interested readers are referred to refs. [129–131] for details. In this chapter, the macroscopic models are classified by the following four types, namely, (i) the 0D (or interface) models that treat the catalyst layer with zero thickness, (ii) macrohomogeneous approach, (iii) film models, and (iv) the agglomerate models.

The majority of the catalyst-layer models in the literature are for the cathode, owing to the fact that the cathode reaction is slower and contributes to the principal losses in the fuel cell. However, the modeling approaches for the cathode catalyst layer are generally applicable to the anode catalyst layer, with only different kinetic expressions and values of properties. Furthermore, the expression for the transport and electrochemical phenomena in the previous sections consist of a general set of governing equations for the catalyst layer, although different models discussed in the following may adopt a different subset of equations.

4.2.2.4.1 Microscopic Models

Some early catalyst models are microscopic single-pore models, which are further divided into gas pore models [132, 133] and flooded-agglomerate models [134–136]. In the first type of models, the catalyst layer consists of a solid matrix with cylindrical gas pores of a defined radius. The gas pores extend the thickness of the catalyst layer, and the electrochemical reaction takes place at the cylindrical surface. For the second type of models, the catalyst layer is made up of a number of porous cylinders flooded with the electrolyte, and the species diffusion and reaction occur within the cylinders [137–139]. Since the two modeling approaches are very similar, detailed discussion is only given on the flooded-agglomerate model in this subsection.

Figure 4.8 shows a schematic of an agglomerate-type structure, which is used in both microscopic and macroscopic models. The gray, black, and white areas in the largest magnification denote the electrolyte, carbon supported catalyst solid, and the gas pore, respectively. During operation of the fuel cell, reactant gas diffuses through the gas pore, dissolves and diffuses in the electrolyte contained in the agglomerates, and reacts on the active sites of the catalyst particles.

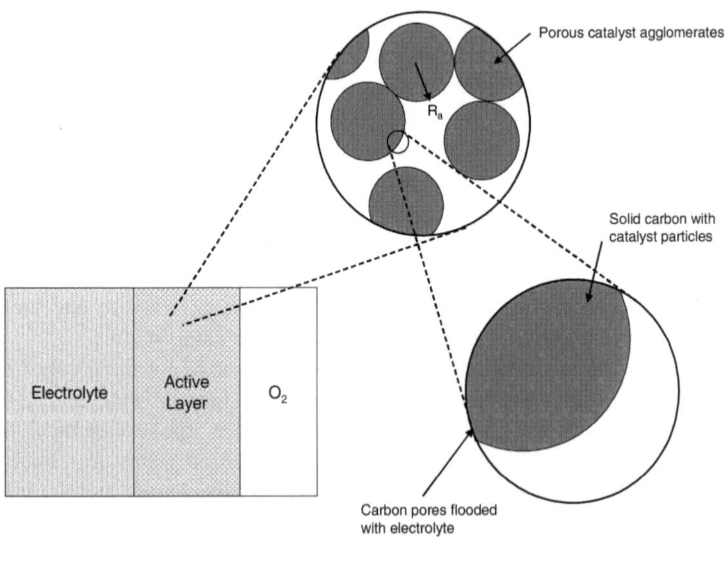

Figure 4.8. Schematic of a flooded-agglomerate structure (redrawn from [152])

The flooded-agglomerate model [137] treats the above working mechanism via studying an idealized porous cylinder perpendicular to the catalyst layer surface with radius r_0 and length h (shown in Figure 4.9), wherein the catalyst particles and electrolyte are homogeneously dispersed as continuous media. The reactant gas arrives at the lateral surface of the cylinder and diffuses radially toward the center, with simultaneous electrochemical reaction in the diffusion path. It is assumed that the diffusion of the dissolved gas occurs only in the radial direction, and the voltage in the cylinder changes only in the axial (x) direction. The diffusion of the reactant gas is governed by Fick's law with a source term accounting for the simultaneous bulk reaction [137]:

$$D_m^{eff} \frac{\partial^2 C(r)}{\partial r^2} + D_m^{eff} \frac{1}{r}\frac{\partial C(r)}{\partial r} - S_k = 0 \tag{4.56}$$

where D_m^{eff} is the effective diffusion coefficient of the dissolved species, $S_k = \frac{\nu_k j}{nF}$ is the consumption rate given by Eq. (4.12). The boundary conditions for the concentration are given by:

$$\frac{dC}{dr} = 0 \text{ at } r = 0; C = C_0 \text{ at } r = r_0 \tag{4.57}$$

The solution for Eq. (4.56) with the boundary conditions, Eq. (4.57), involves an analytical form with a Bessel function [137].

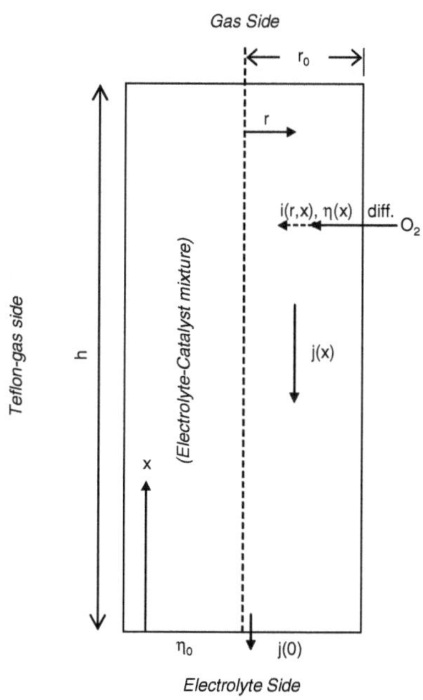

Figure 4.9. Schematic representation of a flooded cylinder used to model diffusion and electrochemical reactions of the reactants (redrawn from [137])

From the model assumption, the activation overpotential, η, in the source term S_k, is a function of x, and may be obtained from Ohm's law:

$$\frac{d^2\eta}{dx^2} = \frac{2nFD_m^{eff}}{\kappa^{eff} r_0}\left(\frac{\partial C}{\partial r}\right)_{r=r_0} \tag{4.58}$$

where κ^{eff} is the effective ionic conductivity. Equation (4.58) may be easily solved numerically to obtain the radial distribution of current density at various locations, which, in turn, may be used to evaluate the performance of the electrode as a function of physical properties such as the intrinsic activity of the catalyst, agglomerate size, internal porosity, and active surface area.

Another microscopic model was developed by Durand and coworkers [140–143], where spherical agglomerate structures are arranged in three-dimensional hexagonal arrays. The region between the agglomerates either consists of gas pores or is flooded with electrolyte, and the exact placement of the agglomerates depends on the interaction among the phases. The solution of the governing equations (*i.e.* Ohm's law and Fick's law with kinetic source term) yields the concentration field near an electrocatalyst particle. The results illustrate the effects of the packing and structure of the agglomerate particles on the overall efficiency of the catalyst layer.

4.2.2.4.2 Interface Models

Most theoretical studies on fuel cells adopt macroscopic models, and a simple approach is to treat the catalyst layer as an infinitely thin interface between the gas diffusion layer and the membrane. The interface models are used in complete fuel-cell simulations when the emphasis is not on the catalyst-layer effect but rather on the membrane, the water balance, or the nonisothermal effects. In some models focusing mainly on water management, the catalyst layer is treated as a location where the reactants are consumed and the water is produced [144–146]. Faraday's law, Eq. (4.12), is used in the boundary conditions for mass balance of each species between the membrane and the gas diffusion layer. The effect of the catalyst interface on the overall polarization behavior of a cell is also reported [35, 89, 147].

In more sophisticated interface models, both Faraday's law and the kinetics expressions are used in the boundary conditions at the gas diffusion layer and membrane interface [19, 119]. These models can account for the multidimensional effects, such as the distribution of the current density and the potential. Overall, the interface models assume that the values of the relevant variables are constant across the thickness of the catalyst layers, which may be justified by the fact that the layers are extremely thin. However, detailed treatment of the catalyst layer is required when the structure parameters of the layer, such as the catalyst loading, need to be optimized.

4.2.2.4.3 Macrohomogeneous and Thin Film Models

Within the framework of macrohomogeneous models, the catalyst layer is assumed to consist of uniformly dispersed carbon supported catalyst solid and the ionomer electrolyte without gas pores. The gas species dissolve and diffuse in the membrane phase, and consequently, the diffusion rate is low. In the thin film model, however, gas pores are assumed to exist, and the catalyst particles are covered by a thin film of electrolyte polymer. Both the macrohomogeneous models and the thin film models ignore microstructural details and share similar governing equations. The major distinction between the two types of models may be the values of the diffusion coefficients, since the reactant gas diffuses in different phases, namely, in the gas phase for the thin film models and in the membrane phase for the macrohomogeneous models.

A one-dimensional macrohomogeneous model for a cathode catalyst layer with thickness δ_{cat} is briefly summarized in the following discussion of the subsection [148]. In the model, the oxygen flux, N_{O_2}, through the catalyst thickness (or x direction) is determined by Fick's law of diffusion:

$$\frac{dC_{O_2}}{dx} = -\frac{N_{O_2}}{D_{O_2}^{eff}} = \frac{i(x) - I_0}{4FD_{O_2}^{eff}} \tag{4.59}$$

where $D_{O_2}^{eff}$ is the effective oxygen diffusion coefficient, $i(x)$ is the local proton current density, and I_0 is the total current density through the cell. The solid phase is considered to be equipotential, *i.e.* $\Phi_s = 0$, since the ohmic losses are negligible in the highly conductive materials. Consequently, the local electrode potential, $\eta(x)$, defined as the potential difference between the membrane and solid phase, is only determined by the local electrolyte potential, namely, $\eta(x) = \Phi_e - \Phi_s = \Phi_e(x)$. Ohm's law for the conduction of protons in the ionomer phase yields:

$$\frac{d^2\eta(x)}{dx^2} = \frac{d^2\Phi_e(x)}{dx^2} = \frac{j_c(x)}{\kappa^{eff}} \tag{4.60}$$

where $j_c(x)$ is the volumetric transfer current (see Eq. (4.16)], and κ^{eff} is the effective proton conductivity. The boundary conditions for Eqs. (4.59) and (4.60) are

$$x = 0 \quad \kappa^{eff} \tfrac{d\Phi_e}{dx} = 0; \; C_{O_2} = C_{O_2}^*$$

$$x = \delta_{cat} \quad \kappa^{eff} \tfrac{d\Phi_e}{dx} = I_0; \; \tfrac{dC_{O_2}}{dx} = 0$$

The solution of Eqs. (4.59) and (4.60) may be used to illustrate the effect of the effective oxygen diffusion coefficient $D_{O_2}^{eff}$ and effective proton conductivity κ^{eff} on the cathode performance. Since both $D_{O_2}^{eff}$ and κ^{eff} are functions of the membrane content in the catalyst layer, the model determines an optimal membrane gradient that significantly improves the cathode performance via optimizing both oxygen diffusion and proton transport [148].

4.2.2.4.4 Agglomerate Models

The macrohomogeneous and thin film models describe the transport processes on the macroscale across the layer thickness, while the variation of physical variable in the local agglomerate-scale is assumed to be negligible. To decide if changes in the agglomerate or pore-scale are required for more accurate modeling of the phenomena, the characteristic length of the pore-agglomerate must be compared with the diffusion lengths, *i.e.* the distances over which the physical variables related to the transport process change significantly [85]. Specifically, when the diffusion lengths are smaller than the pore-scale lengths, the nonhomogeneity of the pore-agglomerate region cannot be ignored and the diffusion equations in the microscale region must be solved. The characteristic pore scale lengths of a PEMFC are in the range 10^{-7} to 10^{-8} m, while the diffusion lengths for the reactants in the electrolyte phase changes from 10^{-7} to 10^{-8} m [149]. Consequently, the diffusion of the reactants within the agglomerates is considered by various modeling studies [85, 124, 149–156].

In the following discussion, a one-dimensional, steady-state, isothermal agglomerate model developed by Wang *et al.* [156] is outlined for the benefit of the reader. The cathode catalyst layer is assumed to consist of uniformly distributed spherical agglomerates with radius R_a and void space as shown in Figure 4.8. The coordinate x is defined in the thickness direction of the catalyst layer, and r is the local radical coordinate for the agglomerates. The ohmic losses within the solid are ignored, and consequently, the potential within the agglomerate is constant. The model considers the diffusion and reaction of oxygen within the spherical agglomerate as follows [156]:

$$D_a^{eff} \frac{1}{r^2}\frac{d}{dr}\left(r^2 \frac{dC_{O_2}}{dr}\right) = -\frac{j_c}{nF} \tag{4.61}$$

where D_a^{eff} is the effective diffusivity of oxygen in the porous agglomerate and j_c is the volumetric transfer current given by Eq. (4.16).

When the concentration of the oxygen on the agglomerate surface is $C_{O_2}^s$, an analytical solution to Eq. (4.61) is given as [156]:

$$C_{O_2} = C_{O_2}^s \frac{R_a}{r}\frac{\sinh(\phi r)}{\sinh(\phi R_a)} \tag{4.62}$$

where the product ϕR_a is commonly known as the Thiele modulus

$$\phi R_a = \sqrt{\frac{A_{av} i_{0,c}^{ref}}{nFD_a^{eff} C_{O_2}^{ref}}} R_a \exp\left(\frac{\alpha_c F}{2RT}\eta\right) \tag{4.63}$$

The current produced in the agglomerate, I_a, is obtained by using Faraday's law and Eq. (4.62):

$$I_a = nF(4\pi R_a^2) N_{O_2}(r = R_a) = -4\pi nFR_a D_a^{eff} C_{O_2}^s [\phi R_a \coth(\phi R_a) - 1] \tag{4.64}$$

where $N_{O_2}(r=R_a)$ is the oxygen flux at the agglomerate surface, and the concentration $C_{O_2}^s$ may be related to the gas concentration in the void space, $C_{O_2,g}$, by Henry's law:

$$C_{O_2}^s = H_{O_2} C_{O_2,g}$$

where H_{O_2} is Henry's constant for the oxygen gas.

Based on the solution of the current in the agglomerate, Eq. [4.64], the current density variation in the macroscopic catalyst scale, $i(x)$, may be obtained from the proton mass balance in the layer:

$$\frac{di}{dx} = -\rho_a I_a \tag{4.65}$$

where ρ_a is the density of the agglomerates in the catalyst layer and is defined by:

$$\rho_a = \frac{1-\varepsilon_c}{(4/3)\pi R_a^3}$$

with ε_c the porosity of the catalyst layer. Furthermore, the overpotential is governed by Ohm's law as:

$$\frac{d\eta}{dx} = \frac{i}{\kappa^{eff}} \tag{4.66}$$

Finally, the governing equation for the oxygen concentration in the gas pores may be obtained from Eq. (4.65) and Fick's law:

$$nFD_{O_2,c}^{eff} \frac{d^2 C_{O_2,g}}{dx^2} = -\rho_a I_a \tag{4.67}$$

where $D_{O_2,c}^{eff}$ is the effective oxygen diffusivity in the gas phase. Note that Eq. (4.64) for the microscale agglomerate and Eq. (4.67) for the macroscale catalyst layer consist of a multiscale model that may be used to evaluate the effects of micro-structure on the overall cathode performance. More elaborate treatment of the catalyst layer may be found in other agglomerate models [85, 124, 149–155]. Overall, the agglomerate-type models agree better with the physical picture, and they are also easy to implement in fuel-cell simulations.

4.2.2.5 Results of the Modeling Studies

Two major groups of parameters are responsible for the overall cell performance, namely, the operating parameters, *e.g.*, temperature, pressure, reactant stoichiometry, and gas composition, and design parameters, *e.g.*, thickness of cell components, electrode porosity, and platinum catalyst loading. A good understanding of the effects of the design and operating conditions on the fuel cell performance is required to reduce the capital cost and improve the reliability. In this subsection, the effects of several critical parameters on the performance of a PEMFC are first presented. Detailed discussion is subsequently presented on four specific aspects of (i) water and thermal management, (ii) electron transport, (iii) transient phenomena, and (iv) flow field design.

4.2.2.5.1 Parametric Studies

Operating temperature is a key variable that directly influences the values of various properties. The exchange current density, $i_{0,c}$, increases dramatically with temperature (see Eq. (4.22)), leading to an enhanced electrochemical reaction rate. The membrane conductivity, κ, and the binary gas diffusivity, D_{ij}, also increase with increasing temperature, as per Eqs. (4.19) and (4.24), respectively. Furthermore, for fully humidified feed streams, the temperature governs the partial pressure of the water vapor, which, in turn, determines the molar fractions of the hydrogen and oxygen species.

The effect of the operating temperature, T_c, on the cell voltage as a function of the current density is shown in Figure 4.10(a), which is obtained from a one-dimensional non-isothermal model by Mishra *et al* [32]. In Figure 4.10(a), a range of operating temperature, T_c, between 293 and 363 K is chosen for typical PEM fuel cell operation. For a fixed value of T_c, the cell potential, E_{cell} decreases monotonically with increasing current density, as seen previously in Figure 4.3. The cell potential increases when the operating temperature increases from 293 K to 353 K, which may be explained by the reduced losses in the cell. The ionic conductivity deceases with increasing temperature, leading to smaller resistive loss in the membrane. The transport losses are also reduced since the diffusivities decrease with increasing T_c. Furthermore, the activation loss in the catalyst layer is smaller at higher operating temperatures [28]. However, further increase in operating temperature may cause increased partial vapor pressure water, leading to enhanced mass transport losses. Consequently, the cell voltage is seen to slightly decrease as T_c increases to 363 K.

Pressure is another important variable that affects the values of various properties, including the exchange current density of the cathode reaction $i_{0,c}$, the reversible open circuit voltage E, the binary diffusivity D_{ij}, and the inlet gas compositions. The dependence of the cathodic exchange current density on the oxygen pressure at a temperature of 50°C is investigated experimentally by Parthasarathy *et al.* [8]:

$$i_{0,c} = 1.27 \times 10^{-8} \exp(2.06 p_{O_2}) \tag{4.68}$$

According to the Nernst equation, Eq. (4.2), the reversible potential E increases monotonically with increasing pressure. Equation (4.24) indicates that the product of the pressure and the binary diffusivity is a constant at a given temperature; consequently, the doubling of the pressure leads to a reduction of D_{ij} by half. Also, the change of pressure may cause condensation/vaporization of saturated vapor, leading to the composition change of the inlet gas mixture.

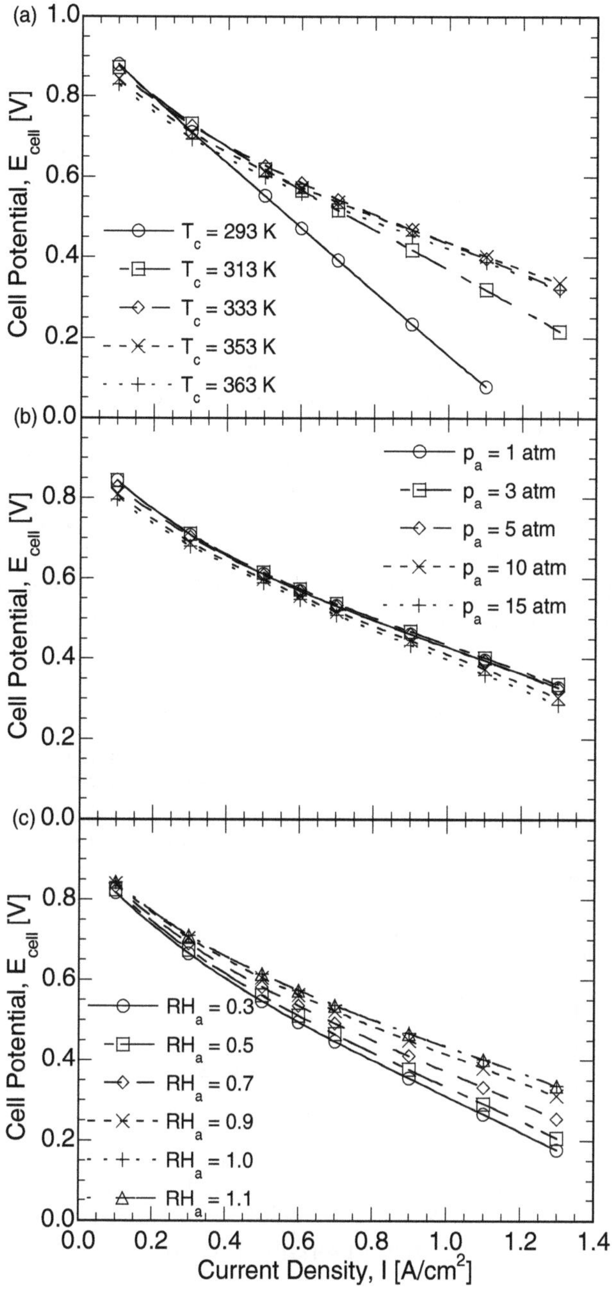

Figure 4.10. Parametric effects of current density I and (a) cell temperature, T_c, (b) anode pressure, p_a, and (c) anode relative humidity, RH_a, on cell potential, E_{cell} (after [32])

Figure 4.10(b) shows the effect of the anode pressure, p_a, on the cell performance, and a non-monotonic trend is again observed for the relation between the cell potential, E_{cell}, and the anode pressure, p_a. When p_a increases from 1 *atm* to 3 *atm*, higher partial pressure (or concentration) of hydrogen is present at the anode catalyst layer, leading to reduced concentration loss or higher cell potential. Further increase in p_a results in reduced water vapor diffusivity, which causes lower membrane hydration. The higher ohmic loss in the membrane, in turn, contributes to the lower cell potential. Overall, the influence of p_a on the cell potential is not as significant as that of the operating temperature in Figure 4.10(a), for the range of parameters considered.

Since PEM fuel cells operated under conditions of low temperature, high current density, and high pressure are prone to membrane dehydration at the anode side, the anode feed stream is commonly humidified with water vapor. Figure 4.10(c) presents the effect of the relative humidity in the anode feed stream, RH_a, on the polarization performance of the fuel cell. In Figure 4.10(c), the cell potential increases monotonically with increasing RH_a, on account of improved hydration and reduced ohmic losses. No apparent improvement in cell potential is observed when RH_a increases from 1.0 to 1.1, which may be explained by the fact that the super-saturated water vapor ($RH_a = 1.1$) will quickly be condensed in the anode electrode, and the actual water vapor concentration in the anode catalyst layer is equivalent to the case of $RH_a = 1.0$.

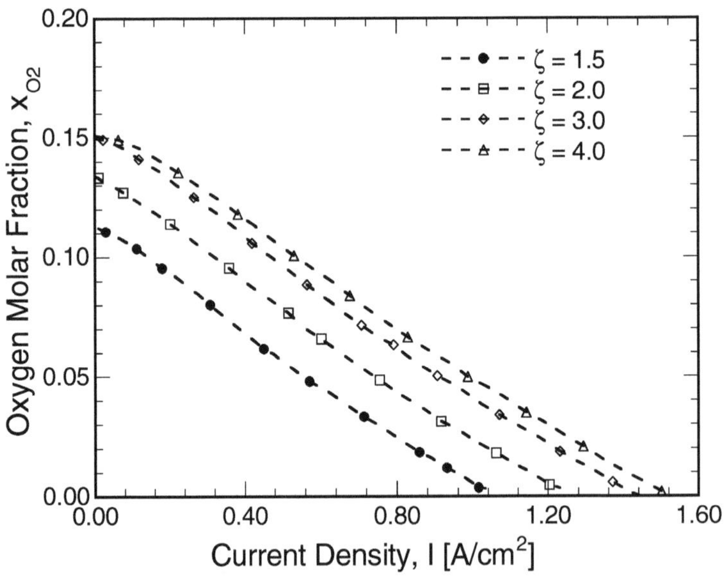

Figure 4.11. Average molar fraction of oxygen at the cathode catalyst layer as a function of current density and stoichiometric ratio (redrawn from [157])

The stoichiometric flow ratio ξ, defined as the ratio of actual flowrate to the flowrate required for the operating current density, is an important parameter affecting the theoretical fuel cell efficiency. In Figure 4.11, the average oxygen molar fraction at the cathode catalyst layer, x_{O_2}, is shown as a function of the current density for various values of ξ [157]. The oxygen concentration is seen to decrease linearly with increasing current density for a fixed ξ, owing to enhanced consumption of the oxidant, and the limiting current density is reached when the oxygen concentration reduces to zero. For a fixed current density, the oxygen concentration at the catalyst layer increases significantly when the dimensionless supply rate ξ increases from 1.5 to 3.0, while a relatively small increase in x_{O_2} is seen when ξ further increases to 4.0. The limiting current density increases significantly from 1.04 A/cm^2 to 1.52 A/cm^2 when ξ increases from 1.5 to 4.0. It must be mentioned that the slope of the lines in Figure 4.11 does not depend on the value of ξ, but rather on parameters such as the porosity of the gas diffusion layer, as seen later in this section.

Figures 4.12(a) and 4.12(b) show the distribution of the local membrane current density, i_m, at two stoichiometric ratios, 2.0 and 4.0, respectively [157]. In general, the current density under the flow channel area is much larger than that under the land area, owing to easier access to the reactants. From the channel inlet to outlet, the concentrations of the oxygen and hydrogen species decrease due to consumption, and a corresponding decrease in i_m, from around 2.2 A/cm^2 to 1.0 A/cm^2, is seen in Figure 4.12(a). A higher stoichiometric ratio results in a more even distribution of the local current density in Figure 4.12(b), where i_m decrease approximately from 1.7 A/cm^2 to 1.2 A/cm^2 from channel inlet to outlet.

The results presented so far pertain to the effects of the operating parameters on the performance of the fuel cell, and it is further illustrative to examine the effects of the design parameters, as shown in Figures 4.13–4.15. The porosity of the gas diffusion layer, ε_0, affects two competing transport mechanisms in the cell, namely, the diffusion of the reactants and the conduction of the electrons. With increase in porosity, more pore space is available for the reactants to diffuse towards the catalyst region, while the electrical resistance, including the contact resistance between the GDLs and the bipolar plates, also increases. Berning and Djilali [157] assumed a linear relation between the interfacial contact resistance and the porosity, and showed that the power density may decrease with increasing porosity due to increased contact resistance. Evidently, an optimum porosity exists for best performance of the cell.

Figure 4.13 presents the average oxygen molar fraction x_{O_2} at the cathode catalyst layer as a function of the current density for three porosity values [157]. For a fixed current density, the oxygen concentration x_{O_2} increases monotonically with increasing ε_0 due to the reduced diffusion resistance. With significant increase in the effective gas diffusivity when ε_0 increases from 0.3 to 0.5 (see the Bruggeman relation, Eq. (4.36)), the limiting current density is more than tripled from around 0.75 A/cm^2 to 2.4 A/cm^2.

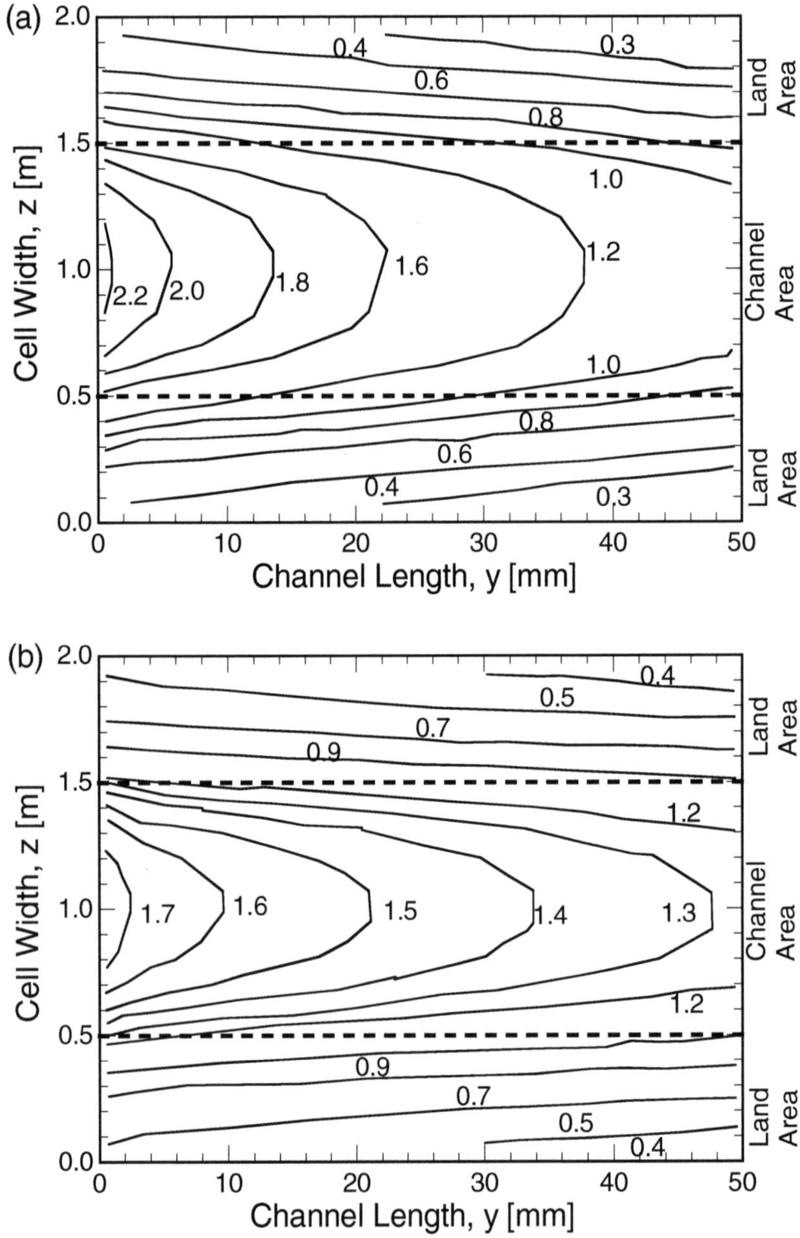

Figure 4.12. The distribution of the local membrane current density at stoichiometric ratio of (*a*) 2.0 and (*b*) 4.0 (redrawn from [157])

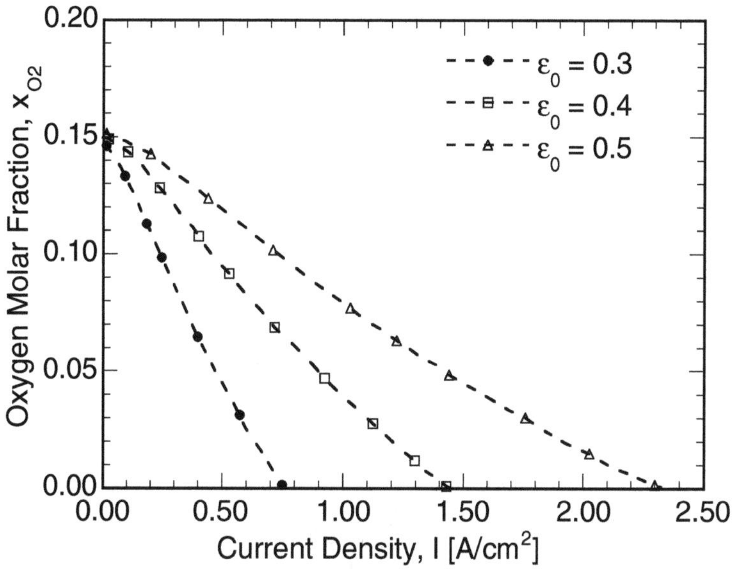

Figure 4.13. Average molar fraction of oxygen at the cathode catalyst layer as a function of current density and porosity of gas diffusion layer (redrawn from [157])

Higher porosity values also facilitate a relatively uniform distribution of the current density, as illustrated in Figures 4.14(a) and 4.14(b) [157]. For the case of small porosity value of 0.4 in Figure 4.14(a), a much higher portion of the total current is generated under the channel area, which, in turn, may cause drying out of the membrane and increased electric resistance. Note that the increase in resistance results in more heat generation and dehydration, leading to the failure of the membrane. When ε_0 increases to 0.6 in 14(b), the maximum and minimum current density are around 1.4 A/cm^2 near the channel inlet and 0.7 A/cm^2 under the land area close to the outlet, respectively, which correspond to a much smaller current density gradient than that seen in Figure 4.14(a).

Other important design parameters include the channel width, the platinum loading, and the membrane thickness. The effects of channel width on the cell performance are similar to those of the porosity. A narrow channel may decrease the contact resistance due to larger land area, however, the resistance for gas diffusion also increases, resulting in an undesired large current gradient as seen in Figure 4.14(a). Platinum catalyst loading is a critical parameter that governs the available sites for the electrochemical reactions. The dependence of the polarization curve on the cathode platinum loading, $m_{c,pt}$, is illustrated in Figure 4.15(a) [32]. When $m_{c,pt}$ increases from 0.32 mg/cm^2 to 3.00 mg/cm^2, the cell potential is observed to have a notable improvement (about 0.1 V for all the current density values considered), owing to the increased reaction area. Further increase in $m_{c,pt}$ to 8.00 mg/cm^2 only yields a small increase in the cell potential, indicating that an optimum platinum loading may be determined by simultaneously evaluating the increases in performance and cost. Figure 4.15(b) presents the effect of the membrane thickness, δ_m, on the cell polarization performance [32]. The increase in the membrane thickness results in a decrease in cell potential, owing to the increased membrane resistivity.

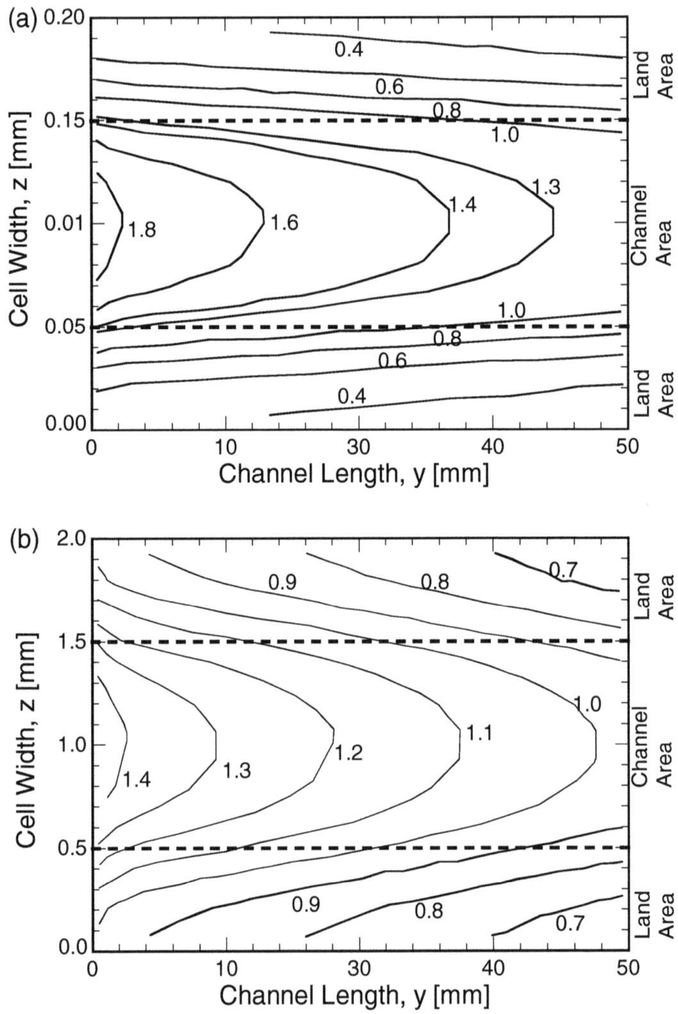

Figure 4.14. The distribution of the local membrane current density at gas diffusion layer porosity of (*a*) 0.4 and (*b*) 0.6 (redrawn from [157])

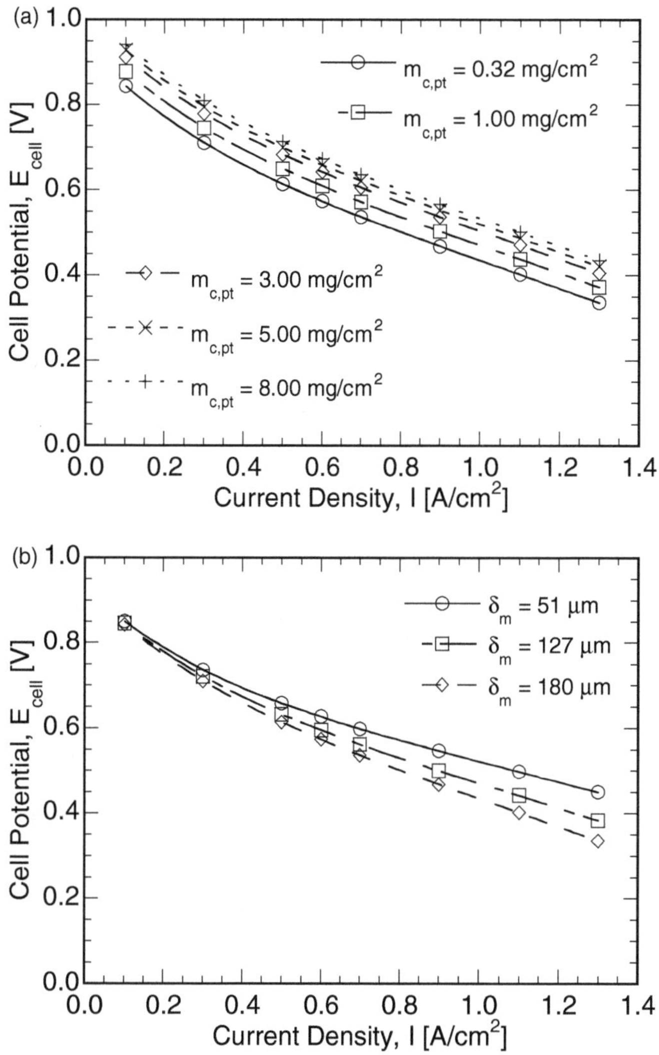

Figure 4.15. Parametric effects of current density I and (a) cathode platinum loading, $m_{c,pt}$, and (b) membrane thickness, δ_m, on cell potential, E_{cell} (after [32])

4.2.2.5.2 Water and Thermal Management

The presentation on the parametric effects is instrumental in the derivation of optimum operating and design conditions, which will be discussed in Section 4.5. In the discussion below, the four specific topics mentioned above are covered, starting with the water and thermal management. It is well known that water and thermal management are critical to the overall cell performance [36, 81, 158, 159]. To maintain ionic conductivity, the membrane in a PEM fuel cell requires adequate humidification, which raises the critical issue of water management. During practical

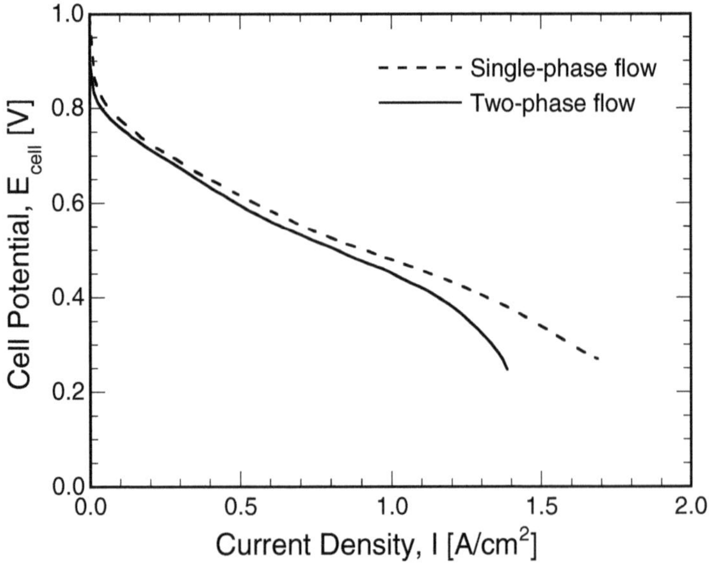

Figure 4.16. Comparison of single-phase and two-phase description of water transport on the polarization curve (redrawn from [160])

operation of fuel cells, both the gas streams are humidified to ensure the proper membrane hydration. However, excessive water will accumulate in the electrode pores and result in electrode flooding, which degrades the cell performance by preventing the reactants from reaching the catalyst sites. In general, a higher operating temperature is desirable due to decreased mass transport limitations and increased electrochemical reaction rates; at the same time, high temperatures may lead to increased mass transport losses due to the increase in water vapor pressure. A careful design of the cell and its operating parameters is therefore imperative to balance such competing constraints.

Figure 4.16 compares the polarization curves with and without liquid water transport, where the solid line corresponds to the two-phase flow modeling results with liquid water accumulation, and the dashed line denotes the single-phase transport without liquid [160]. Since the two-phase model accounts for the flooding effect on the gaseous reactant transport, the cell potential is lower than the corresponding single-phase case due to higher concentration losses. It is seen that the water accumulation effect becomes apparent even at relatively low current densities. Furthermore, the single-phase model does not predict a significant drop of the potential when the current density approaches a limiting value. The relatively large gradient of the polarization curve in the vicinity of the limiting current density is demonstrated by experimental data [160]. This trend is better predicted by the two-phase flow model than the single phase model in Figure 4.16.

The effect of the liquid water on the current density variation along the flow channel is illustrated in Figure 4.17, where the two modeling approaches in Figure 4.16 are adopted again for comparison [160]. For the case of the single-phase flow without electrode flooding, the catalyst sites may be easily accessed by the reactant

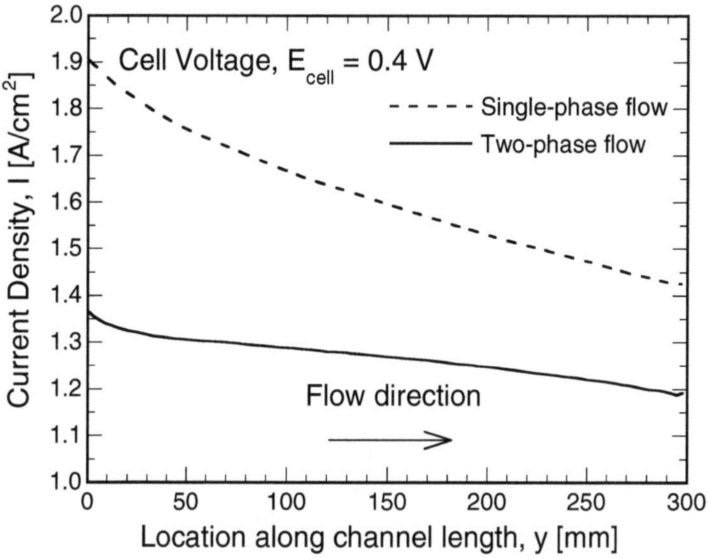

Figure 4.17. Comparison of single-phase and two-phase description of water transport on the current density distribution along the cathode flow channel (redrawn from [160])

gases. Consequently, the current density predicted by the single-phase model is higher than that of the two-phase model. Furthermore, due to the reactant depletion, the current density predicted by both models decreases along the flow channel direction. For the case of two-phase flow, the liquid saturation prevents fast reaction at the channel inlet, which reduces the rate of reactant depletion in the flow direction and leads to a more uniform distribution of the current density along the channel.

Figure 4.18 shows the distribution of liquid saturation in the cathode GDL and catalyst layer at an average current density of 1.18 A/cm^2 [160]. The level of saturation decreases along the flow direction, which may be explained by two contributing factors. As seen in Figure 4.17, the reaction rate is reduced in the flow direction, leading to less liquid water product. Furthermore, the reduced current density causes less water being dragged across the membrane from the anode to the cathode. Note that the electro-osmotic drag may be detrimental to the cell performance by causing dehydration of the anode, which decreases the ionic conductivity, and flooding of the cathode, which decreases the effective permeability of the cathode gas diffusion layer. Consequently, the reduced current density near the channel outlet may be desirable for elevated electro-osmotic drag effects.

In Figure 4.19, the water content, λ, in the membrane electrode assembly (MEA) is shown as a function of the location in the thickness direction and the current density [160]. At the low current density of 0.05 A/cm^2, the hydration λ has a nearly uniform distribution across the thickness, owing to identical humidification conditions adopted for both anode and cathode layers. With the increase in current density, the water concentration decreases at the anode due to electro-osmotic drag, and increases at the cathode from the water production, leading to a

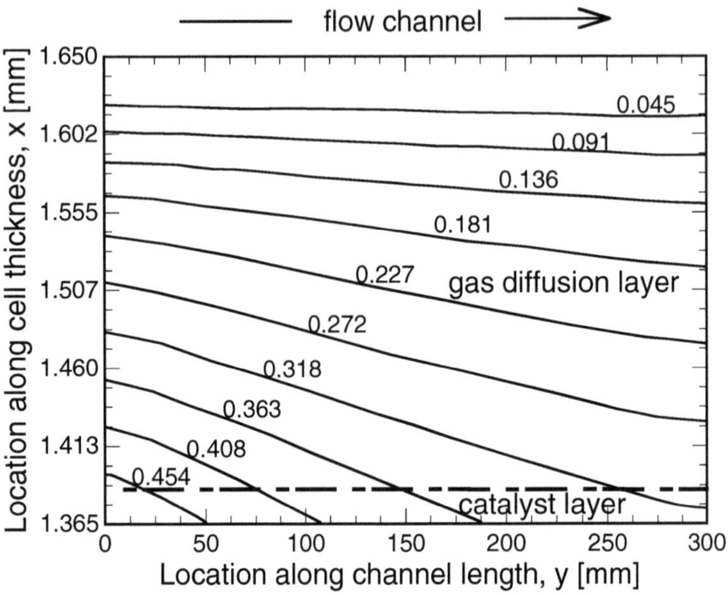

Figure 4.18. Contour plot of liquid saturation in the cathode GDL and catalyst layer at an average current density of 1.18 A/cm^2 (redrawn from [160])

larger gradient of membrane hydratio/cmn in the thickness direction. It must be mentioned that the geometric area under the hydration curve decreases with increasing current density, indicating that the total amount of water contained in the MEA decreases due to the electro-osmotic removal of the water species.

Since the membrane hydration and most transport properties (*e.g.* diffusivity and the ionic conductivity) are strong functions of the temperature, it is imperative to predict the temperature distribution in an accurate modeling of PEM fuel cells. Figure 4.20 presents the three-dimensional nonisothermal temperature distribution inside the cathode gas diffusion layer [128]. The temperature has the highest value near the channel inlet and decreases along the flow channel direction, which may be explained by the higher reaction rate near the inlet as shown in Figure 4.14. Due to the high thermal conductivity of the bipolar plates and slow reaction rate, the temperature under the land area is also lower than that under the channel. Depending on the length of the channel, boundary conditions, and the operating conditions, the largest temperature variation may be either in the thickness direction or along the channel. In Figure 4.20, the temperature gradient is biggest through the cell thickness; however, under the condition of dry-gas feed, the inlet region becomes susceptible to membrane dehydration, and the temperature gradient along the channel may become larger [25]. Overall, the nonisothermal effects influence the water balance and current distribution, which, in turn, affect the water content in the membrane. Evidently, a two-way coupling exists between the water and thermal management issues.

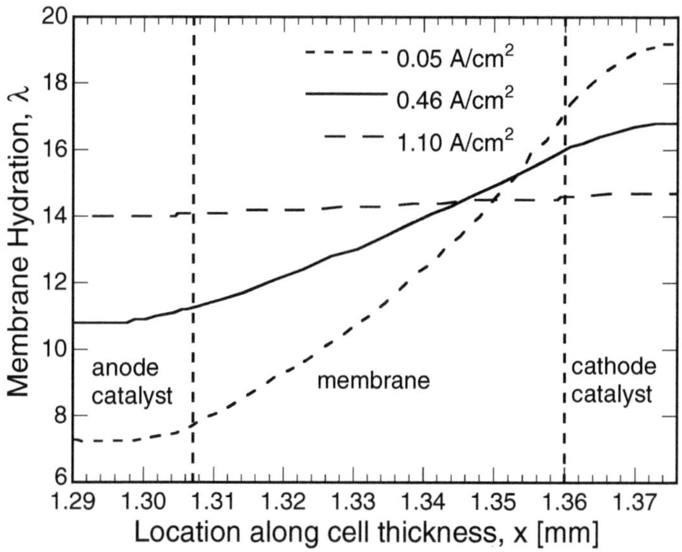

Figure 4.19. The water content, λ, in the membrane electrode assembly as a function of the location in the thickness direction and the current density (redrawn from [160])

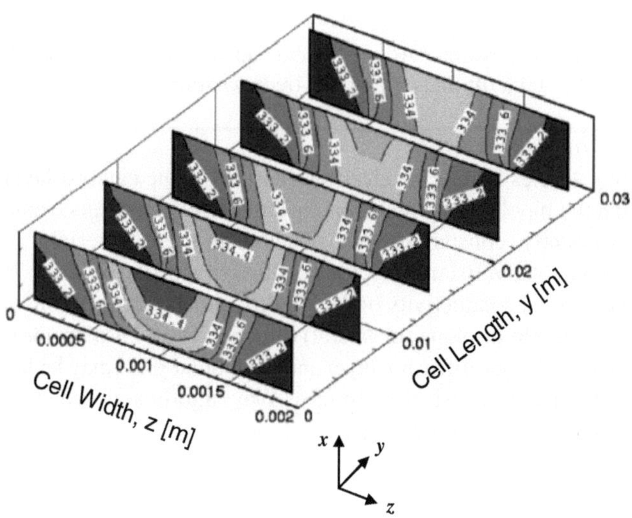

Figure 4.20. Three-dimensional temperature distribution inside the cathode gas diffusion layer (reproduced with permission from [160])

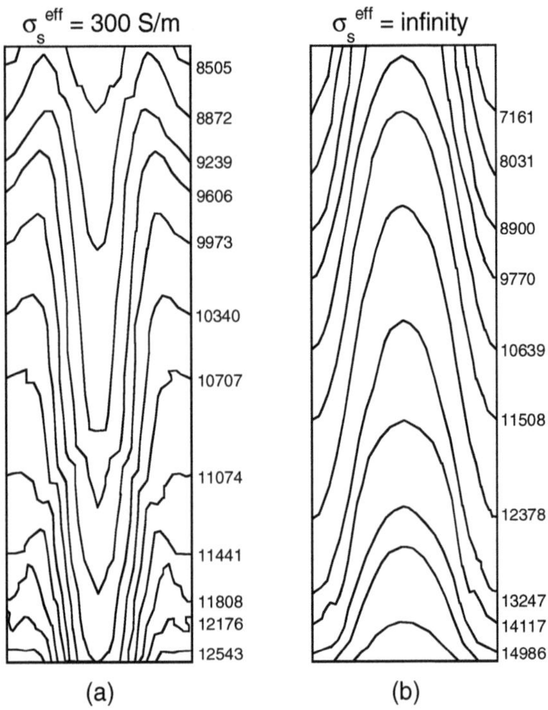

Figure 4.21. Current distribution (A/cm^2) in the middle of membrane for GDL electronic conductivity of (a) 300 S/m and (b) infinity (redrawn from [50])

4.2.2.5.3 Electron Transport

Most modeling studies ignore the electron transport in the catalyst layer, gas diffusion layer, and the bipolar plate by assuming sufficiently large electronic conductivity, and consequently, constant electric potential in these materials. Although the bipolar plate has a relatively large electric conductivity, on the order of 2×10^4 S/m, the effective electrical conductivity of the gas diffusion layer, σ_s^{eff} , only ranges from 300 to 500 S/m in the in-plane (or lateral) direction [50]. Furthermore, the contact resistance between the gas diffusion layer and the bipolar plate may be large [42], and the lateral and contact electrical resistances may significantly change the current distribution and the overall cell voltage.

Figures 4.21(a) and 4.21(b) compare the current distributions in the membrane with and without the lateral electron transport effects [50]. In Figure 4.21(b), the current density is highest in the middle of the channel when the electrical conductivity is assumed to be infinite, since the area has the easiest access to the reactants. For the case of finite lateral electrical resistance in Figure 4.21(a), the region with the highest current density shifts towards the land areas, owing to the desirable combination of easy access to reactants and shorter path for the electron transport to the current-collecting land. The results reveal the possibility that the current distribution may be controlled by tailoring the GDL electrical property.

The electrical contact resistance between gas diffusion layers and bipolar flow channel plates is one of the important factors contributing to the operational voltage loss in PEM fuel cells. Despite its significance, relatively little work is reported in the open literature on the measurement and modeling of the contact resistance in fuel cell systems. Mishra *et al.* [42] reported experimental data to show the effects of different gas diffusion layer materials and contact pressure on the electrical contact resistance. A fractal asperity based model was also adopted to predict the contact resistance as a function of pressure, material properties, and surface geometry. Good agreement was demonstrated between the data and the model predictions for a wide range of contacting pressures and materials.

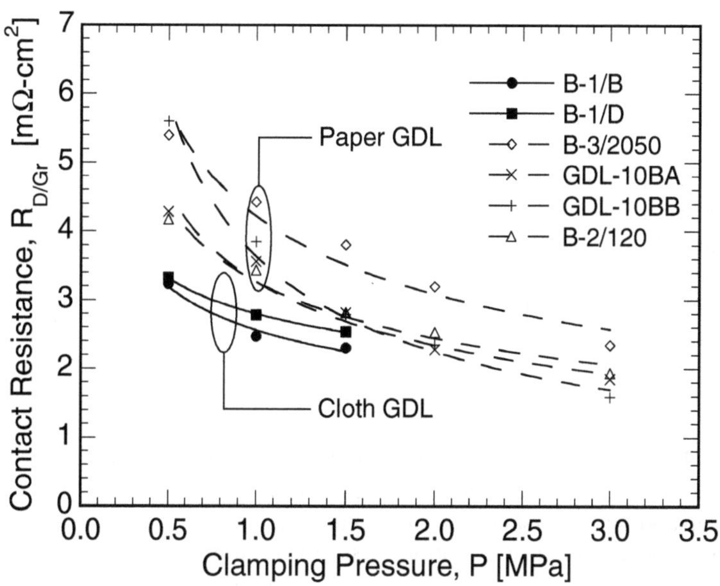

Figure 4.22. Variation of the measured contact resistance between bipolar plate and gas diffusion layers, over a range of pressure for all the gas diffusion layer samples evaluated (after [42])

Experimental measurements of the contact resistance between various GDLs and the XM9612 bipolar plate from SGL Carbon Group are summarized in Figure 4.22 [42]. The contact resistance, $R_{D/Gr}$, decreases monotonically with increasing clamping pressure for all the GDL materials (denoted by the legend in the figure), as physically expected. Of the two paper-based GDLs with wetproofing, GDL-10BA had a hydrophobic treatment of a plain substrate, whereas the GDL-10BB sample was coated on one side with a customized microporous layer. The other two paper-based gas diffusion layers, B-3/2050 and B-2/120, do not have any hydrophobic treatment. When the clamping pressure $P < 1.5\ MPa$, the contact resistance for GDL-10BB is observed to be higher than that of GDL-10BA; however, for $P > 1.5\ MPa$, relative magnitudes are seen to be reversed. Among all the paper-based GDLs, the contact resistance for B-3/2050 is the highest when

$P = 1.5\ MPa$, while that for B-2/120 is the lowest for the entire pressure range studied.

In the case of cloth-based gas diffusion layers, B-1/B is a plain weave carbon cloth without any wetproofing, whereas B-1/D has a knitted weave architecture and is not wetproofed. Since both materials are relatively soft, the measurements were limited to $P < 1.5\ MPa$. Evidently, all the cloth-based gas diffusion layers have lower values of contact resistance than the paper-based samples. Since the cloth-based GDLs have relatively smaller compressive moduli, the materials are easily deformed upon contact with the bipolar plate, leading to the reduced contact resistance. The data presented in Figure 4.22 and the asperity based model in ref. [42] can be used in a comprehensive simulation model of a PEM fuel cell to account for the effect of interfacial contact resistance on the cell performance.

4.2.2.5.4 Transient Phenomena

The dynamic response of fuel cells to load change is important for cell start up and automotive applications. Three major transient processes are responsible for the dynamic characteristic of a PEM fuel cell, namely, (1) formation and discharging of the charge double layer, (2) reactant gas transport in the channel and porous media, and (3) membrane hydration and dehydration [26]. The liquid water transport in the membrane, with a time constant of about 15 seconds, is the time limiting process, since the time constant for both charge double layer and the gas transport are typically below 1 second [26]. A simple stack-level model is developed by Amphlett *et al.* [161], where the transient behavior in a stack is found to last a few minutes. Detailed transient models are also developed to study the response of a single cell to dynamic load change [30].

Figure 4.23 shows an example time response of the average current density in a single cell to a step change of the cell voltage from $0.6\ V$ to $0.7\ V$ [26]. Significant undershoot and overshoot are observed, which are attributed to the transient gas transport and membrane hydration, respectively. It is seen that the current density drops instantaneously when the cell voltage is increased from 0.6 to $0.7\ V$, owing to the quick response of the charge double layer. The undershoot in the current density is caused by the low oxygen concentration resulting from the fast consumption rate at a cell voltage of $0.6\ V$. When the reaction rate becomes less than $0.7\ V$, the oxygen concentration gradually increases, leading to the increase in current density seen in Figure 4.23. The rise in the current density experiences an overshoot, since the membrane still maintains a higher water content corresponding to $0.6\ V$, and it takes about $15\ s$ for the membrane to achieve the lower steady-state level of hydration. Although selected studies have focused on the transient response of PEM fuel cells, this area remains relatively less explored than steady state operation of fuel cells.

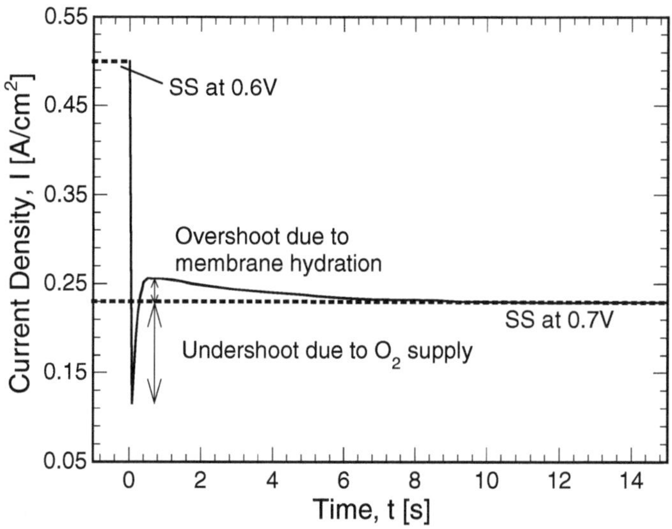

Figure 4.23. Dynamic response of current density to a step change of cell voltage from 0.6 V to 0.7 V (redrawn from [26])

4.2.2.5.5 Flow-field Design

A major benefit of detailed modeling studies is the possibility to examine the effects of various flow-field features, such as the counterflow and co-flow configurations, size and shape of flow channels, and serpentine and interdigitated flow

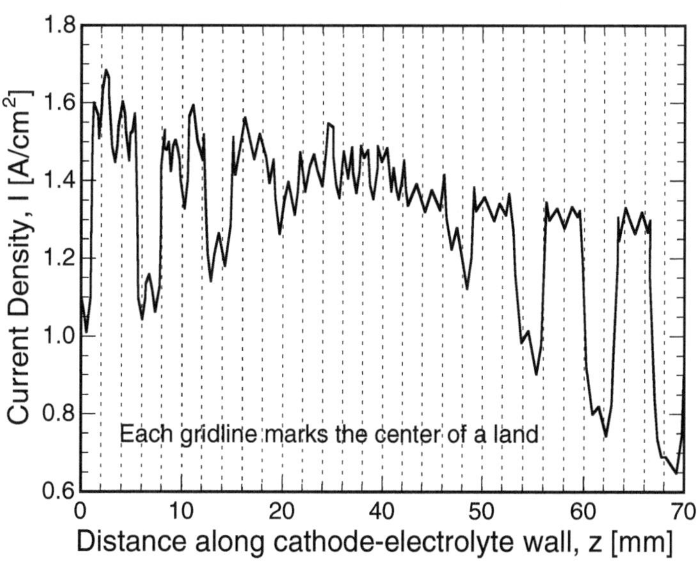

Figure 4.24. Current density profile in the middle cross-section of a 36-channel fuel cell (redrawn from [26])

fields. Fuel cells typically consist of bipolar plates with dozens of flow channels as shown in Figure 4.5. Numerical simulations for the complicated geometries may involve millions of computational cells and tremendous CPU time, and such large scale simulations may be conducted using the parallel computing techniques mentioned previously [51]. Figure 4.24 shows the current distribution in a 50 cm^2 fuel cell with a counterflow serpentine flow field [26]. Such a flow field design is intended to promote moisture exchange between a dry gas channel near the inlet pass and its neighboring moist gas channel, and thus to enhance the feed humidification. The current density shows both sharp local variation from channel to land, and gradual macro-scale gradient in the whole computational domain.

Figure 4.25 compares the polarization behavior between a conventional straight flow-field design (see Figure 4.1) and an interdigitated design shown in Figure 4.7 [5]. Note that diffusion is the dominant transport process in the gas diffusion layer for the conventional design, while gases are forced by convection through the diffusion media in the interdigitated design. For a cell current density below 0.8 A/cm^2, the voltage losses are dominated by the cathode activation polarization and the ohmic resistance through the membrane, and little difference in cell voltage is observed between the two flow-field designs. When the current density exceeds 0.8 A/cm^2, the mass transport loss in the cathode becomes significant, and the advantage of the interdigitated design is apparent. Due to enforced convective transport of oxidant to the catalyst layer, the interdigitated flow-field improved the current density by 40% at a cell voltage of 0.3 V, as compared to that of the conventional design. Detailed presentation on the flow-field design is beyond the scope of the Chapter, and interested readers are referred to refs. [38, 39, 162, 163] for relevant discussion.

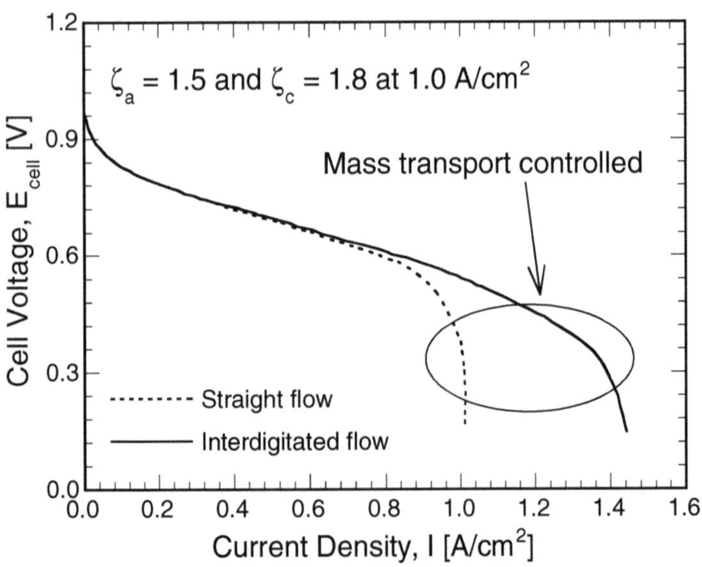

Figure 4.25. Effect of the interdigitated flow-field design on the polarization curve (redrawn from [5])

The PEM fuel cells covered in this section require an operating temperature below 100°C, since water evaporation at higher temperatures leads to dehydration of the polymer membrane. However, higher operating temperature in a fuel cell system is desirable for several advantages: (i) the electrochemical reactions proceed quickly with lower activation losses, (ii) the need for expensive noble metal catalysts may be eliminated, (iii) the exit gases with high temperature are a valuable source of heat for facilities such as buildings, generators, and reforming reactors. A promising high temperature cell, the solid oxide fuel cell, is discussed in the next section.

4.3 Modeling of Solid Oxide Fuel Cells

A solid oxide fuel cell (SOFC) consists of three main components: a porous cathode (or air electrode), a porous anode (or fuel electrode), and an ion-conducting ceramic membrane, as shown in Figure 4.26. The operation temperature of SOFCs are high, typically between 700°C and 1000°C, to ensure sufficient ionic conductivity in the membrane and to promote electrochemical reactions. Similar to the PEM fuel cells, the advantages of the SOFCs include high-efficiency, high energy density, and low pollution. Furthermore, a major constraint of the PEM fuel cells, *i.e.*, the requirement of hydrogen fuel, is relaxed in SOFCs. Although hydrogen is taken as a future alternative to the depleting fossil fuel, a number of issues related

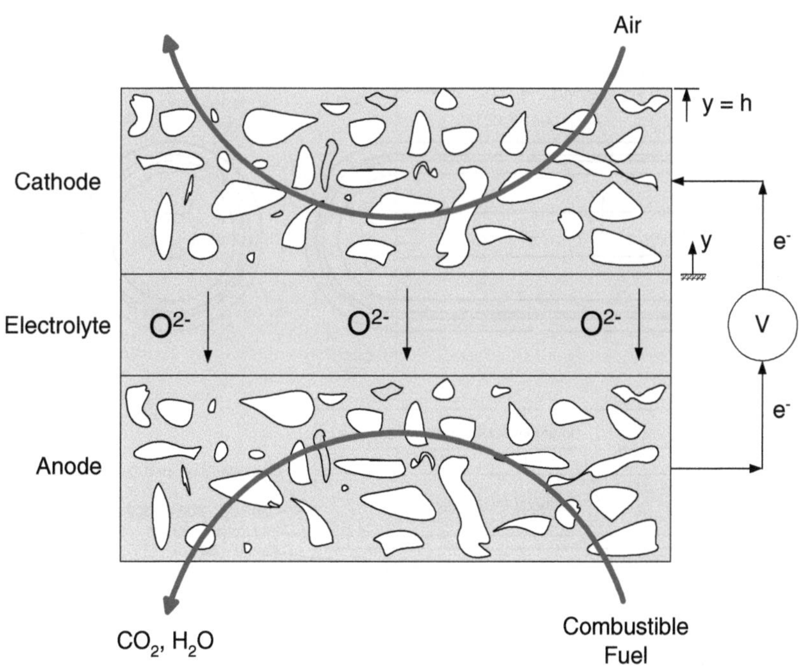

Figure 4.26. Schematic of a solid oxide fuel cell (redrawn from [178])

to hydrogen generation and storage limit the large-scale usage of hydrogen. About 96% of hydrogen is currently obtained by reforming hydrocarbons, and 20 to 30 percent of the chemical energy is wasted during the process [164]. Since the energy density of hydrogen gas is low compared to that of hydrocarbon liquid, hydrogen storage may be difficult and expensive.

Two types of geometries are commonly used for SOFCs, namely, tubular and planar configurations as shown in Figures 4.27(a) and 4.27(b), respectively. Some model power plants manufactured by Siemens-Westinghouse adopt the tubular design, which has demonstrated high efficiency and long term stability (3). However, two technical problems are associated with the tubular design, namely, high fabrication cost and relatively large ohmic losses. To address these issues, cell/stack designs with planar configurations have been developed by companies such as Ztek, Ceramatec SOFCO, and Siemens. In the planar design, any one of the three components, *i.e.* the cathode, anode, or electrolyte, may be used as structural support. Traditional design adopts a thick membrane as the support structure, which requires high operating temperature; electrode-supported design [*e.g.*, the cathode-supported configuration in Figure 4.27(b)] allows the use of a thinner membrane and lower temperature.

Due to the fact that the O^{2-} anion is the species transported in the membrane, SOFCs may operate, in principle, on any combustible fuel. Another advantage of SOFC is the possibility of efficient utilization of waste heat generated at high operating temperatures. A number of comprehensive reviews are available in the literature on the issues such as the component materials, stack designs, and transport properties [165–180]. The present discussion focuses on a fundamental understanding of

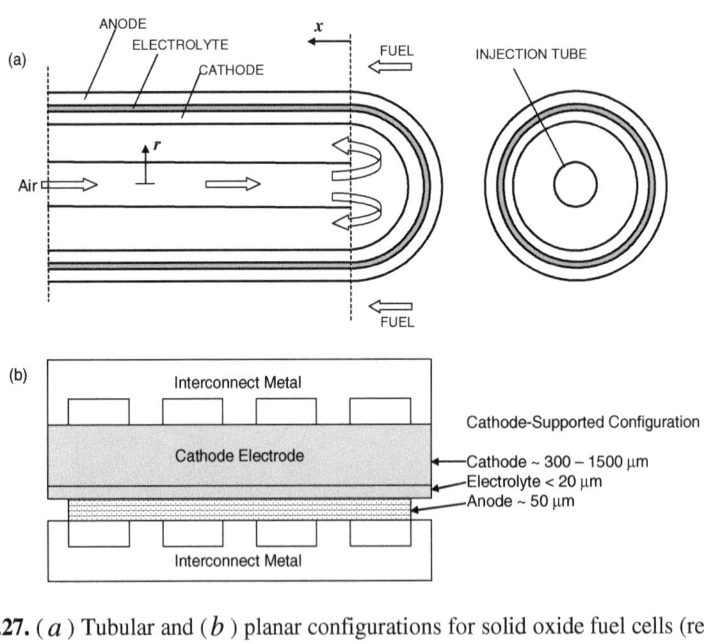

Figure 4.27. (*a*) Tubular and (*b*) planar configurations for solid oxide fuel cells (redrawn from [26, 206])

the transport and electrochemical kinetics toward optimal design and operation of a SOFC. A brief presentation on the cell components and electrochemistry is given in Section 4.3.1. Modeling approaches for a single cell and the cell electrodes are subsequently discussed in Section 4.3.2, and simulation results and discussion are presented in Section 4.3.3.

4.3.1 Component Materials and Electrochemistry

Figure 4.26 shows the basic transport and electrochemical processes involved in a SOFC. The ceramic electrolyte typically consists of 85–90% zirconia with 10–15% yttria, referred to as yttria-stabilized zirconia (YSZ). The fuel and air compartments of the cell are separated by the YSZ, which exhibits high ion conductivity and low electronic conductivity over a wide range of oxygen pressures, from about 1 atm on the cathode side to as low as 10^{-20} atm at the anode. The operating temperature is determined by the requirement for high ionic conductivity in the electrolyte. Since lower operating temperatures simplify the material requirements in other parts of the cell, the development of the substrate concept (*i.e.* a thin electrolyte layer supported on a thick electrode) is aimed at a reduction of temperature below 800°C.

During operation of a cell, an oxygen molecule is reduced to two oxygen anions at the cathode/electrolyte interface by combining with four electrons from an external circuit [178]:

$$O_2(g) + 4e^- \longrightarrow 2O^{2-} \tag{4.69}$$

To facilitate the cathode half-cell reaction, the material of the air electrode must have two necessary properties: (1) catalyzing the dissociation of O_2, and (2) conducting the electrons. Ionic conductivity is also desirable for extending the two-dimensional reaction interface further into the cathode. Thus, in most cases, the cathode is a composite of electronically conductive perovskite-based ceramic, typically Sr-doped $LaMnO_3$ (LSM), and the electrolyte material, YSZ. Note that the chemical species are in the gas state at the cathode owing to high operating temperature, and consequently, the flooding problem for the cathode seen in the case of a PEM fuel cell does not exist in SOFCs.

Driven by the concentration gradient, the oxygen anions migrate through the electrolyte from the cathode side to the anode, where the oxidation of the fuel takes place in the following general form [178]:

$$C_nH_{2n+2} + (3n+1)O^{2-} \longrightarrow nCO_2 + (n+1)H_2O + (6n+2)e^- \tag{4.70}$$

The electrons produced by the anode reaction are conducted through an external load to the cathode to complete the electric circuit, as shown in Figure 4.26. The anode material must also be catalytically active for the oxidation reaction, and possess sufficient electronic and ionic conductivities. The present state-of-the-art anode material is Ni-YSZ, a ceramic-metallic (cermet) composite, where Ni provides the catalytic activity and electronic conductivity, while the YSZ component gives ionic conductivity. The various components in a SOFC must have chemical compatibility and similar thermal expansion coefficient, and a large amount of research has been devoted to address the two issues [168, 169, 171, 172].

For a direct natural gas (CH_4) fueled SOFC, the generic oxidation reaction in Eq. (4.70) may be further divided into the endothermic reforming reaction (Eq. (4.71)), the exothermic shift reaction (Eq. (4.72)), and the electrochemical reaction (Eq. (4.73)):

reforming: $$CH_4 + H_2O \leftrightarrow CO + 3H_2 \quad (4.71)$$

shift: $$CO + H_2O \leftrightarrow CO_2 + H_2 \quad (4.72)$$

electrochemical: $$H_2 + O^{2-} \rightarrow H_2O + 2e^- \quad (4.73)$$

It must be mentioned that the reforming/shift reactions do not involve the transport of the oxygen anions, and may take place at the entire anode half-cell (*i.e.*, fuel channel and anode electrode); while the electrochemical reaction in Eq. (4.73) can only occur at the three-phase boundary (TPB) as shown schematically in Figure 4.28.

The performance of a SOFC depends strongly on the microscale structure of the three-phase boundary, where the electrolyte, the electron-conducting electrode, and the gas species meet in the space (see Figure 4.28). When a breakdown in connectivity is present in any of the three phases, the electrochemical reaction

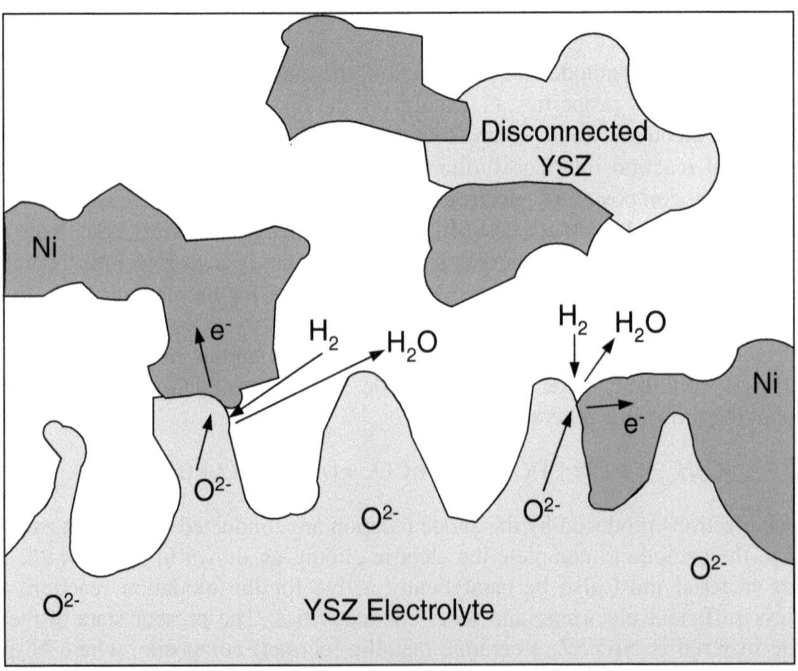

Figure 4.28. Schematic of a Ni/YSZ anode three-phase boundary (redrawn from [178])

cannot occur. Various experimental and theoretical studies have indicated that the TPB occupies a thin layer of space (around 10 μm) between the bulk electrode and bulk membrane [178]. Note that the three-phase boundary region in SOFC is similar to the catalyst layer in the PEM fuel cells.

4.3.2 Performance Models for SOFC

Similar to the discussion in Section 4.2.1, the voltage of a SOFC, E_{cell}, is related to the open cell voltage, E, and various losses, η, [181]:

$$E_{cell} = E - (\eta_{Ohm} + \eta_{conc,a} + \eta_{conc,c} + \eta_{act,a} + \eta_{act,c}) \tag{4.74}$$

where the subscripts Ohm, $conc$, and act imply ohmic, concentration, and activation losses, respectively, and a and c denote anode and cathode. The open circuit potential E depends on the temperature and gas composition at the electrodes, and may be given by the Nernst equation:

$$E = E^0 - \frac{RT}{2F}\ln\left(\frac{p_{H_2O,f}}{p_{H_2,f}\,p_{O_2,air}^{1/2}}\right) \tag{4.75}$$

where E^0 stands for the open circuit potential at standard temperature and pressure, p is pressure, and the subscript f and air denote fuel and air channels, respectively. The voltage losses are associated with the electrochemical reactions at the three-phase boundary, and are affected by the temperature, pressure, gas flowrate and composition, electrode/membrane materials, and cell designs. The losses consist of part the heat generation in an operating fuel cell. Furthermore, since both electrodes typically have high electrical conductivity, the cell voltage may be considered constant throughout the cell [182].

The ohmic loss, η_{Ohm}, is caused by the resistance to the ion conduction through the electrolyte and electron conduction through the electrodes and current collectors, and by the contact resistance between cell components. Ohm's law may be applied to relate η_{Ohm} to the current density, i, and the internal resistance, R_{Ohm}, of the cell:

$$\eta_{Ohm} = iR_{Ohm} \tag{4.76}$$

The internal resistance may be obtained experimentally or be estimated from conductivity data and thickness of each layer [183]:

$$R_{Ohm} = \frac{\delta_a}{\sigma_a} + \frac{\delta_m}{\sigma_m} + \frac{\delta_c}{\sigma_c} \tag{4.77}$$

where the subscript m denotes the membrane, and δ and σ represent the thickness and conductivity, respectively. Note that Eq. (4.77) ignores the contact resistance, which may be determined via the experimental or theoretically methods discussed previously for the PEM fuel cells [42].

Under open circuit conditions, the reactant and product concentrations at the three-phase boundaries are equal to those in the bulk channel flow, and the Nernst equation, Eq. (4.75), predicts the open circuit potential. During cell operation with nonzero current density, concentration gradients develop across the cell, resulting in lower concentration at the three-phase boundaries and concentration losses. The cathode and anode concentration losses, $\eta_{conc,c}$ and $\eta_{conc,a}$, may be expressed as [181]:

$$\eta_{conc,c} = \frac{RT}{4F}\ln\left(\frac{p_{O_2,air}}{p_{O_2,TPB}}\right) \tag{4.78}$$

$$\eta_{conc,a} = \frac{RT}{2F}\ln\left(\frac{p_{H_2O,TPB}\,p_{H_2,f}}{p_{H_2O,f}\,p_{H_2,TPB}}\right) \tag{4.79}$$

where the subscript TPB denotes the three-phase boundary. To calculate the concentration losses, the relationship between the partial pressures at the three-phase boundary and the current density i must be determined. Considering the gas diffusion in the porous electrodes, the following expressions are obtained [184–187]:

$$p_{H_2,TPB} = p_{H_2,f} - \frac{RT\delta_a}{2FD_{eff,an}}i \tag{4.80}$$

$$p_{H_2O,TPB} = p_{H_2O,f} + \frac{RT\delta_a}{2FD_{eff,a}}i \tag{4.81}$$

$$p_{O_2,TPB} = P-(P-p_{O_2,air})\exp\left(\frac{RT\delta_c}{4FPD_{eff,c}}i\right) \tag{4.82}$$

where the $D_{eff,a}$ and $D_{eff,c}$ are the effective diffusivities at the anode and cathode, respectively, and P is the total pressure.

Slow reaction kinetics at the three-phase boundary leads to the activation losses, which are governed by the Butler-Volmer equation [181]:

$$i = i_{0,elec}\left[\exp\left(\frac{\alpha nF}{RT}\eta_{act,elec}\right)-\exp\left(-\frac{(1-\alpha)nF}{RT}\eta_{act,elec}\right)\right] \tag{4.83}$$

where the subscript $elec$ denotes the electrodes, and may represent either a (anode) or c (cathode), α is the transfer coefficient, n is the number of electrons transferred in a single elementary rate-limiting reaction, and the exchange current density is expressed as [181]:

$$i_{0,elec} = \frac{RT}{nF}k_{elec}\exp\left(-\frac{E_{elec}}{RT}\right) \tag{4.84}$$

The kinetic constants k_{elec} and E_{elec} depend on the detailed electrode reaction mechanism and microstructure, and are reported by refs. [188–190]. The discussion in this subsection pertains to the gross polarization behavior of SOFCs fed on the pure hydrogen fuel, and analogous relationships may be developed for other fuels such as CH_4 and CO [178]. For the purpose of design and optimization, it is paramount to understand the in-depth description of the transport and electrochemical processes, which are covered in the next subsection.

4.3.3 Mechanistic Models for SOFC

In the earlier stage of development, focus had been put on introducing new materials for various component of a SOFC [165–180]. Active research efforts have been made recently to improve the understanding of the fundamental transport and electrochemical processes, aiming at optimal design and operation of the solid oxide fuel cells [191–210]. With the availability of affordable and effective computational hardware and software, the trends of SOFC modeling moved toward multiphysics and multiscale descriptions. Prinkey *et al.* [200] developed a computational fluid dynamics model to describe the reactant flow, transport, and electrochemical reaction in a SOFC. Recknagle *et al.* [207] simulated the operation of a SOFC with three flow configurations: co-flow, counter-flow, and cross flow. A self-consistent SOFC model based on the single-domain framework was developed by Pasaogullari and Wang [209] to solve the conservation equations for mass, momentum, species, thermal energy, and electric charge along with the electrochemical kinetics. The modeling studies at the single cell level and the component level are summarized in Sections 4.3.2.1 and 4.3.2.2, respectively.

4.3.3.1 Modeling Considerations at a Single Cell Level

The basic conservation laws embedded in Eqs. (4.3–4.8) may also be applied to various layers of a SOFC, and consequently, the presentation of the entire set of governing equations is omitted in this subsection for the purpose of brevity. The discussion is focused on two specific features during the solid oxide cell operation, namely, (i) the radiation effect and (ii) the modeling of the internal reforming. Most modeling studies neglect the radiation heat transfer, which may significantly influence the temperature distribution and electrochemical kinetics [194]. To account for the radiation effect, an additional source term must be added in the energy equation, Eq. (4.8), as follows [194]:

$$\text{energy}: \frac{\partial[(\rho c_p)_m T]}{\partial t} + \nabla\cdot(\rho c_p \mathbf{u} T) = \nabla\cdot(k^{eff}\nabla T) + S_T - \nabla\cdot\mathbf{q}_R \qquad (4.85)$$

where $\mathbf{q}_R$ is the radiation heat flux from an semi-transparent body, and is a function of the radiative properties (*e.g.*, the optical thickness τ) of the material and the temperature.

The optical thickness, τ, is defined as the product of the absorption coefficient, κ, and the thickness, δ, of the medium:

$$\tau = \kappa\delta$$

For optically thick slabs ($\tau >> 1$) including the Ni-YSZ anode and LSM cathode, the radiative heat flux may be estimated with the Rosseland approximation as [211]:

$$\mathbf{q}_R = -k_R \nabla T = -\frac{16n^2\sigma T^3}{3\beta_R}\nabla T \tag{4.86}$$

where k_R represents the radiative conductivity, n denotes the refractive index of the media, σ is the Stefan-Boltzmann constant, and β_R refers to the Rosseland-mean extinction coefficient (equal to the constant absorption coefficient for the gray and non-scattering medium). It must be mentioned that the diffusion approximation in Eq. (4.86) is not valid near a boundary, and may be modified with the Schuster-Schwartzchild two-flux approximation for locally optically thin regions near the boundaries of the overall optically thick material [194].

For the case of optically thin media such as the YSZ membrane, the Schuster-Schwartzchild or two-flux approximation provides a simple solution for one-dimensional radiation in a thin slab [194]. Assuming a gray, non-scattering medium with thickness δ confined between two isothermal black plates at temperatures T_{top} and T_{bot}, the two-flux model gives the radiative heat flux at a thickness location z as [211]:

$$q_R(z) = -\sigma(T_{top}^4 - T^4)e^{2\kappa(z-\delta)} + \sigma(T_{bot}^4 - T^4)e^{-2\kappa z} \tag{4.87}$$

The radiative fluxes given in Eqs. (4.86) and (4.87) may be substituted into the energy equation, Eq. (4.85) to determine the radiation effect.

An important advantage of SOFC is the possibility of direct utilization of hydrocarbon fuels such as CH_4, which eliminates the need for complicated external reforming unit and therefore saves capital costs. The internal reforming reactions, given in Eqs. (4.71) and (4.72), must be appropriately treated in the species conservation equation, Eq. (4.5). The generation/consumption rates (i.e, S_k in Eq. (4.5)) of the species involved (*i.e.*, CH_4, CO, H_2, H_2O, and CO_2) must be determined from the corresponding reaction kinetics. For the hydrogen species, the generation rates from the reforming and shift reactions, S_{r,H_2} and S_{s,H_2}, may be given as [212, 213]:

reforming: $$S_{r,H_2} = k_r^+ p_{CH_4} p_{H_2O} - k_r^- p_{CO} p_{H_2}^3 \tag{4.88}$$

shift: $$S_{s,H_2} = k_s^+ p_{CO} p_{H_2O} - k_s^- p_{CO_2} p_{H_2} \tag{4.89}$$

where k denotes the rate constant, p is the partial pressure, the subscript r (s) indicates the reforming (shift) reaction, and the superscripts + and – refer to the forward and backward chemical reactions, respectively. Note that the production/ consumption rates of other species may be obtained from Eqs. (4.88) and (4.89) through the stoichiometric ratios indicated in Eqs. (4.71) and (4.72).

The forward reaction constants, k_r^+ and k_s^+, may be determined by the following Arrhenius functions [212, 213]:

reforming: $$k_r^+ = 2395 \exp\left(-\frac{231266}{RT}\right) \tag{4.90}$$

shift: $$k_s^+ = 0.0171 \exp\left(-\frac{103191}{RT}\right) \tag{4.91}$$

The backward reaction constants, k_r^- and k_s^-, are obtained from the equilibrium constants, K_{pr} and K_{ps} for the two reactions:

reforming: $$K_{pr} = \frac{k_r^+}{k_r^-} = \frac{p_{CO} p_{H_2}^3}{p_{CH_4} p_{H_2O}} \tag{4.92}$$

shift: $$K_{ps} = \frac{k_s^+}{k_s^-} = \frac{p_{CO_2} p_{H_2}}{p_{CO} p_{H_2O}} \tag{4.93}$$

which, in turn, are defined as functions of temperature by the following empirical relations [214]:

$$K_{pr} = 1.0267 \times 10^{10} \times \exp(-0.2513Z^4 + 0.3665Z^3 + 0.5810Z^2 - 27.134Z + 3.2770) \tag{4.94}$$

$$K_{ps} = \exp(-0.2935Z^3 + 0.6351Z^2 + 4.1788Z + 0.3169)(Pa^2) \tag{4.95}$$

$$Z = \frac{1000}{T(K)} - 1 \tag{4.96}$$

The modeling approaches for the radiation heat transfer and the internal reforming kinetics presented in the subsection may be incorporated with the conservation laws in Eqs. (4.3–4.8) to predict the detailed transport and reactions and the overall performance of a SOFC. In the next subsection, mechanistic models for the cell electrodes are discussed, and the simulation results for SOFCs are presented in Section 4.3.4.

4.3.3.2 Modeling Approaches for the SOFC Electrodes

To date, the modeling studies of SOFC components have focused on the anode or cathode electrodes, while no work seems to be available to specifically treat the membrane region. The performance of electrodes depends on the relevant transport and reaction parameters, as well as the material properties and geometry (*e.g.*, ionic conductivity, porosity, particle size). Depending on the length scale involved, three types of models are reported in the literature, namely, (i) macroscopic continuum electrode theories, (ii) many particle or Monte Carlo methods, and (iii) models for local current density distribution within individual particles [215, 216].

The continuum electrode approach disregards the actual geometric details, and the porous electrode is characterized by homogeneous effective parameters (*e.g.*, effective ionic and electronic conductivities, effective electrochemical reaction rate, and effective gas diffusion coefficient). Despite the simplifications, valuable insights can be obtained from the continuum simulation studies [216]. To illustrate the mathematical procedure employed in these studies, consider a composite cathode of height h made from an electrode/electrolyte mixture, with high electronic conductivity (*i.e.* constant electronic potential in the electrode material), fast gas diffusion in the pores (*i.e.*, zero oxygen partial pressure gradient), and an effective ionic conductivity σ_{ion}. A voltage U is applied on the cathode such that the current only flows in the thickness y direction (see Figure 4.26). On the continuum level, oxygen is reduced at any location of the cathode and the current density of the oxide ions is given by [216]:

$$i_{O^{2-}} = -\frac{\sigma_{ion}}{2e}\frac{d\Phi_{O^{2-}}}{dy} \tag{4.97}$$

where e represents the electric charge of an electron and $\Phi_{O^{2-}}$ denotes the electrochemical potential of oxide ions. The rate of the electrochemical reaction reads [216]:

$$\frac{di_{O^{2-}}}{dy} = \frac{Y_{vol}}{2e}(\Phi_{O^{2-},eq} - \Phi_{O^{2-}}) \tag{4.98}$$

In Eq. (4.98), Y_{vol} is a rate constant for the volumetric reaction and $\Phi_{O^{2-},eq}$ denote the electrochemical potential of O^{2-} at zero current density. The combination of Eqs. (4.97) and (4.98) yields the governing equation for the potential $\Phi_{O^{2-}}$:

$$\frac{d^2\Phi_{O^{2-}}}{dy^2} = \frac{Y_{vol}}{\sigma_{ion}}(\Phi_{O^{2-}} - \Phi_{O^{2-},eq}) \tag{4.99}$$

which may be solved analytically with the boundary conditions $\Phi_{O^{2-}} = \Phi_{O^{2-},eq} - 2eU$ at $y = 0$ and $d\Phi_{O^{2-}}/dy = 0$ at $y = h$, and the solution has the form [217]:

$$\Phi_{O^{2-}} = \Phi_{O^{2-},eq} - 2eU\frac{\cosh\left[(h-y)\sqrt{Y_{vol}/\sigma_{ion}}\right]}{\cosh\left[h\sqrt{Y_{vol}/\sigma_{ion}}\right]} \tag{4.100}$$

The effective parameters, σ_{ion} and Y_{vol} in Eq. (4.100), depend on the material properties and geometrical parameters such as the porosity and contact geometry between particles, and may be evaluated by the microstructural modeling studies discussed in the following.

The many particle or Monte Carlo models [216] treat an electrode as a random mixture of particles of three phases, namely, (1) electrode, (2) electrolyte, and (3) gas; and the electrical and electrochemical properties of the electrode are obtained from various types of resistors. The current between two adjacent solid particles of phase k is governed by a resistor r_k, which is related to the conductivity of phase σ_k and a geometric factor f_{geo} as [216]:

$$r_k = f_{geo}/\sigma_k$$

where f_{geo} accounts for the shape of the particles and the contact necks between adjacent particles. An electrochemical reaction takes place at sites where the electrode, electrolyte, and gas particles meet, and the heterojunctions of the three phases are represented by transfer resistors r_{ech}. The assembly of electrode particles consists of a network of the elementary resistors, and the desired electrical and electrochemical properties (*e.g.*, σ_{ion} and Y_{vol}) are calculated by applying Kirchhoff's law to the resistor network [216].

A critical assumption made in the Monte Carlo calculations is uniform current density within individual particles. Several modeling studies [216, 218] have shown that the current density distribution on the particle length scale is inhomogeneous, and may strongly influence the polarization behavior. It is found that the reaction zones may be confined to the three-phase boundaries, resulting in larger resistance. Modeling studies on the local current density distributions on the particle length scale have been performed by finite element method, finite difference method or resistor network simulations with numerous resistors in one particle [219, 220].

4.3.4 Results and Discussion

Numerical results on the operation of SOFCs are widely available in the literature, including the polarization behavior, the distributions of temperature, pressure, velocity, species concentration, and current density for various cell designs (planar/tubular) and flow field configurations [191–210]. The present discussion focuses only on the two specific issues considered in Section 4.3.2.1 for SOFCs, namely, the radiation heat transfer and the internal reforming, which are less studied in the literature. Figures 4.29(a) and 4.29(b) show the temperature variation in the flow channel direction along the (a) anode/electrolyte and (b) cathode/electrolyte interfaces, respectively [194]; the solid (dashed) lines correspond to results with (without) inclusion of radiation effects. Evidently, the cooling enhancement due to radiation results in an overall decrease of the temperature within the cell, as well as a smaller temperature gradient along the flow channel direction.

The cell voltage is proportional to the change of in Gibbs free energy of the electrochemical reaction ΔG, which is significantly changed by the radiation effect in Figure 4.29. The thermodynamic relation for the Gibbs free energy is given by [221]:

$$\Delta G = \Delta H - T\Delta S \tag{4.101}$$

where the reaction enthalpy ΔH corresponds to the total thermal energy available in the system, ΔS represents the change in entropy and is a nonlinear function of temperature [221]

$$\Delta S = \Delta S^0 + a \ln\left(\frac{T}{298}\right) + b(T - 298) + 0.5c(T - 298)^2 \qquad (4.102)$$

where a, b, and c are empirical constants, and ΔS^0 is the standard entropy of formation. With the additional heat transfer path provided by radiation, a significant decrease (around 180 K) in fuel cell temperature is seen in Figsures 4.29(a) and 4.29(b), which, in turn, results in a 20% drop in the $T\Delta S$ value in light of

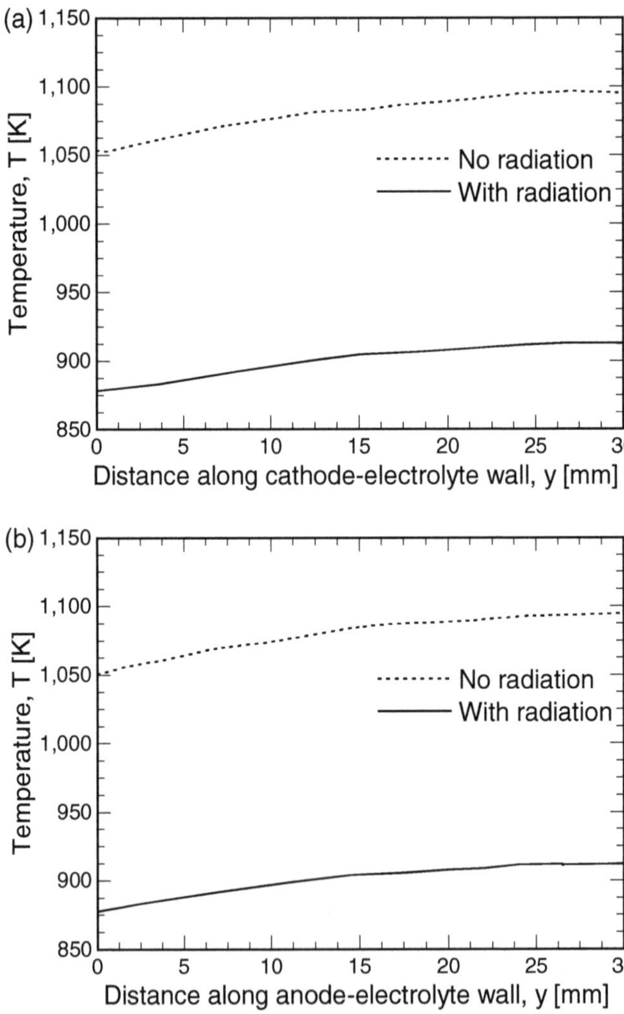

Figure 4.29 Effect of radiation on the temperature variation in the flow channel direction along the (*a*) anode/electrolyte and (*b*) cathode/electrolyte interfaces (redrawn from [194])

Eq. (4.102). From Eq. (4.101), the Gibbs free energy ΔG increases with decreasing $T\Delta S$, resulting in a significantly increase in cell voltage from 0.65 to 0.74 V [194].

Most fuel cells require the conversion of the primary fuel into a hydrogen-rich gas via an external steam reformer, whereas direct reforming is possible in a SOFC due to the high operating temperatures. Both indirect and direct internal reforming designs are developed for SOFCs. In the indirect approach, a separate reformer section is in close thermal contact with the fuel cell anode, as shown in Figure 4.30 [192]. For the case of direct internal reforming design, methane is fed directly into the cell and reforming takes place inside the anode electrode [192]. A common problem associated with both internal reforming designs is the mismatch between the amount of heat required for the endothermic reforming reaction and the heat released from the fuel cell electrochemical reactions. The rates of the reforming reactions may be orders of magnitude larger than those of the fuel cell electrochemical reactions, leading to significant local cooling and thermally induced fractures of the ceramic components [192].

To reduce the local cooling rate in an indirect internal reforming design, a catalyst with significantly smaller activity is adopted [222]. However, a less active reforming catalyst can not fully convert all the methane to synthesis gas in the reformer section, and additional methane reforming (or methane slippage) is performed within the anode, resulting in a combination of indirect and direct internal reforming [192]. Figure 4.31(a) presents the averaged temperature profiles of the reformer, fuel channel, and air channel in Figure 4.30 for a less active reformer catalyst [192]. Local temperature decreases are seen near both ends of the fuel cell, owing to the endothermic reforming reactions at the inlet of the reformer ($z=0$) and the inlet of the fuel channel ($z=1$). For the case considered in Figure 4.31(a), the ratio between the rate of steam reforming reaction and the rate of electrochemical reaction is approximately 13 at the fuel channel inlet, and consequently, relatively large temperature gradients are seen at $z=1$.

When a more active catalyst is used in the reformer, large temperature gradients only appear near the reformer inlet, as shown in Figure 4.31(b). Since all the methane is converted within the reformer, no local decrease in temperature is seen near the

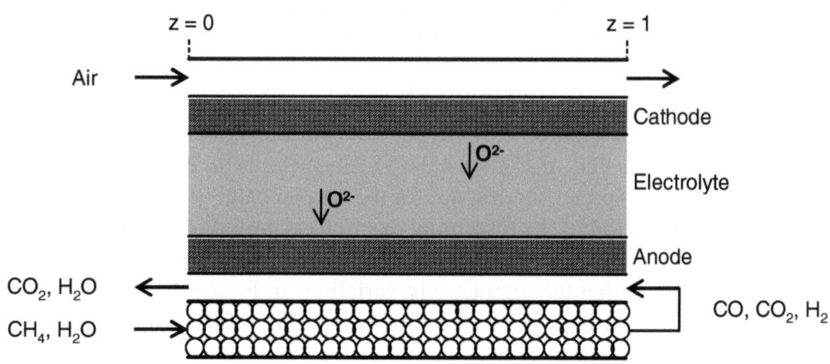

Figure 4.30. Schematic diagram of a solid oxide fuel cell with an indirect internal reformer (redrawn from [192])

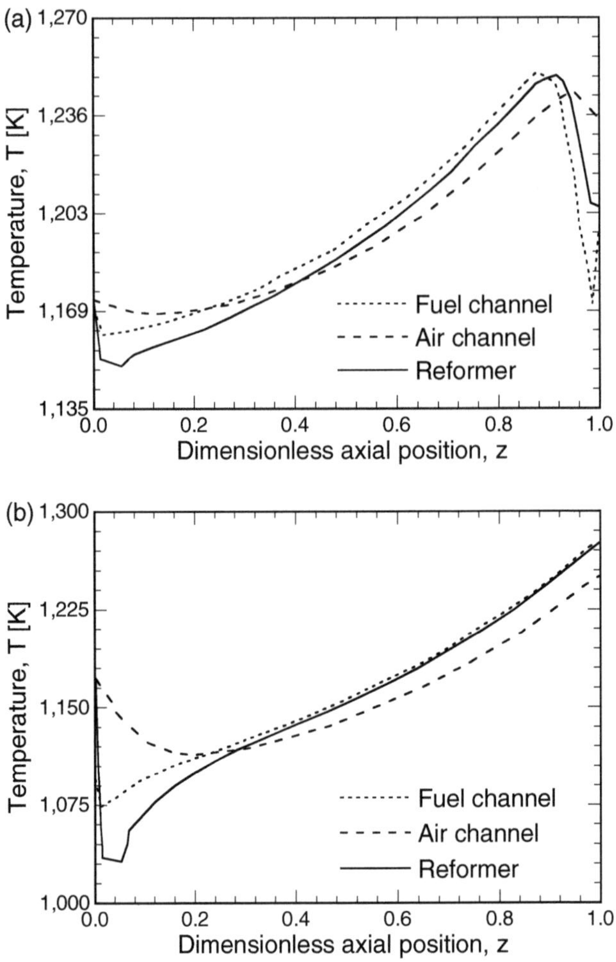

Figure 4.31. Temperature profiles in the reformer, fuel channel, and air channel for relative catalyst activity of (*a*) 0.008% and (*b*) 0.04% within the reformer region (redrawn from [192])

inlet of fuel channel in Figure 4.31(b). The overall temperature variation in Figure 4.31(a) is less than that in Figure 4.31(b), since the heat-absorbing reforming reactions are relatively spread out in space for the smaller catalyst activity case. Since uniform space distribution of the reforming reactions is preferred, the anode catalyst activity is reduced in Figure 4.32 to spread the reforming reactions within the anode region [192]. It is seen that the temperature variations in Figure 4.32 are further reduced when compared with the corresponding cases in Figures 4.31(a) and 4.31(b).

The solid oxide fuel cell described in this section may be an ideal candidate for stationary power generation. For applications such as portable electronic devices, the direct methanol fuel cell discussed next is appropriate since it allows for a simpler fuel delivery system and a more compact design.

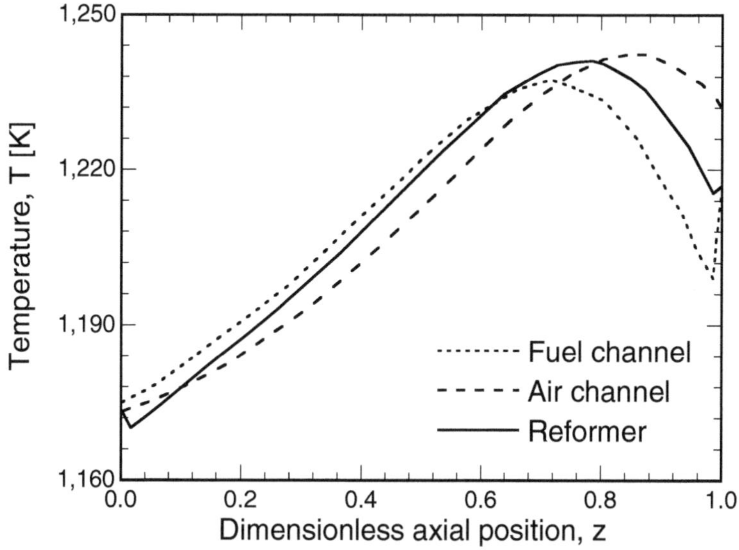

Figure 4.32. Temperature profiles in the reformer, fuel channel, and air channel for relative catalyst activity of 0.008% within the anode region (redrawn from [192])

4.4 Direct Methanol Fuel Cells

In comparison with PEM fuel cells, the technical and economical advantages of direct methanol fuel cells (DMFCs) include (1) elimination of the fuel reforming unit in the power plant and (2) easy storage of the methanol liquid than the hydrogen gas [3, 223–227]. The component materials of DMFCs are similar to those of PEM fuel cells with the main difference being the adoption of platinum-ruthenium as the anode catalyst. An aqueous methanol solution of low molarity, introduced in the fuel channel, diffuses through the backing layer to the anode catalyst layer. The global methanol oxidation reaction at the platinum-ruthenium catalyst sites generates carbon dioxide, protons, and electrons, as shown in Table 4.1. Air is fed to the cathode side flow channel, and the oxygen species react with the protons transported through the polymer membrane and the electrons from the external circuit to form water.

The flow field at the anode and cathode sides of DMFCs involves gas-liquid two-phase flow with different characteristics. A distinctive flow pattern in DMFCs is caused by the formation of CO_2 bubbles on the anode side [228–230]. Carbon dioxide bubbles resulting from the anode electrochemical reaction nucleate at certain locations and grow with time to form discrete gas slugs in the flow channel. Small bubbles tend to cover the surface of the backing layer by strong surface tension, and thus decrease the effective mass transfer area. With the growth of the bubble size, detachment of large bubbles takes place along the backing layer surface, resulting in the sweeping and cleaning of the small bubbles. The gas-liquid two-phase flow then restarts from the nucleation of small bubbles and proceeds in the periodic sequence of nucleation-growth-detachment. The flow of the oxidant on the cathode side is characterized by the formation of liquid water droplets and the flooding phenomenon, as discussed previously for the PEM fuel cells.

The key challenges in the development of a DMFC are (1) the slow electro-oxidation rate of the methanol, (2) large rate of methanol crossover through the Nafion electrolyte, and (3) water and heat management [3, 223–227]. Owing to the combination of slow anode kinetics and methanol crossover, the power density in a DMFC is about three to four times lower than that in a hydrogen fuel cell. With the electrocatalyst (Pt-Ru) loading ten-fold higher than that in PEM fuel cells, the electro-oxidation rate of methanol may be still orders of magnitude lower than the corresponding value of hydrogen. Efforts have been made to develop different anode catalyst structures and to explore different materials [231–234]. A higher reaction rate enables the reduction of activation overpotential, leading to enhanced energy efficiency and power density.

Unlike hydrogen, methanol is completely soluble in the perfluorosulfonic acid (PFSA) membrane and the crossover rate is much higher than that of hydrogen. Diffusion, electro-osmotic drag and pressure-driven advection are the prime mechanisms for the methanol transport through the electrolyte. The chemical reaction of methanol on the cathode side causes a mixed potential or decreased cell voltage, resulting in reduced fuel efficiency and energy density. Extensive experimental and theoretical studies have been conducted on the methanol crossover in DMFCs [235–239]. Two approaches are reported to control the crossover via increasing the mass transfer resistance namely, (1) development of novel membranes with low methanol permeability and (2) adoption of a microporous layer between the membrane and the backing layer [26]. A higher concentration of fuel may be used for a cell design with lower methanol crossover rate, resulting in much higher power density of the DMFC system.

Constrained by the methanol crossover, the fuel concentration adopted in typical DMFCs is low, and consequently, a large amount of water is present in the fuel stream. Note that the electrochemical reactions of DMFC consume water at the anode side while producing water at the cathode side. The electro-osmotic drag effect further increases the water consumption (generation) at the anode (cathode). Water management in DMFCs refers to the dual tasks of (i) effectively removing water from the cathode to prevent flooding and (ii) supplying H_2O to the anode to replenish the water losses.

The issue of thermal management is strongly coupled with the water and methanol transport processes. Due to the much lower energy efficiency (only 20–25% when operating between 0.3 and 0.4 V), a large amount of heat is generated in a DMFC; *e.g.*, 60–80 W of waste heat may be produced for a 20 W system. The heat is removed by (i) the liquid fuel flow on the anode side and (ii) water evaporation on the cathode side. Note that the second heat removal route influences the water management issue. Furthermore, higher operating temperature promotes the anode oxidation reaction and thus decreases the activation losses, while the adverse effect is the increase in methanol crossover rate. Fundamental modeling of the processes discussed so far is needed for the design and operation optimization of the DMFC system.

4.4.1 Performance Models

Various models have been reported in the literature on the polarization performance of a direct methanol fuel cell [240–243]. Two distinctive processes affect the efficiency and voltage of DMFCs, namely, the multistage nature of the overall anode electrochemical reaction (see Table 4.1) and the methanol transport. During operation of a DMFC, CO may be produced as stable and adsorbed intermediates at the anode side. The adsorbants occupy the active catalyst sites, leading to considerably high anodic overpotential and low cell voltage. In general, high methanol molarity is desirable to reduce the concentration overpotential. However, the increased methanol permeation at higher fuel concentration enhances the methanol oxidation at the cathode, forming a mixed potential that reduces the overall cell voltage.

The voltage of a DMFC, E_{cell}, is given by the difference between the open circuit voltage, E, and three types of overpotentials [243]:

$$E_{cell} = E - \eta_a - \eta_c - \eta_{ohm} \tag{4.103}$$

where η_a and η_c denote the anode and cathode overpotentials, respectively, and η_{ohm} is the ohmic loss. The open circuit potential may be calculated using Nernst's equation, and has a similar value ($E = 1.21\ V$) to the hydrogen feed PEM fuel cells under room temperature and standard atmosphere pressure [243]. The determination of the three overpotentials is discussed in the following.

The anodic overpotential, η_a, depends on the temperature, methanol concentration, and the kinetic rates of the following four elementary electro-oxidation steps [244–246]:

$$CH_3OH + Site_1 \rightarrow CH_3OH_{ad,1} \tag{4.104}$$

$$CH_3OH_{ad,1} \leftrightarrow CO_{ad,1} + 4H^+ + 4e^- \tag{4.105}$$

$$H_2O + Site_2 \leftrightarrow OH_{ad,2} + H^+ + e^- \tag{4.106}$$

$$CO_{ad,1} + OH_{ad,2} \rightarrow CO_2 + H^+ + e^- + Site_1 + Site_2 \tag{4.107}$$

The first two steps in Eqs. (4.104) and (4.105) correspond to the adsorption and dissociation of the methanol species, and Eq. (4.106) denotes the ionization of the water molecule at a catalyst site. The final step, Eq. (4.107), involves the reaction of two adsorbed species and is the rate-determining step at relatively low current densities; when the methanol concentration is small, the transport of methanol to the catalyst surface (Eq. (4.104)) limits the overall reaction. Other possible elementary steps of the anodic methanol oxidation are discussed in the literature [247].

Based on the multistep analyses of the anode electrochemical reaction, a physical model predicts the anode overpotential, η_a, as a function of the current density i [245]:

$$i = \frac{i_{0,a} c_M a \exp(2.303\eta_a/b)}{c_M + Ka\exp(2.303\eta_a/b)} \tag{4.108}$$

where $i_{0,a}$ denotes the exchange current density at the anode, c_M is the methanol concentration in the catalyst layer, b represents the Tafel slope, and K and a are constants. For large methanol concentration c_M and small anode overpotential, the exponent term in the denominator of Eq. (4.108) may be neglected, a zero-order kinetics is obtained with the current density i being independent of the methanol concentration:

$$i = i_{0,a} a \exp(2.303\eta_a / b) \tag{4.109}$$

In the case of high overpotential, the concentration term in the denominator can be neglected, indicating a transition to first-order kinetics.

To calculate the cathode overpotential, η_c, the methanol permeation across the membrane must be known. In the absence of a pressure gradient between anode and cathode, the methanol permeation depends on the combined effects of the diffusion and electro-osmotic processes. Dohle and Wippermann [243] assumed the methanol permeation, i_{perm}, to be a linear function of the current density, i, with an offset m_1 and a slope m_2:

$$i_{perm} = m_1 + m_2 i \tag{4.110}$$

The offset m_1 pertains to the diffusion contribution, and is related to the membrane thickness δ_m, methanol concentration c_M, and temperature T as [243]:

$$m_1 = 4.65V - 2\frac{\delta_m}{\delta^{ref}} \frac{c_M}{c_M^{ref}} \exp\left[0.01329(T - 273.15)\right] \tag{4.111}$$

where the superscript, ref, denotes a reference value. The slope m_2 represents the effect of the electro-osmotic drag, and is a function of the methanol concentration, c_M, and the drag coefficient, n_d [243]:

$$m_2 = -0.4 + 0.27\frac{n_d}{n_d^{ref}} \frac{c_M}{c_M^{ref}} \tag{4.112}$$

The cathode overpotential, η_c, is influenced by the temperature, current density, and the methanol permeation effect, and may be given by [243]:

$$\eta_c = \frac{RT}{\alpha F} \ln\left[\frac{i}{\delta_c i_{0,c}} \frac{c_{O_2}^{ref}}{c_{O_2}}\right] + k(i) i_{perm} \tag{4.113}$$

The Tafel equation is employed to derive the first term in Eq. (4.113), where α is the transfer coefficient, δ_c denotes the catalyst layer thickness, $i_{0,c}$ represents the exchange current density at the cathode, c_{O_2} is the oxygen molar concentration, and $c_{O_2}^{ref}$ is a reference oxygen concentration. The second term in Eq. (4.113) represents the methanol permeation effect, and $k(i)$ is an empirical factor depending on the current density.

Finally, the ohmic loss, η_{ohm}, is related to the conductivity σ and thickness δ of various layers:

$$\eta_{ohm} = (\frac{\delta_m}{\sigma_m} + \frac{\delta_a}{\sigma_a} + \frac{\delta_c}{\sigma_c})i \tag{4.114}$$

Note that the electrical resistances are assumed to be in series in Eq. [4.114], and the influence of the contact resistance is neglected.

4.4.2 Mechanistic Models

Due to the great complexities of transport and electrochemical processes discussed so far in Section 4.4, detailed simulations on the operation of DMFCs are less reported in the literature. Scott and coworkers [229, 241] presented several simplified single-phase models for the transport and electrochemical processes in DMFC. Baxter *et al.* [248] developed a one-dimensional model for a liquid-feed DMFC anode, where the carbon dioxide is assumed to be dissolved in the liquid and no gas phase is present. Kulikovsky and coworkers [249, 250] performed numerical studies on both vapor-feed and liquid-feed DMFCs; however, the important issue of methanol crossover was ignored. A thermodynamic framework for a multicomponent membrane was developed by Meyers and Newman [244–246], where the concentrated-solution theory was adopted for the species transport. The study neglected the effects of pressure-driven flow and the transport of gaseous carbon dioxide at the anode.

A few studies are presented to attempt the two-phase flow problems involved during the operation of DMFCs. Murgia *et al.* [251] developed a one-dimensional, two-phase model for a liquid-feed DMFC based on phenomenological transport equations for the backing, catalyst, and membrane layers. Wang and Wang [252] presented a two-phase multicomponent model that accounts for the capillary effects in both anode and cathode backings. The model describes the coupled processes of the electrochemical reactions at the catalyst layers, and the diffusion/convection of both gas and liquid phase species in the backing layers and the flow channels. The simulation results indicate that the gas-phase transport is critical to the cell performance. A comprehensive two-phase model for DMFCs was presented by Divisek *et al.* [247] where the capillary effects of hydrophilic and hydrophobic pores were taken into account. Mass transport of various species occurs in parallel in the gas and liquid phases, and exchange between the phases is governed by condensation and evaporation. Multistep electrochemical reactions are described in ref. [247] to illustrate the effect of blocking the reaction sites with various intermediate species.

Although much effort has been made to simulate the DMFC system, considerable work remains to be done to effectively assist design and operation optimization. Few studies illustrate the important effects of two-phase flow, and no simulation result has been reported to predict the two-phase flow patterns observed in experimental studies. Likewise, no model is available in the literature to provide a microfluidic theory for portable DMFC systems [26]. Furthermore, comprehensive studies are needed to clearly illustrate the coupling among various processes, including the methanol transport, CO_2 gas dynamics, elementary electrochemical reactions, and water and thermal management.

4.4.3 Results and Discussion

Detailed two/three dimensional simulation results on the transport processes in DMFCs are still less reported in the literature, while most modeling studies focus on the prediction of the overall polarization behavior. In this subsection, a brief discussion on the experimental validation of the models and the effects of the methanol crossover are presented. Validation of the theoretical models is commonly carried out by comparison with experimental data on cell voltage as a function of current density [240–243].

Figure 4.33(a) demonstrates the validation of a two-phase DMFC model by Wang and Wang [252] at two operating temperatures; the lines represent the model predictions while the symbols denote the experimental data. Good agreement is seen for the entire range of current density and temperature considered. A smaller limiting current density is seen for the case of temperature $T = 50^o$ C, which may be explained by the lower diffusion coefficients in both gas and liquid phases and lower saturation methanol vapor concentration in the gas phase at lower temperature. Figure 4.33(b) presents the model validation at two different methanol feed concentrations. In agreement with experimental data, the model prediction for the higher concentration of 2 M shows a slightly lower voltage for current density $I < 0.36 A/cm^2$ and relatively larger limiting current density.

The rate of methanol crossover (in mol/cm^2s) was predicted by the model of Jeng and Chen [253] as a function of the current density I and methanol feed concentration in Figure 4.34(a). Recall that methanol permeation is influenced by the diffusion and electro-osmotic drag processes. For fixed current density, the crossover rate is seen to increase monotonically with increasing feed concentration, owing to an enhanced diffusion process. With increasing current density, the methanol concentration at the anode catalyst/membrane interface decreases due to the higher consumption rate, and consequently, the diffusion rate of methanol through the membrane decreases. When the methanol feed molarity is low, the methanol flux caused by the electro-osmotic drag also decreases with increasing I, since most methanol molecules are consumed by the electrochemical reaction. Consequently, the overall rate of methanol crossover decreases monotonically with increasing current density for low feed concentrations, *e.g.*, $\leq$ 3 M in Figure 4.34(a). For the case of a 4 M feed concentration, however, an initial increase in methanol crossover is seen at relatively small current densities (< 0.1 A/cm^2), owing to the ehanced electro-osmotic drag effect with increasing current density.

Figure 4.34(b) shows the methanol crossover as a percentage of the total methanol flux transported from the feed stream to the electrode for various current densities and feed concentrations [253]. As physically expected, the crossover mechanism consumes the entire methanol feed when the current density equals zero. For the feed concentration of 1 M, zero crossover may be obtained when the current density reaches a limiting value of 0.22 A/cm^2. The limiting current density is seen to increase monotonically with the feed concentration. For feed concentrations higher than 2M and current density lower than 0.1 A/cm^2, the crossover is more than 60% of the total methanol flux, which corresponds to an overall efficiency of less than 20%.

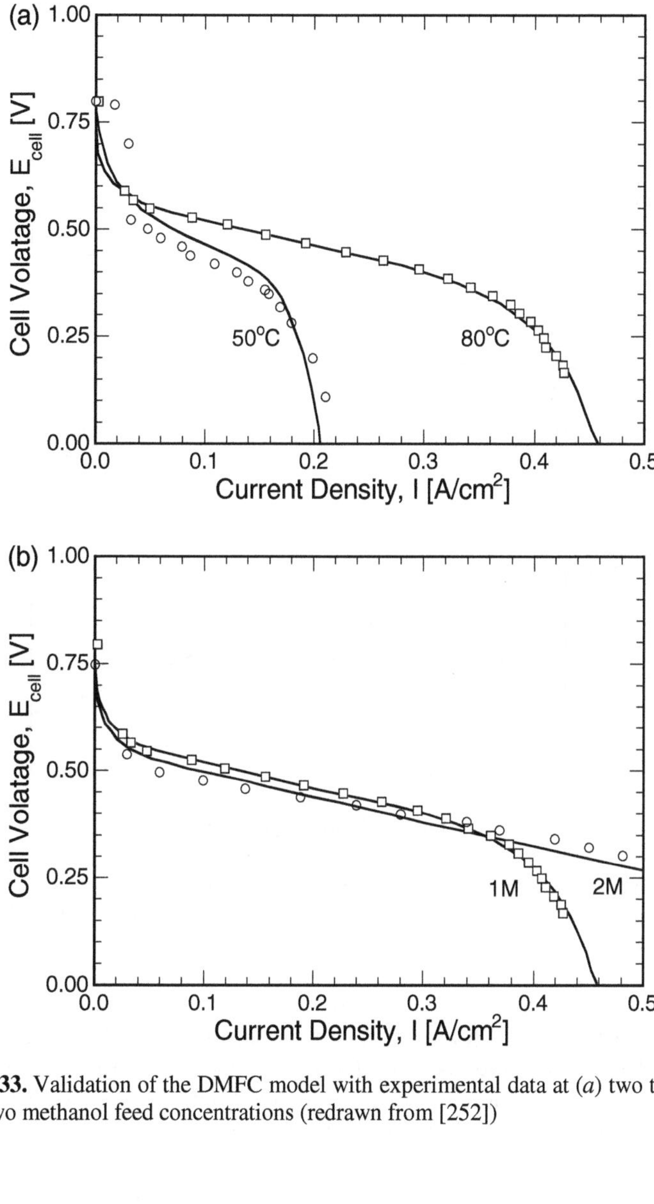

Figure 4.33. Validation of the DMFC model with experimental data at (*a*) two temperatures and (*b*) two methanol feed concentrations (redrawn from [252])

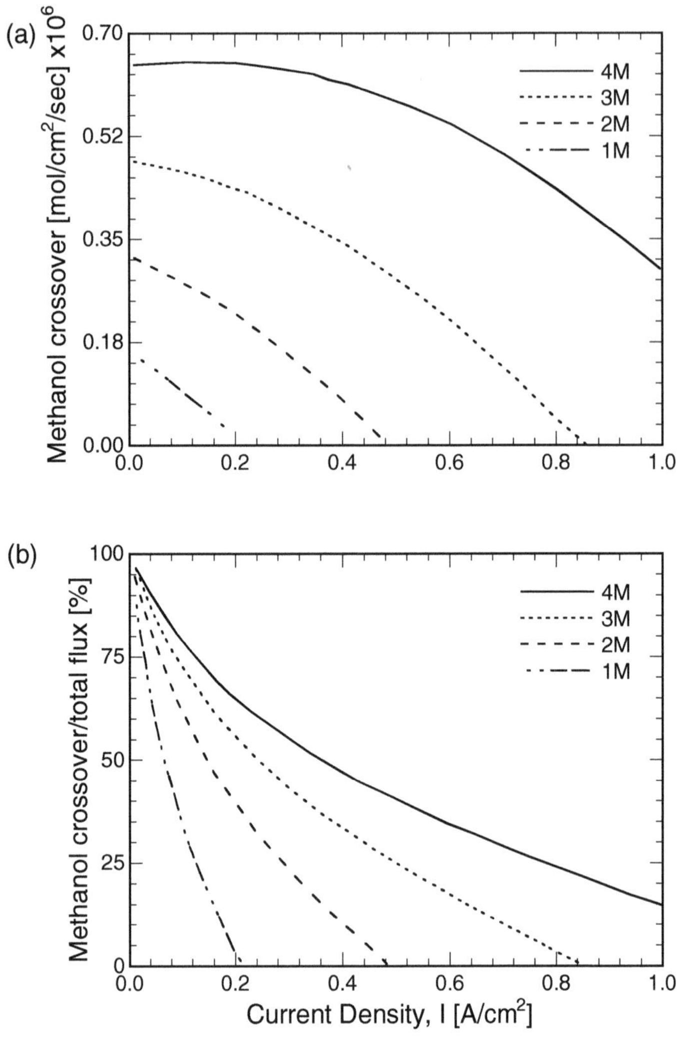

Figure 4.34. (a) Absolute rate of methanol crossover and (b) percentage of methanol crossover with respect to the total methanol feed flux as functions of current density and feed concentration (redrawn from [253])

The effects of methanol feed concentration on the polarization curve are summarized in Figure 4.35 [252]. For small current densities, the cell voltage increases with lower feed concentrations since the rate of methanol crossover is limited. However, low feed concentration suffers from low limiting current density. Operation with current density $I > 0.22 A/cm^2$ requires a relatively higher feed concentration (> 0.5 M). When the methanol feed concentration is greater than 1 M, the cell may experience substantial methanol crossover, resulting in a significant decrease in cell voltage. In the case of 6 M feed, a significant amount of oxygen in

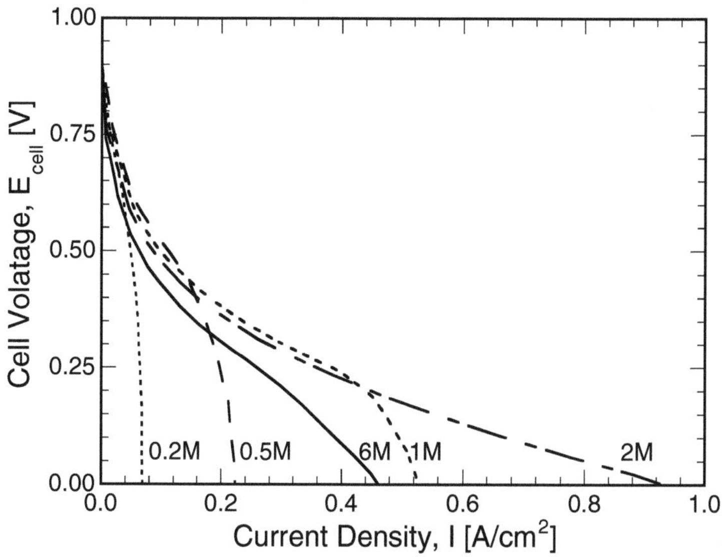

Figure 4.35. Prediction of polarization curves at different methanol feed concentration (redrawn from [252])

the cathode is consumed by the reaction with crossed methanol, and the maximum current is limited by the oxygen transport.

Detailed theoretical analyses on three types of fuel cells are reviewed in Sections 4.2–4.4, along with presentation of the significant physical insights gained from the description on the relevant transport and electrochemical processes. It is of interest to use the physics-based models for practical engineering considerations, which forms the focus of the next section.

4.5 Application Considerations

The modeling studies presented in the previous sections may be used to address several practical considerations such as fuel cell optimization and design under uncertainty. Despite their significances, however, little effort has been made to systematically address these issues for fuel cell systems. Two approaches were adopted by Pitchumani and coworkers to seek the optimal design and operating conditions for PEM fuel cells [32, 254]. In the first approach, a methodology is illustrated for model-based design and optimization of the operating and design parameters using a comprehensive parametric analysis on the various physical and electrochemical phenomena [32]. Specific optimization solutions were obtained by changing one of the operating or design parameters while fixing the values of the remaining ones. Alternatively, Mawardi *et al.* [254] provided an optimization framework to derive more general optimum solutions. Furthermore, to analyze the effects of uncertainty in the operating and design parameters on the fuel cell performance, a sampling based stochastic model was developed by Mawardi *et al.* [255] to predict the variability of power density of the fuel cell. The two optimization approaches and the stochastic modeling framework are discussed in Sections 4.5.1–4.5.3, respectively.

4.5.1 Optimization Based on Parametric Studies

In the approach adopted by Mishra *et al.* [32], the optimization of a PEM fuel cell is based on systematic parametric studies and consideration of a few illustrative constraints on cell operation. A comprehensive numerical model was developed by combining a one-dimensional version of the transport equations, Eqs. (4.3–4.8) in section 4.2 [28], and the CO poisoning kinetics from Springer *et al.* [31]. The numerical model, validated with the available experimental and numerical results in the literature, was used to conduct a parametric exploration in terms of the operating and design parameters. Based on the parametric studies and specified constraints, operating windows were identified on the current density as a function of the various parameters. As a working example, the procedure for the derivation of operating window for the cell operating temperature is outlined below.

The numerical model was first used in a parametric study to evaluate the performance of the fuel cell over a wide range of design and operation conditions. In the example, four quantities, namely, the cell potential, E_{cell}, the power density, P, the maximum temperature rise, ΔT_{max}, and the minimum hydration λ_{min} are adopted to evaluate the overall fuel cell performance. Figures 4.36(a)–4.36(d) present the performance quantities as functions of the current density, I, for different values of operating temperature [32]. In Figure 4.36(a), a range of operating cell temperature, T_c, between 293 and 363 K is chosen for typical PEM fuel cell operation. Note that Figure 4.36(a) is reproduced from Figure 4.10(a), and the reader is referred to the corresponding discussion in Section 4.2.2.5.1 for the trends seen.

Figure 4.36(b) presents the power density, P_d, as a function of the current density, I, and the operating temperature, T_c. The effect of the operating temperature on the power density is similar to that on the cell potential, *i.e.*, P_d increases initially with increasing T_c, and starts to decrease when $T = 363K$. The maximum temperature difference in Figure 4.36(c) decreases as T_c increases from 293 K to 353 K, owing to the decreased losses mentioned above; a slight increase in ΔT_{max} is caused by the increased transport losses when $T_c = 363K$. The minimum hydration in the membrane, λ_{min}, is shown as a function of I for different values of T_c in Figure 4.36(d). For fixed value of T_c, λ_{min} is seen to decrease monotonically with increasing I, which may be attributed to the loss of the water species at the anode catalyst/membrane interface caused by the increased osmotic drag effect. The decrease in λ_{min} is more pronounced at lower temperatures, since water species can not be effectively transferred to the membrane due to reduced diffusivities with temperature. The minimum hydration is seen to increase monotonically with the operating temperature for fixed values of I, owing to increased water vapor pressure at higher temperatures.

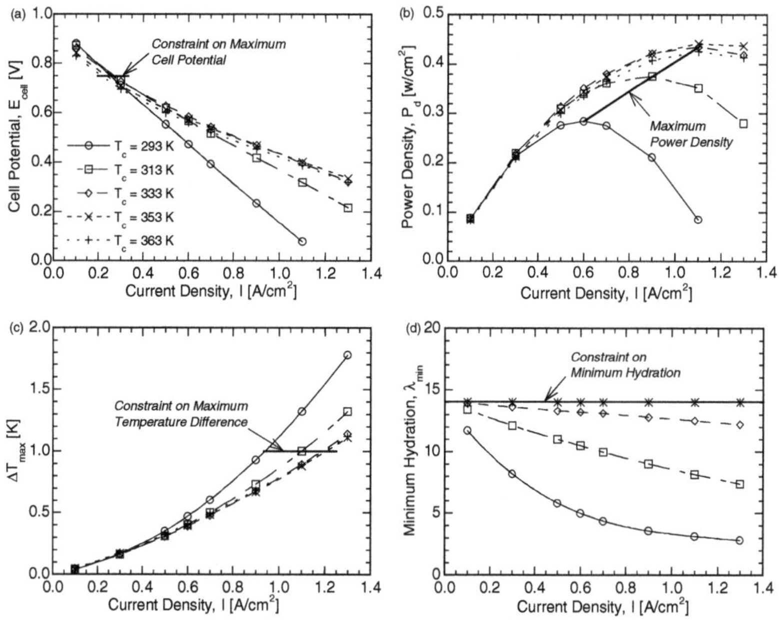

Figure 4.36. Parametric effects of I and T_c on (a) cell potential, (b) power density, (c) maximum temperature difference, and (d) minimum hydration (after [32])

It is evident from the parametric effects presented in Figures 4.36(a)–4.36(d) that increasing the operating temperature can generally reduce ohmic, transport, and activation losses. However, an excessive increase in T_c leads to higher water vapor pressure, which, in turn, results in enhanced transport loss. An optimization of T_c is therefore necessary to balance these competing considerations. In the study by Mishra *et al.* [32], the objective of the optimization problem is considered to be that of maximizing the power density, subject to constraints on the maximum temperature difference, the minimum membrane hydration, and the maximum cell potential. The goal of the optimization problem is to develop the ranges of feasible current density, I, as function of the operating parameters such as temperature.

The optimization problem is written mathematically as:

$$\text{Maximize } P_d \tag{4.115}$$

subject to:

$$\Delta T_{max} - \Delta T_{crit} \leq 0$$

$$\lambda_{min} - \lambda_{crit} \geq 0$$

$$E_{cell} - E_{crit} \leq 0 \tag{4.116}$$

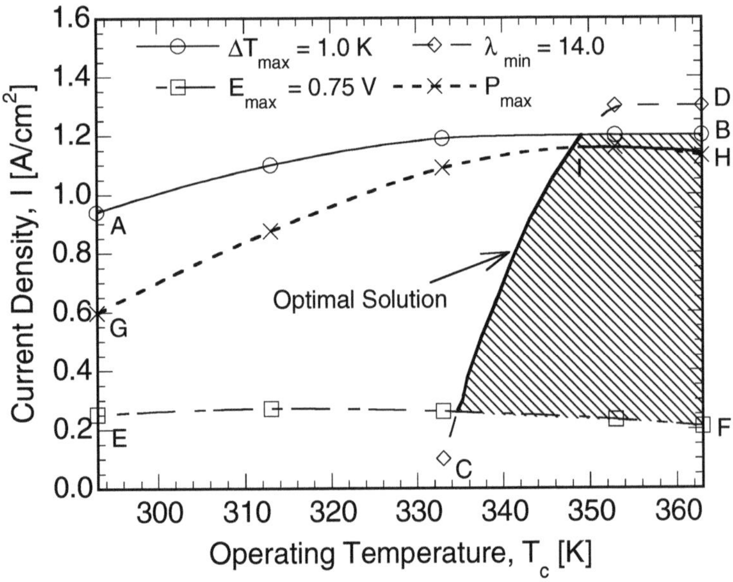

Figure 4.37 Example operating window as function of cell operating temperature (after [32])

where the critical values in this example are chosen as: $\Delta T_{crit} = 1.0K$, $\lambda_{crit} = 14$, and $E_{crit} = 0.75V$. The physical significance of the constraints are as follows [32]: The consideration of membrane degradation requires that the maximum temperature rise, ΔT_{max}, must be lower than a critical value, ΔT_{crit}. Since the ionic conductivity of the membrane material increases with increasing membrane hydration, the minimum hydration in the membrane, λ_{min}, must therefore be larger than a lower limit λ_{crit}. Furthermore, the size and the capital cost of the fuel cell system increase for the higher cell potential, owing to the decreased current density with increasing cell potential. The cell potential must, therefore, have an upper bound, Φ_{crit}, determined based on the capital cost consideration [32].

Based on the parametric study shown in Figure 4.36, the construction of an example operating window on the current density, I, is illustrated in Figure 4.37 as a function of the operating temperature, T_c [32]. Corresponding to the constraint on the maximum allowable temperature, shown by the thick straight line in Figure 4.36(c), the upper limits of the current density for different values of T_c are indicated by the solid line **AB** in Figure 4.37. Similarly, the long-dashed line **CD** and the chain-dashed line **EF** in Figure 4.37, are determined based on the constraints on the minimum membrane hydration and the maximum cell potential, shown in Figures 4.36(d) and 4.36(a), respectively. The shaded region in Figure 4.37, enclosed among the three constraints, is identified as the operating window. Values of the current density that lie inside the operating window will ensure that the maximum temperature difference, the minimum membrane hydration, and the maximum cell potential are within the prescribed limits that correlate to desired overall cell performance.

While satisfying the specified constraints is a necessary condition, maximizing the power density simultaneously is imperative for affordable fuel cells with small size. The variation of the peak value of the power density with T_c [shown by the thick solid curve in Figure 4.36(b)] is plotted as the short-dashed line **GH** in Fig. 4.37. This line represents the unconstrained current density as a function of operating temperature, T_c, required for maximizing the power density. The superposition of the unconstrained solution on the processing window is used to identify the optimal variation of the current density as a function of the operating temperature so as to maximize the power density, P. It is seen in Figrue 4.37 that in the region **GI**, since the unconstrained current density values for maximizing the power density fall above the upper bound corresponding to the λ_{min} constraint, the upper bound constitutes the constrained optimal solution. In the segment between **I** and **H** the unconstrained optimum current density profile passes within the feasible window, and as a result, satisfies all the specified constraints. The constrained optimum current density variation with the operating temperature is indicated in the plot by the thick solid line. Since the constraints on λ_{min} and E_{max} can not be satisfied simultaneously when T_c is less than 333 K, neither a feasible operating window nor an optimal solution exist for $T_c < 333 \ K$. The constrained solution in Figure 4.37 indicates that operating conditions for fuel cell systems in practice mostly correspond to off-peak values of power density. Following the foregoing procedure, operating windows and optimal solution of the current density can be obtained for the other operating and design parameters as well, and the reader is referred to ref. [32] for detailed results.

4.5.2 Optimization Based on a Numerical Optimizer

Evidently, the optimal conditions obtained from parametric studies are specific to the parameter combinations. In the optimization study by Mawardi *et al.* [254], the optimization problem represented by Eqs. (4.115) and (4.116) was solved numerically using a numerical optimizer. The same numerical simulation model for a PEM fuel cell discussed in the previous subsection is adopted to obtain the information on objective function and constraints, which, in turn, is combined with a simplex search based simulated annealing optimizer to obtain the optimum operating conditions. The decision variables are the following nine operating parameters in the fuel cell: the cell operating temperature T_c, the anode pressure p_a, the cathode pressure p_c, the relative humidity at the anode RH_a, the relative humidity at the cathode RH_c, the anode stoichiometry ζ_a, the cathode stoichiometry ζ_c, and dry gas mole fraction at the anode, x_{CO_2/H_2} (defined as the ratio of the mole fraction of the CO_2 to that of H_2 in the fuel stream), and the dry gas mole fraction at the cathode x_{N_2/O_2} (defined as the ratio of the mole fraction of the N_2 to that of O_2 in the oxidant stream).

The numerical optimizer combines the Nelder-Mead simplex method [256] with a simulated annealing technique [257] to improve the effectiveness of the search [258]. The Nelder-Mead simplex method is an algorithm that performs a continuous search for selecting a new point during an optimization iteration that guarantees

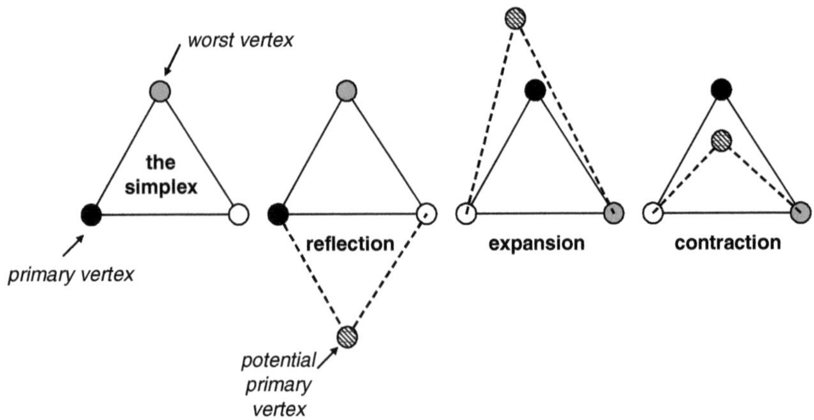

Figure 4.38. Schematic diagram of the possible simplex reconfigurations during a Nelder-Mead simplex search (after [254])

objective function improvement. A simplex is defined as a convex hull of $m+1$ vertices in an m-dimensional space, representing the m decision variables that govern the objective function evaluation. The vertices are ranked, from best to worst, based on the corresponding objective function evaluations, and the best vertex is defined as the *primary vertex*. Since a primary vertex represents a set of decision variables that corresponds to the lowest objective function evaluation, the finding of a new primary vertex constitutes an improvement to the objective function evaluation. The algorithm searches for a new simplex by replacing the worst vertex by a potential new primary vertex, through a set of predefined simplex movements, which include reflection, expansion, and contraction, illustrated in Figure 4.38 for a two-dimensional simplex. Further details on the simplex movements in the Nelder-Mead algorithm, including parameter selection for the various movement and tie-breaking rules (if two vertices of equal objective function value are evaluated), may be found in ref. [259].

In the subsection, optimal solutions of the power density at five current densities, namely, $I = 0.3, 0.5, 0.7, 0.9$ and 1.1 A/cm^2, are presented for the following example fuel cell design case: membrane thickness $\delta_m = 180 \mu m$, electrode thickness $\delta_{el} = 200$ μm, and the critical values in Eq. [4.116] for cell potential $E_{crit} = 0.75V$, temperature difference $\Delta T_{crit} = 1.0K$, and minimum hydration $\lambda_{crit} = 14$ [254]. Optimization runs are carried out by initially supplying trial values of the nine decision variables. Each optimization run was repeated for ten different initial trial solutions, and the optimum result corresponding to the maximum power density among the ten trial results was taken to be the global optimum solution. In Figure 4.39(a), the solid line presents the optimum power density P_d^* as a function of the current density I for the above-mentioned design case [254], while the dashed line corresponds to the results of a reference case in Rowe and Li [28]. The power density is seen to increase monotonically as the current density increases from 0.3 A/cm^2 to 1.1 A/cm^2. For all the current densities, the optimized power density shows improvement with respect to the corresponding base case

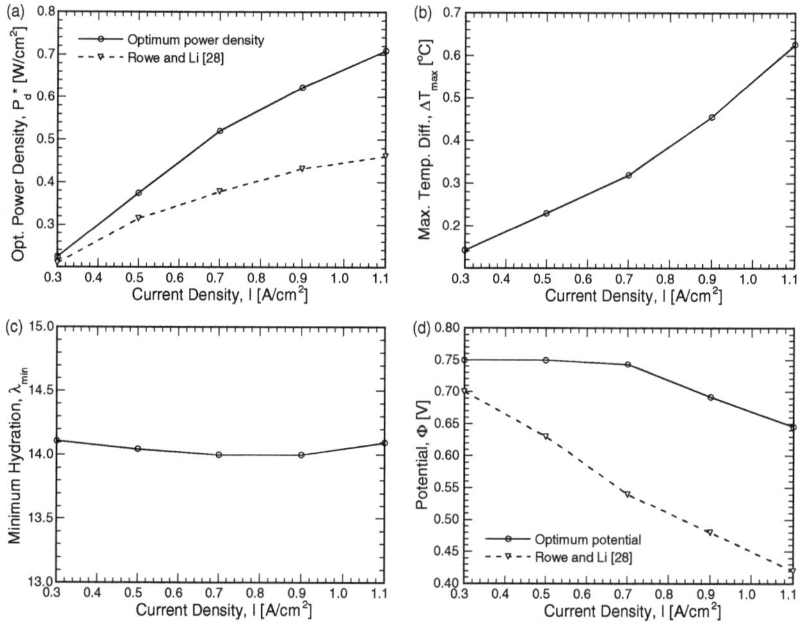

Figure 4.39.(*a*) Optimum power density and the corresponding constraints: (*b*) maximum temperature difference, (*c*) minimum hydration, and (*d*) cell potential as functions of the current density, for the example cell design case (after [254])

value in Rowe and Li [28]. When $I = 1.1 A/cm^2$, the optimum P_d^* is seen to increase to 0.71 A/cm^2 relative to that of 0.46 A/cm^2 for a typical combination of parameters reported in the literature [28], indicating the significance of the present optimization study.

Figures 4.39(b) and 4.39(c) present the two constraints, the maximum temperature difference ΔT_{max} and the minimum hydration λ_{min}, as functions of the current density under the optimum operating conditions for the present fuel cell design case considered. Since the voltage losses and the resulting heat generation increases with increasing current density, ΔT_{max} in Figure 4.39(b) increases at larger I. Note that the constraint on the maximum temperature difference is satisfied, and all the values of ΔT_{max} are well below the critical value of $1.0\ K$. In Figure 4.39(c), the requirement for membrane hydration is satisfied with the minimum hydration, λ_{min}, being just above the critical value of 14 for all current densities. Figure 4.39(d) compares the cell potential under the optimum operating conditions (solid line) with the polarization curve predicted by the base case in ref. [28] [dashed line]. It is seen that the adoption of the optimum design variables boosts the cell potential for the entire range of current density considered. When the current density $I \le 0.7\ A/cm^2$, the optimized cell potential reaches the upper bound value of 0.75 V. The results indicate that the optimum solution is governed by the minimum hydration constraint, and for $I \le 0.7 A/cm^2$ additionally by the cell potential constraint.

The corresponding nine optimum decision variables are presented in Figures 4.40(a)–4.40(f) as functions of the current density. In Figure 4.40(a), the optimum operating temperature, T^*, reaches the upper temperature bound of $100^o C$ when $I \geq 0.7 \;\; A/cm^2$, since higher T^* reduces the losses and thereby enhances the cell performance [32]. For the lower current densities ($I < 0.7 \;\; A/cm^2$), the optimum cell potential assumes the upper bound of $0.75 \; V$ [see Figure 4.39(d)], and consequently, T^* is maintained to be less than $100^o C$ to satisfy the potential constraint. Since a smaller dry gas mole fraction is desirable to reduce the concentration loss, optimum $x^*_{N_2/O_2}$ at the cathode [Figure 4.40(b)] is seen to decrease with increasing current density and reaches the lower bound value of 0 for $I \geq 0.7 \;\; A/cm^2$. Again, the relatively large values of $x^*_{N_2/O_2}$ for $I < 0.7 \;\; A/cm^2$ may be attributed to the constrained cell potential [Figure 4.39(d)]. Based on the same consideration of reducing the concentration loss, the optimum $x^*_{CO_2/H_2}$ at the anode [Figure 4.40(c)] decreases monotonically with increasing I. Note that $x^*_{CO_2/H_2}$ is at the upper bound value of 0.4 for $I < 0.7 A/cm^2$ and gradually drops to the lower bound value of 0 at $I = 1.1 A/cm^2$.

Figure 4.40(d) presents the optimum cathode and anode pressures, p^*_c and p^*_a, for the range of current density considered in the example design case. Since the increase in cathode pressure improves the cell performance by decreasing the concentration loss, the optimum p^*_c is observed to increase with increasing I and hits the upper bound of 15 atm when $I \geq 0.5$. The optimum solution for the anode pressure p^*_a, however, does not reach the upper bound at larger current densities, which may be explained by the non-monotonic effect of p^*_a on the cell performance as follows. Similar to p^*_c, the initial increase in p^*_a from 1 atm results in enhanced cell potential due to reduced concentration loss. However, further increase in p^*_a may decrease the water diffusivity in the anode region, causing membrane dehydration and lower performance [32]. Consequently, the optimum anode pressure always assumes medium values between the lower and upper bounds, as seen in Figure 4.40(d). When $I \geq 0.7 \;\; A/cm^2$, the optimum p^*_a is observed to be about 10 atm. It must be mentioned that the cathode pressure does not have the above-mentioned non-monotonic effect since the water product at the cathode keeps the membrane fully hydrated.

To satisfy the membrane hydration constraint in Figure 4.39(c), the optimum relative humidity at the anode, RH^*_a, assumes the upper bound value of 1.1 for all the current densities, as seen in Figure 4.40(e). The optimum relative humidity at the cathode, RH^*_c, is observed to decrease to the lower bound with increasing I. Since larger stoichiometry of the oxygen is desirable to reduce the concentration loss, the optimum stoichiometry at the cathode ζ^*_c is seen to increase with increasing I, and reaches the upper bound value of 7 when $I \geq 0.5 A/cm^2$ [Figure 4.40(f)]. The half-cell electrochemical reaction at the anode side is much faster than that at the cathode side, and is less sensitive to the reactant concentration. Consequently, a lower bound value of 1.1 for the anode stoichiometry is sufficient for the optimized performance of the cell for all the current densities, as shown in Figure 4.40(f).

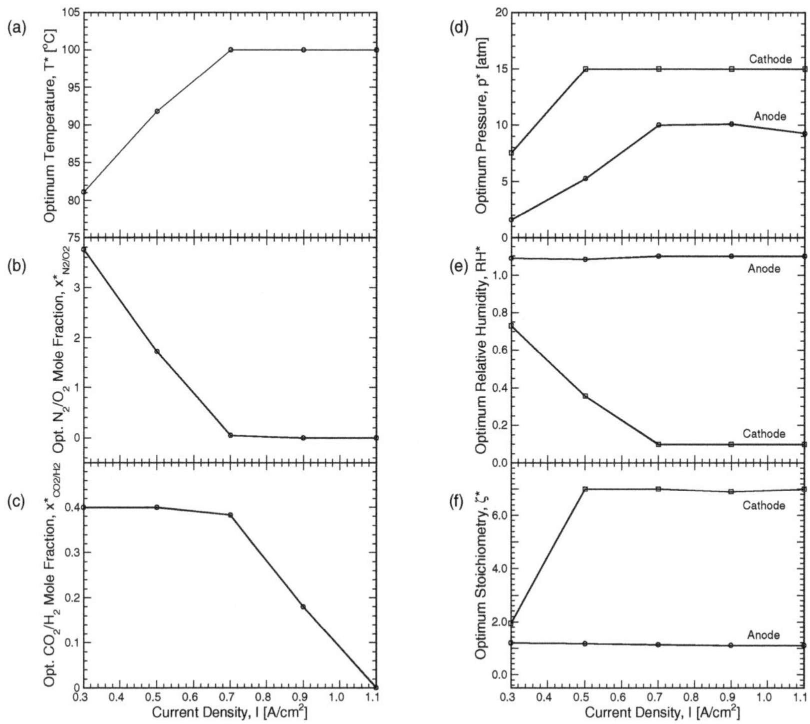

Figure 4.40. Optimum decision variables: (*a*) temperature, (*b*) NO_2/O_2 mole fraction, (*c*) CO_2/H_2 mole fraction, (*d*) anode and cathode pressures, (*e*) relative humidities in anode and cathode, and (*f*) anode and cathode stoichiometries as functions of the current density, for the example cell design case (after [254])

The presentation in the subsection pertains to the optimum solutions for an example design case, and the readers are referred to ref. [254] for detailed discussion on other PEM fuel cell design cases.

4.5.3 Stochastic Modeling of Fuel Cell Performance under Uncertainty

It must be pointed out that actual fuel cell design and operation are conducted under uncertainty in the involved parameters, while typical theoretical analyses consider the parameters to be deterministic. The uncertainty in the operating parameters arises from inaccuracies in parameter setting, monitoring, and control, while the uncertainty in the design parameters arises from the sources associated with manufacturing and characterization of transport/electrochemical properties. The interactive effects of uncertainty in the operating and design parameters potentially cause large variability in the performance of PEM fuel cells. To achieve a more realistic fuel cell analysis, it is imperative to account for the parameter uncertainty in the modeling. A sampling-based stochastic analysis of fuel cells was conducted by Mawardi *et al.* to address the issue [255].

Figure 4.41 schematically illustrates the sampling-based stochastic model, which principally consists of *(a)* representation, quantification, and sampling of the parameters under uncertainty, and *(b)* propagation of the uncertainty through a deterministic fuel cell model to shape the output parameter distributions. In ref. [255], the cell temperature, T_c, the anode and cathode pressures, p_a and p_c, anode and cathode relative humidity, RH_a and RH_c, the anode and cathode stoichiometries, ζ_a and ζ_c, the dry gas mole fractions, x_{N_2/O_2} and x_{CO_2/H_2}, and the anode and cathode reaction rates were considered to be uncertain. Each of the parameters under uncertainty was assumed to be a random variable represented by a probability distribution (typically Gaussian), instead of a fixed value in deterministic model. The uncertain process parameters, which are represented by Gaussian distributions in ref. [255], are collectively referred to as the input parameters, ξ_i, in Figure 4.41. Note that the probability distribution can typically be quantified by its mean value (μ) and standard deviation (σ). The mean, μ, denotes the nominal value of the parameter, while the standard deviation, σ, is proportional to the uncertainty in the parameter. A degree of uncertainty can be expressed in terms of the coefficient of variance (COV), defined as σ/μ; thus, a deterministic parameter (*i.e.* one with no uncertainty in its value) corresponds to the coefficient of variance of zero ($\sigma/\mu = 0$).

The combinations of input parameters are selected from their respective distributions using an appropriate sampling technique such as the Latin hypercube sampling [260] or the Monte Carlo method. The deterministic PEM fuel cell model is invoked for each of the resulting sets of input parameters (samples) to calculate the outputs, from which output distributions are constructed as shown in Figure 4.41 [255]. Note that the number of samples used in the stochastic model and the number of input parameters are denoted by N and m, respectively in Figure 4.41. A typical power density distribution for input parameters with COV of 2% is

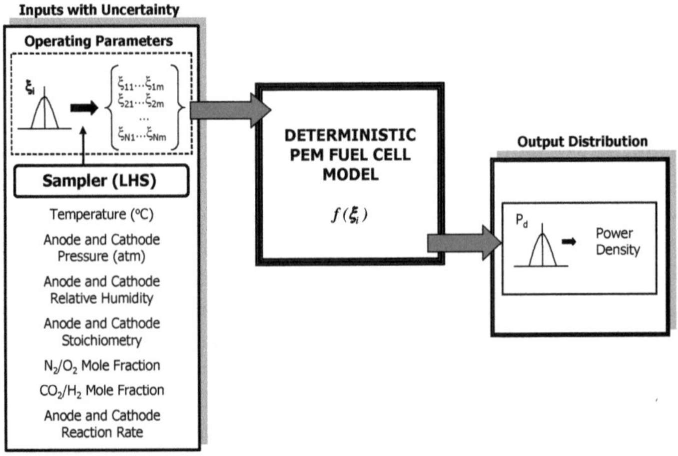

Figure 4.41. Schematic diagram of a sampling-based stochastic analysis (after [255])

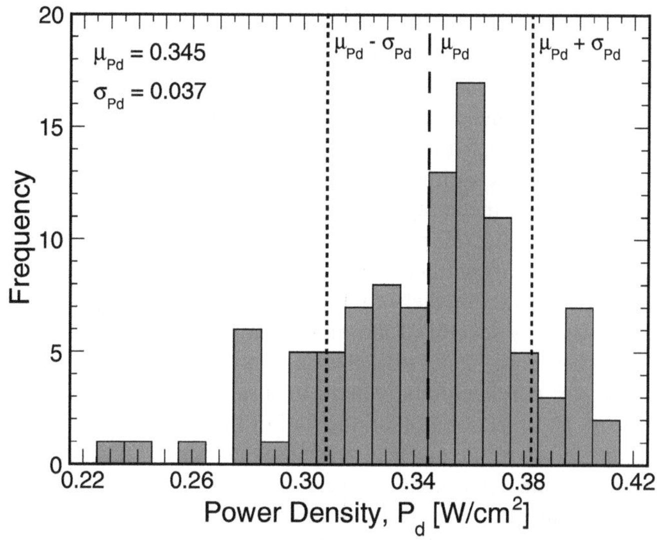

Figure 4.42. Typical distribution of power density obtained from stochastic analysis using 100 samples (after [255])

presented in Figure 4.42, which shows the histogram of the power density from the stochastic analysis using 100 sampling sets [255]. The dashed lines in the figure correspond to the mean value of the power density, $\mu_{P_d} = 0.345$, while the dotted lines indicate the $\mu_{P_d} - \sigma_{P_d}$ and $\mu_{P_d} + \sigma_{P_d}$ values, where $\sigma_{P_d} = 0.037$ is the standard deviation of the power density in the present example. The figure shows that the distribution of the power density corresponding to input parameters with Gaussian uncertainty is not symmetric about the mean value, which is expected due to the nonlinearities in the fuel cell model. Note that the COV of the power density has a relatively large value (more than 10%), which may significantly influence the performance of the fuel cell. Systematic studies were conducted in ref. [255] to determine the distribution of power density as a function of the input parameters and their variations, and the results were used to determine the fuel cell design and operating conditions that reduce the variation of the output power density. It must be mentioned that although the discussion in this section is specific to PEM fuel cells, the adopted approaches are generally valid for other types of fuel cells.

4.6 Concluding Remarks

Modeling studies for three types of fuel cells were reviewed in this chapter with particular focus on the transport and electrochemical processes within the cells. The physical phenomena and their governing equations for the overall cell as well as individual cell components were elucidated. Example results were presented on the performance of fuel cells under various design and operating conditions, and the use of the theoretical models for application considerations of cell optimization and design under uncertainty was demonstrated. The Chapter provides a compendium of advances in the modeling of fuel cells, to form the foundation for future developments. While several significant advances have been made with regard to the theoretical description and computational modeling, several challenges remain to be addressed, some of which are noted below.

Model validation remains an important issue toward improving the fidelity and the predictive capability of the computational developments. Presently, the models are routinely compared to overall polarization data, whereas more detailed validation with regard to the concentration and temperature profiles is less common. Moreover, comparisons of the models with experimental data are made mostly in a correlation sense, whereby several of the model parameters are fitted to match the experimental data. It is well known that the same polarization curve may be fitted by multiple parameter combinations, which, in turn, may correspond to dramatically different flow and reaction conditions within the cell. A related issue is the large scatter in the values of the material properties and model parameters reported in the literature. Accurate determination of the properties and parameters is, therefore, imperative. An allied requirement is that of systematically quantifying the uncertainty in the properties and the parameters of the models. Advanced diagnostics tools are needed to generate benchmark-quality data for detailed model validation and for accurate measurement of the parameters and properties.

Some issues specific to the various types of fuel cells considered in this chapter are as follows:

- In the case of PEM fuel cells, the mechanism for the liquid water transport and the related flooding phenomena needs further computational and experimental investigations. Experimental techniques, such as optical diagnostics and 3-D neutron tomography, offer viable opportunities in this regard.
- Comprehensive models have been available to describe the electrochemistry, gas dynamics, heat transfer, and multicomponent transport in SOFC. In addition to the prediction of voltage-current behavior, the models provide detailed distributions of the reactant and product, current density, and temperature. The temperature distribution can be coupled with structural stress analysis, thereby providing an integrated tool for SOFC design and operation. Other future topics of SOFC modeling include (1) internal reforming at the anode, (2) radiation heat transfer, and (3) direct simulation of the SOFC electrodes to establish the relationship between cell performance and electrode microstructure.
- Models for the simultaneous transport of methanol, water, and heat are developed for DMFCs. The modeling studies, however, are less elaborate due to the complicated two-phase flow patterns and multistep electrochemistry at the anode, which need to be considered in future work. Improving the performance of the DMFC is the focus of research and development, and the possible approaches include (1) find alternate electrocatalysts with higher activities and (2) develop membrane materials with a low methanol crossover rate.

Reliability of the operation of fuel cells remains one of the foremost impediments to their widespread commercialization. Quantification of parameter uncertainty, and modeling and analysis of the effects of uncertainty on performance variability offer a viable means of extracting reliability and robustness measures.

Furthermore, combined with numerical optimization schemes, optimal material and operational designs under uncertainty may be derived using physics-based simulation models as a basis. A framework for such implementation was discussed in this chapter. Further work is needed to develop a comprehensive paradigm for physics-based optimal and robust design of fuel cells.

The focus of the chapter was on individual cell-level modeling and analysis. Integration of the cell-level modeling with a stack-level and system-level description remains a fertile ground for future contributions. Effective strategies for bridging the phenomena across the cascade of length and time scales are needed to bring the detailed physical understanding at the cell-level for stack-level and system-level modeling. As model rigor and sophistication increase, the computational requirements for such a large-scale simulation and optimization under uncertainty will be tremendous. Parallel efforts in addressing the computing challenges are needed to efficiently bridge the science and the practice of fuel cells.

4.7 Acknowledgement

Portions of the authors' work cited in this chapter were funded by the U.S. Army RDECOM through Contract no. DAAB07-03-3-K-415. Sincere appreciation is extended to Dr. Andryas Mawardi and Ms. Yanyan Zhang for all their assistance in the preparation of the manuscript.

4.8 References

1. Larminie J, Dicks A. Fuel Cell Systems Explained. Wiley, West Sussex England, 2000.
2. Francis M. Modeling: driving fuel cells. *Materials Today.* 2002;5(5):34–39.
3. Srinivasan S, Mosdale R, Stevens P, Yang C. Fuel Cells: Reaching the Era of Clean and Efficient Power Generation in the Twenty-First Century. *Annu. Rev. Energy Environ.* 1999;24:281–328.
4. Shah R K, Desideri U, Hsueh K-L, Vikar A V. Research Opportunities and Challenges in Fuel Cell Science and Engineering. Proceedings of the 4th Baltic Heat Transfer Conference Kaunas, August 25–27, Lithuania, 2003.
5. Um S, Wang C Y. Three-Dimensional Analysis of Transport and Electrochemical Reactions in Polymer Electrolyte Fuel Cells. *J. Power Sources.* 2004;125:40–51.
6. Paganin V A, Ticianelli E A, Gonzalez E R. Development and Electrochemical Studies of Gas Diffusion Electrodes for Polymer Electrolyte Fuel Cells. *J. Appl. Electrochem.* 1996;26:297–304.
7. Parthasarathy A, Srinivasan S, Appleby A J, Martin C R. Temperature Dependence of the Electrode Kinetics of Oxygen Reduction at the Platinum/Nafion Interface – a Microelectrode Investigation. *J. Electrochem. Soc.* 1992;139:2530–2537.
8. Parthasarathy A, Srinivasan S, Appleby A J, Martin C R. Pressure Dependence of the Oxygen Reduction Reaction at the Platinum Microelectrode/Nafion Interface: Electrode Kinetics and Mass Transport. *J. Electrochem. Soc.* 1992;139(10):2856–2862.
9. Ticianelli E A, Derouin C R, Redondo A, Srinivasan S. Methods To Advance Technology Of Proton Exchange Membrane Fuel Cells. *J. Electrochem. Soc.* 1988;135:2209–2213.

10. Kim J, Lee S M, Srinivasan S, Chamberlin C E. Modeling of Proton Exchange Membrane Fuel Cell Performance with an Empirical Equation. *J. Electrochem. Soc.* 1995; 142:2670–2674.
11. Mosdale R, Srinivasan, S. Analysis of Performance and of Water and Thermal Management in Proton Exchange Membrane Fuel Cells. *Electrochim. Acta.* 1995;40: 413–421.
12. Newman J. Optimization of Potential and Hydrogen Utilization in an Acid Fuel Cell. *Electrochim. Acta.* 1979;24:223–229.
13. Beattie P D, Basura V I, Holdcroft S. Temperature and Pressure of Oxygen Reduction at a Pt/Nafion 117 and a Pt/BAM3G Interface. *J. Electronanal. Chem.* 1999;468: 180–192.
14. Sena D R, Ticianelli E A, Paganin V A, Gonzalez E R. Effect of water transport in a PEFC at low temperatures operating with dry hydrogen. *J. Electronanal. Chem.* 1999;477:164–170.
15. Lee S J, Mukerjee S, McBreen J, Rho Y W, Kho Y T, Lee T H. Effects of Nafion Impregnation On Performances of PEMFC Electrodes. *Electrochim. Acta.* 1998; 43:3693–3701.
16. Liebhafsky H A, Cairns E J, Grubb W T, Niedrach L W. In Fuel Cell Systems; Young, G. J., Linden, H. R., Eds.; American Chemical Society: Washington, DC, 1965; Vol. 47.
17. Passalacqua E, Lufrano F, Squadrito G, Patti A, Giorgi L. Nafion Content in the Catalyst Layer of Polymer Electrolyte Fuel Cells: Effects on Structure and Performance. *Electrochim. Acta.* 2001;46:799–805.
18. Squadrito G, Maggio G, Passalacqua E, Lufrano F, Patti A. An Empirical Equation for Polymer Electrolyte Fuel Cell (PEFC) Behaviour. *J. Appl. Electrochem.* 1999;29:1449–1455.
19. Amphlett J C, Baumert R M, Mann R F, Peppley B A, Roberge P R. Performance Modeling of the Ballard Mark Iv Solid Polymer Electrolyte Fuel Cell. *J. Electrochem. Soc.* 1995;142(1):1–15.
20. Pisani L, Murgia G, Valentini M, D'Aguanno B. A New Semi-Empirical Approach to Performance Curves of Polymer Electrolyte Fuel Cells. *J. Power Sources.* 2002;108: 192–203.
21. Kulikovsky A A. The Voltage-Current Curve of a Polymer Electrolyte Fuel Cell: "Exact" and Fitting Equations. *Electrochem. Commun.* 2002;4(11):845–852.
22. Gurau V, Barbir F, Liu H. An Analytical Solution of a Half-Cell Model for PEM Fuel Cells. *J. Electrochem. Soc.* 2000;147:2468–2477.
23. Standaert F, Hemmes K, Woudstra N. Analytical Fuel Cell Modeling. *J. Power Sources.* 1996;63:221–234.
24. Standaert F, Hemmes K, Woudstra N. Analytical Fuel Cell Modeling; Non-Isothermal Fuel Cells. *J. Power Sources.* 1998;70:181–199.
25. Weber A Z, Newman J. Modeling Transport in Polymer-Electrolyte Fuel Cells. *Chem. Rev.* 2004;104:4679–4726.
26. Wang C Y. Fundamental Models for Fuel Cell Engineering. *Chem. Rev.* 2004;104: 4727–4766.
27. Yao K Z, Karan K, McAuley K B, Oosthuizen P, Peppley B, Xie T. A Review of Mathematical Models for Hydrogen and Direct Methanol Polymer Electrolyte Membrane Fuel Cells. *Fuel Cells.* 2004;4(1–2):3–29.
28. Rowe A, Li X. Mathematical Modeling of Proton Exchange Membrane Fuel Cells. *J. Power Sources.* 2001;102:82–96.
29. Gurau V, Liu H, Kakac S. Two-Dimensional Model for Proton Exchange Membrane Fuel Cells. *AIChE J.* 1998;44(11):2410–2422.

30. Um S, Wang C Y, Chen K S. Computational Fluid Dynamics Modeling of Proton Exchange Membrane Fuel Cells. *J. Electrochem. Soc.* 2000;147:4485–4493.
31. Springer T E, Rockward T, Zawodzinski T A, Gottesfeld S. Model for Polymer Electrolyte Fuel Cell Operation on Reformate Feed. *J. Electrochem. Soc.* 2001;148(11): A11–A23.
32. Mishra V, Yang F, Pitchumani R. Analysis and Design of PEM Fuel Cells. *J. Power Sources.* 2005;141:47–64.
33. Bernardi D M, Verbrugge M W. A Mathematical Model of the Solid Polymer Electrolyte Fuel Cell. *J. Electrochem. Soc.* 1992;139(9):2477–2745.
34. Gottesfeld S, Zawodzinski T A. In Advances in Electrochemical Science and Engineering; Tobias C, Ed.; Wiley and Sons: New York, 1997; Vol. 5.
35. Springer T E, Zawodzinski T A, Gottesfeld S. Polymer Electrolyte Fuel Cell Model. *J. Electrochem. Soc.* 1991;138(8):2334–2342.
36. Zawodzinski T A, Springer T A, Davey J, Jestel R, Lopez C, Valero J, Gottesfeld S. A Comparative Study of Water Uptake and Transport Through Iononomeric Fuel Cell Membranes. *J. Electrochem. Soc.* 1993;140(7):1981–1985.
37. Zawodzinski T A, Derouin C, Radzinski S, Sherman R J, Smith V T, Springer T A, Gottesfeld S. Water Uptake By and Transport Through Nafion 117 Membranes. *J. Electrochem. Soc.* 1993;140(4):1041–1047.
38. Dutta S, Shimpalee S, Van Zee J W. Three-Dimensional Numerical Simulation of Straight Channel PEM Fuel Cells. *J. Appl. Electrochem.* 2000;30:135–146.
39. Dutta S, Shimpalee S, Van Zee J W. Numerical Prediction of Mass-Exchange between Cathode and Anode Channels in a PEM Fuel Cell. *Int. J. Heat Mass Transfer.* 2001;44:2029–2042.
40. Parthasarathy A, Srinivasan S, Appleby A J, Martin C R. Temperature Dependence of the Electrode Kinetics of Oxygen Reduction at the Platinum/Nafion Interface-A Microelectrode Investigation. *J. Electrochem. Soc.* 1992;139(9):2530–2537.
41. Gasteiger H A, Gu W, Makharia R, Mathias M F, Sompalli B. In Handbook of Fuel Cells; Vielstich W, Gasteiger H A, Lamm A, Eds.; John Wiley and Sons Ltd.: England, 2003; Vol. 3, p 594.
42. Mishra V, Yang F, Pitchumani R. Measurement and Prediction of Electrical Contact Resistance between Gas Diffusion Layers and Bipolar Plate for Applications to PEM Fuel Cells. *ASME J. Fuel Cell Sci. Tech.* 2004;1(1):2–9.
43. Pitchumani R, Yao S C. Correlation of Thermal Conductivities of Unidirectional Fibrous Composites Using Local Fractal Techniques. *ASME J. Heat Transfer.* 1991; 113:788–796.
44. Pitchumani R. Evaluation of Thermal Conductivities of Disordered Composite Media Using a Fractal Model. *ASME J. Heat Transfer.* 1999;121:163–166.
45. Mathias M F, Roth J, Flemming J, Lehnert W. In Handbook of Fuel Cells; Vielstich W, Gasteiger H A, Lamm A, Eds.; John Wiley and Sons Ltd.: England, 2003; Vol. 3, p 517.
46. Prausnitz J M, Reid R C, Sherwood T K. The Properties of Gases and Liquids. McGraw-Hill book Company, 3rd edn, 1977.
47. Patankar S. V. Numerical Heat Transfer and Fluid Flow; Hemisphere Publishing Corp.: New York, 1980.
48. Newman J. Electrochemical Systems, 2nd edn.; Prentice Hall: Englewood Cliffs, NJ, 1991.
49. Wu J, Srinivasan V, Xu J, Wang C Y. Newton-Krylov-Multigrid Algorithms for Battery Simulation. *J. Electrochem. Soc.* 2002;149:A1342–A1348.
50. Meng H, Wang C Y. Electron Transport in PEMFCs. *J. Electrochem. Soc.* 2004;151: A358–A367.

51. Meng H, Wang C Y. Large-Scale Simulation of Polymer Electrolyte Fuel Cells by Parallel Computing. *Chem. Eng. Sci.* 2004;59:3331–3343.
52. Hohenberg P, Kohn W. Inhomogeneous Electron Gas. *Phys. Rev. B* 1964;136:B864–B871.
53. Kohn W, Sham L. Self-Consistent Equations Including Exchange and Correlation Effects. *Phys. Rev.* 1965;140:A1133–A1138.
54. Lee C, Yang W, Parr R G. Development of the Colle-Salvetti Correlation-Energy Formula into a Functional of the Electron Density. *Phys. Rev. B* 1988;37:785–789.
55. Perdew J P. Density-Functional Approximation for the Correlation Energy of the Inhomogeneous Electron Gas. *Phys. Rev. B* 1986:33;8822–8824.
56. Perdew J P. In Electronic Structure of Solids; Ziesche P, Eschrig H, Eds.; Academie Verlag: Berlin, 1991.
57. Becke A D. Density-Functional Exchange-Energy Approximation with Correct Asymptotic Behavior. *Phys. Rev. A.* 1988;38:A3098–3100.
58. Johnson B G, Gill P M W, Pople J A. The Performance of a Family of Density Functional Methods. *J. Chem. Phys.* 1993;98:5612–5626.
59. Allen M P, Tildesley D J. Computer Simulation of Liquids; Clarendon Press: Oxford, U.K., 1987.
60. Gray C G, Gubbins K E. Theory of Molecular Fluids; Clarendon Press: Oxford, U.K., 1984; Vol. 1.
61. Harned H S, Owen B B. The Physical Chemistry of Electrolyte Solutions; Reinhold: New York, 1958.
62. Kay R L. In Water, A Comprehensive Treatise; Franks F, Ed.; Plenum Press: New York, 1973; Vol. 3.
63. Frauenheim T, Seifert G, Elstner M, Niehaus T A, Köhler T, Amkreutz M, Sternberg M, Hajnal Z, di Carlo A, Suhai S. Atomistic Simulations of Complex Materials: Ground-State and Excited-State Properties. *J. Phys. Condens. Matter.* 2002;14(11):3015–3047.
64. Münch W, Kreuer K D, Seifert G. Maier. Proton Diffusion in Perovskites: Comparison between $BaCeO_3$, $BaZrO_3$, $SrTiO_3$, and $CaTiO_3$ Using Quantum Molecular Dynamics. *Solid State Ionics* 2000;136–137:183–189.
65. Vishnyakov A, Neimark A V. Molecular Dynamics Simulation of Nafion Oligomer Solvation in Equimolar Methanol-Water Mixture. *J. Phys. Chem. B* 2001;105:7830–7834.
66. Mologin D A, Khalatur P G, Kholhlov A R. Structural Organization of Water-Containing Nafion: A Cellular-Automaton-Based Simulation. *Macromol. Theory Simul.* 2002;11:587–607.
67. Spohr E, Commer P, Kornyshev A A. Enhancing Proton Mobility in Polymer Electrolyte Membranes: Lessons from Molecular Dynamics Simulations. *J. Phys. Chem. B.* 2002;106:10560–10569.
68. Vishnyakov A, Neimark A V. Molecular Simulation Study of Nafion Membrane Solvation in Water and Methanol. *J. Phys. Chem. B.* 2000;104(18):4471–4478.
69. Vishnyakov A, Neimark A V. Molecular Dynamics Simulation of Microstructure and Molecular Mobilities in Swollen Nafion Membranes. *J. Phys. Chem. B.* 2001;105:9586–9594.
70. Paddison S J, Paul R, Zawodzinski T A. A Statistical Mechanical Model of Proton and Water Transport in a Proton Exchange Membrane. *J. Electrochem. Soc.* 2000;147: 617–626.
71. Khalatur P G, Talitskikh S K, Khokhlov A R. Structural Organization of Water-Containing Nafion: The Integral Equation Theory. *Macromol. Theory Simul.* 2002;11: 566–586.

72. Krueger J J, Simon P P, Ploehn H. J. Phase Behavior and Microdomain Structure in Perfluorosulfonated Ionomers via Self-Consistent Mean Field Theory. *Macromolecules* 2002;35:5630–5639.
73. Kreuer K-D, Paddison S J, Spohr E, Schuster M. Transport in Proton Conductors for Fuel-Cell Applications: Simulations, Elementary Reactions, and Phenomenology. *Chem. Rev.* 2004;104:4637–4678.
74. Weber A Z, Newman J. Transport in Polymer-Electrolyte Membranes I. Physical Model. *J. Electrochem. Soc.* 2003;150:A1008–A1015.
75. Weber A Z, Newman J. Transport in Polymer-Electrolyte Membranes II. Mathematical Mode. *J. Electrochem. Soc.* 2004;151:A311–A325.
76. Weber A, Darling R, Meyers J, Newman J. In Handbook of Fuel Cells: Fundamentals, Technology, and Applications; Vielstich W, Lamm A, Gasteiger H A, Eds.; John Wiley & Sons: New York, 2003; Vol. 1.
77. Bennion D N. Mass Transport of Binary Electrolyte Solutions in Membranes; Water Resources Center Desalination Report No. 4; Department of Engineering, University of Californias Los Angeles: 1966.
78. Pintauro P N, Bennion D N. Mass Transport of Electrolytes in Membranes. 1. Development of Mathematical Transport Model. *Ind. Eng. Chem. Fundam.* 1984;23: 230–234.
79. Auclair B, Nikonenko V, Larchet C, Metayer M, Dammak L. Correlation between Transport Parameters of Ion-Exchange Membranes. *J. Membr. Sci.* 2002;195:89–102.
80. Koter S, Hamann C H. Characteristics of lon-Exchange Membranes for Electrodialysis on the Basis of Irreversible Thermodynamics. *J. Non-Equilib. Thermodyn.* 1990;15:315–330.
81. Bernardi D M, Verbrugge M W. Mathematical Model of a Gas Diffusion Electrode Bonded to a Polymer Electrolyte. *AIChE J.* 1991;37(8):1151–1163.
82. Verbrugge M W, Hill R F. Analysis of Promising Perfluorosulfonic Acid Membranes for Fuel-Cell Electrolytes. *J. Electrochem. Soc.* 1990;137:3770–3776.
83. Singh D, Lu D M, Djilali N. A Two-Dimensional Analysis of Mass Transport in Proton Exchange Membrane Fuel Cells. *Int. J. Eng. Sci.* 1999;37:431–452.
84. Baschuk J J, Li X. Modelling of Polymer Electrolyte Membrane Fuel Cells with Variable Degrees of Water Flooding. *J. Power Sources.* 2000;86:181–196.
85. Pisani L, Murgia G, Valentini M, D'Aguanno B. A Working Model of Polymer Electrolyte Fuel Cells: Comparisons Between Theory and Experiments. *J. Electrochem. Soc.* 2002;149:A898–A904.
86. Murgia G, Pisani L, Valentini M, D'Aguanno B. Electrochemistry and Mass Transport in Polymer Electrolyte Membrane Fuel Cells I. Model. *J. Electrochem. Soc.* 2002;149:A31–A38.
87. Berning T, Lu D M, Djilali N. Three-Dimensional Computational Analysis of Transport Phenomena in a PEM Fuel Cell. *J. Power Sources.* 2002;106:284–294.
88. Thampan T, Malhotra S, Tang H, Datta R. Modeling of Conductive Transport in Proton-Exchange Membranes for Fuel Cells. J. Electrochem. Soc. 2000;147:3242–3250.
89. Costamagna P. Transport Phenomena in Polymeric Membrane Fuel Cells. *Chem. Eng. Sci.* 2001;56:323–332.
90. Thampan T, Malhotra S, Zhang J X, Datta R. PEM Fuel Cell as a Membrane Reactor. *Catal. Today* 2001;67:15–32.
91. You L, Liu H. A Two-Phase Flow and Transport Model for the Cathode of PEM Fuel Cells. *Int. J. Heat Mass Tran.* 2002;45:2277–2287.
92. Hertwig K, Martens L, Karwoth R. Mathematical Modelling and Simulation of Polymer Electrolyte Membrane Fuel Cells. Part I: Model Structures and Solving an Isothermal One-Cell Model. *Fuel Cells.* 2002;2:61–77.

93. Yi J S, Nguyen T V. An Along-the-Channel Model for Proton Exchange Membrane Fuel Cells. *J. Electrochem. Soc.* 1998;145:1149–1159.
94. Eikerling M, Kharkats Y I, Kornyshev A A, Volfkovich Y M. Phenomenological Theory of Electro-osmotic Effect and Water Management in Polymer Electrolyte Proton-Conducting Membranes. *J. Electrochem. Soc.* 1998;145:2684–2699.
95. Weber A Z, Newman J. In Advanced Materials for Fuel Cells and Batteries; Ehrlich G, Arora P, Ofer D, Savinell R F, Zawodinski T, D'Souza F, Eds.; The Electrochemical Society Proceeding Series: Pennington, NJ, 2003.
96. De La Rue R E, Tobias C W. On the Conductivity of Dispersions. *J. Electrochem. Soc.* 1959;106:827–833.
97. Bruggeman D A G. Berechnung verschiedener physikalischer Konstanten von heterogenen Substanzen. *Ann. Physik. (Leipzig).* 1935;24:636–679.
98. Li G, Pickup P P. Ionic Conductivity of PEMFC Electrodes Effect of Nafion Loading. *J. Electrochem. Soc.* 2003;150:C745–C752.
99. Bernardi D M. Ph.D. Dissertation, University of California, 1986.
100. Dullien F A L. Porous Media: Fluid Transport and Pore Structure, 2nd edn.; Academic Press: New York, 1992.
101. Bird R B, Stewart W E, Lightfoot E N. Transport Phenomena, 2nd edn.; John Wiley & Sons: New York, 2002.
102. Nam J H, Kaviany M. Effective Diffusivity and Water-Saturation Distribution in Single- and Two-Layer PEMFC Diffusion Medium. *Int. J. Heat Mass Transfer* 2003;46: 4595–4611.
103. Ramakrishnan B, Pitchumani R. Fractal Permeation Characteristics of Preforms Used in Liquid Composite Molding. *Polymer Composites.* 2000;21(2):281–296.
104. Knudsen M. The Kinetic Theory of Gases; Methuen: London, 1934.
105. Mason E A, Malinauskas A P. Gas Transport in Porous Media: The Dusty-Gas Model; Elsevier: Amsterdam, 1983.
106. Arnost D, Schneider P. Dynamic Transport of Multicomponent Mixtures of Gases in Porous Solids. *Chem. Eng. J.* 1995;57:91–99.
107. Ackmann T, de Haart L G J, Lehnert W, Stolten D. Modeling of Mass and Heat Transport in Planar Substrate Type SOFCs. *J. Electrochem. Soc.* 2003;150(6):A783–A789.
108. Wohr M, Bolwin K, Schnurnberger W, Fischer M, Neubrand W, Eigenberger G. Dynamic Modelling and Simulation of a Polymer Membrane Fuel Cell Including Mass Transport Limitation. *Int. J. Hydrogen Energy* 1998;23:213–218.
109. Siegel N P, Ellis M W, Nelson D J, von Spakovsky M R. Single Domain PEMFC Model Based on Agglomerate Catalyst Geometry. *J. Power Sources* 2003;115:81–89.
110. Bradean R, Promislow K, Wetton B. Transport Phenomena in the Porous Cathode of a Proton Exchange Membrane Fuel Cell. *Numer. Heat Transfer, Part A.* 2002;42: 121–138.
111. Natarajan D, Nguyen T V. Two-Dimensional, Two-Phase, Multicomponent, Transient Model for the Cathode of a Proton Exchange Membrane Fuel Cell Using Conventional Gas Distributors. *J. Electrochem. Soc.* 2001;148:A1324–1335.
112. Wang Z H, Wang C Y, Chen K S. Two-Phase Flow and Transport in the Air Cathode of Proton Exchange Membrane Fuel Cells. *J. Power Sources.* 2001;94:40–50.
113. Weber A Z, Newman J. Transport in Polymer-Electrolyte Membranes III. Model Validation in a Simple Fuel-Cell Model. *J. Electrochem. Soc.* 2004;151:A326–A339.
114. Yi J S, Nguyen T V. Multicomponent Transport in Porous Electrodes of Proton Exchange Membrane Fuel Cells Using the Interdigitated Gas Distributors. *J. Electrochem. Soc.* 1999;146:38–45.
115. He W, Yi J S, Nguyen T V. Two-Phase Flow Model of the Cathode of PEM Fuel Cells Using Interdigitated Flow Fields. *AIChE J.* 2000;46:2053–2064.

116. Hu M, Zhu X, Wang M, Gu A, Yu L. Three Dimensional, Two Phase Flow Mathematical Model for PEM Fuel Cell: Part II. Analysis and Discussion of the Internal Transport Mechanisms. *Energy Conversion and Management.* 2004;45:1883–1916.
117. Chu H-S, Yeh C, Chen F. Effects of Porosity Change of Gas Diffuser on Performance of Proton Exchange Membrane Fuel Cell. *J. Power Sources.* 2002;123:1–9.
118. Springer T E, Wilson M S, Gottesfeld S. Modeling and Experimental Diagnostics in Polymer Electrolyte Fuel Cells. *J. Electrochem. Soc.* 1993;140(12):3513–3526.
119. Nguyen T V, White R E. A Water and Heat Management Model for Proton Exchange Membrane Fuel Cells. *J. Electrochem. Soc.* 1993;140(8):2178–2186.
120. Shimpalee S, Dutta S. Numerical Prediction of Temperature Distribution in PEM Fuel Cells. *Numer. Heat Transfer, Part A.* 2000;38:111–128.
121. Miller C T, Christakos G, Imhoff P T, McBride J F, Pedit J A, Trangenstein J A. Multiphase Flow and Transport Modeling in Heterogeneous Porous Media: Challenges and Approaches. *Adv. Water Resources* 1998;21:77–120.
122. Bear J. Dynamics of Fluids in Porous Media; Dover Publications: New York, 1988.
123. You L. The Two Phase Flow, Transport Mechanism and Performance Studies for PEM Fuel Cells. Ph. D. dissertation, University of Miami, 2001.
124. Mazumder S, Cole J V. Rigorous 3-D Mathematical Modeling of PEM Fuel Cells II. Model Predictions with Liquid Water Transport. *J. Electrochem. Soc.* 2003;150: A1510–A1517.
125. Natarajan D, Nguyen T V. Three-Dimensional Effects of Liquid Water Flooding in the Cathode of a PEM Fuel Cell. *J. Power Sources.* 2003;115:66–80.
126. Wang C Y, Cheng P. Multiphase Flow and Heat Transfer in Porous Media. *Adv. Heat Transfer.* 1997;30:93–196.
127. Weber A Z, Darling R M, Newman J. Modeling Two-Phase Behavior in PEFCs. *J. Electrochem. Soc.* 2004;151(10):A1715–A1727.
128. Berning T, Djilali N. A 3D, Multiphase, Multicomponent Model of the Cathode and Anode of a PEM Fuel Cell. *J. Electrochem. Soc.* 2003;150:A1589–A1598.
129. Kinoshita K. Electrochemical Oxygen Technology; John Wiley & Sons: New York, 1992.
130. Vielstich W, Lamm A, Gasteiger H A, Eds.; Handbook of Fuel Cells: Fundamentals, Technology, and Applications, Vol. 2s Electrocatalysis; John Wiley & Sons: New York, 2003.
131. Markovic N M, Schmidt T J, Stamenkovic V, Ross P N. Oxygen Reduction Reaction on Pt and Pt Bimetallic Surfaces: A Selective Review. *Fuel Cells.* 2001;1:105–116.
132. Grens E A, Turner R M, Katan T. A Model for Analysis of Porous Gas Electrodes. *Adv. Energy Convers.* 1964;4:109–119.
133. Viitanen M, Lampinen M J. A Mathematical Model and Optimization of the Structure for Porous Air Electrodes. *J. Power Sources.* 1990;32:207–231.
134. Grens E A. Analysis of Operation of Porous Gas Electrodes with Two Superimposed Scales of Pore Structure. *Ind. Eng. Chem. Fundam.* 1966;5:542–547.
135. Yang S C, Cutlip M B, Stonehart P. Simulation and Optimization of Porous Gas-diffusion Electrodes Used in Hydrogen/oxygen Phosphoric Acid Fuel Cells-I. Application of Cathode Model Simulation and Optimization to PAFC Cathode Development. *Electrochim. Acta.* 1990;35:869–878.
136. Yang S C, Cutlip M B, Stonehart P. Further Development of an Approximate Model for Mass Transfer with Reaction in Porous Gas-Diffusion Electrodes to Include Substrate Effects. *Electrochim. Acta.* 1989;34:703–705.
137. Giner J, Hunter C. The Mechanism of Operation of the Teflon-Bonded Gas Diffusion Electrode: A Mathematical Model. *J. Electrochem. Soc.* 1969;116(8):1124–1130.

138. Cutlip M B, Yang S C, Stonehart P. Simulation and Optimization of Porous Gas-diffusion Electrodes Used in Hydrogen Oxygen Phosphoric Acid Fuel Cells-II Development of a Detailed Anode Model. *Electrochim. Acta.* 1991;36:547–553.
139. Iczkowski R P, Cutlip M B. *J. Electrochem. Soc.* 1980;127:1433.
140. Antoine O, Bultel Y, Durand R, Ozil P. Electrocatalysis, Diffusion and Ohmic Drop in PEMFC: Particle Size and Spatial Discrete Distribution Effects. *Electrochim. Acta.* 1998;43:3681–3691.
141. Bultel Y, Ozil P, Durand R. Modelling the Mode of Operation of PEMFC Electrodes at the Particle Level: Influence of Ohmic Drop within the Active Layer on Electrode Performance. *J. Appl. Electrochem.* 1998;28:269–276.
142. Bultel Y, Ozil P, Durand R. Modified Thin Film and Agglomerate Models for Active Layers of P.E. Fuel Cells. *Electrochim. Acta.* 1998;43:1077–1087.
143. Bultel Y, Ozil P, Durand R. Modelling of Mass Transfer within the PEM Fuel Cell Active Layer: Limitations at the Particle Level. *J. Appl. Electrochem.* 1999;29:1025–1033.
144. Janssen G J M. A Phenomenological Model of Water Transport in a Proton Exchange Membrane Fuel Cell. *J. Electrochem. Soc.* 2001;148:A1313–A1323.
145. Kulikovsky A A. Quasi-3D Modeling of Water Transport in Polymer Electrolyte Fuel Cells. *J. Electrochem. Soc.* 2003;150:A1432–A1439.
146. Bernardi D M. Water-Balance Calculations for Solid Polymer Electrolyte Fuel Cells. *J. Electrochem. Soc.* 1990;137(11):3344–3351.
147. Mann R F, Amphlett J C, Hooper M A I, Jensen H M, Peppley B A, Roberge P R. Development and Application of a Generalised Steady-State Electrochemical Model for a PEM Fuel Cell. *J. Power Sources.* 2000;86:173–180.
148. Wang Q, Eikerling M, Song D, Liu Z, Navessin T, Xie Z, Holdcroft S. Fuctionally Graded Cathode Catalyst Layers for Polymer Electrolyte Fuel Cells I. Theoretical Modeling. *J. Electrochem. Soc.* 2004;151(7):A950–A957.
149. Pisani L, Valentini M, Murgia G. Analytical Pore Scale Modeling of the Reactive Regions of Polymer Electrolyte Fuel Cells. *J. Electrochem. Soc.* 2003;150(12):A1549–A1559.
150. Rho Y W, Srinivasan S, Kho Y T. Mass Transport Phenomena in Proton Exchange Membrane Fuel Cells Using O_2/He, O_2/Ar, and O_2/N_2 Mixtures. II. Theoretical Analysis. *J. Electrochem. Soc.* 1994;141:2089–2096.
151. Mazumder S, Cole J V. Rigorous 3-D Mathematical Modeling of PEM Fuel Cells I. Model Predictions without Liquid Water Transport. *J. Electrochem. Soc.* 2003;150: A1503–A1509.
152. Perry M L, Newman J, Cairns E J. Mass Transport in Gas-Diffusion Electrodes: A Diagnostic Tool for Fuel-Cell Cathodes. *J. Electrochem. Soc.* 1998;145:5–15.
153. Jaouen F, Lindbergh G, Sundholm G. Investigation of Mass-Transport Limitations in the Solid Polymer Fuel Cell Cathode I. Mathematical Model. *J. Electrochem. Soc.* 2002;149:A437–A447.
154. Dannenberg K, Ekdunge P, Lindbergh G. Mathematical model of the PEMFC. *J. Appl. Electrochem.* 2000;30:1377–1387.
155. Maja M, Tosco P, Vanni M. A One-Dimensional Model of Gas-Diffusion Electrodes for O_2 Reduction. *J. Electrochem. Soc.* 2003;148:A1368–A1376.
156. Wang Q, Song D, Navessin T, Holdcroft S, Liu Z. A Mathematical Model and Optimization of the Cathode Catalyst Layer Structure in PEM Fuel Cells. *Electrochim. Acta.* 2004;50:725–730.
157. Berning T, Djilali N J. Three-Dimensional Computational Analysis of Transport Phenomena in a PEM Fuel Cell-a Parametric Study. *J. Power Sources.* 2003;124:440–452.
158. Marr C, Li X. An Engineering Model of Proton Exchange Membrane Fuel Cell Performance. *ARI.* 1998;50:190–200.

159. Fuller T F, Newman J. Water and Thermal Management in Solid Polymer Electrolyte Fuel Cells. *J. Electrochem. Soc.* 1993;140(5):1218–1225.
160. Siegel N P, Ellis M W, Nelson D J, von Spakovsky M R. A Two-Dimensional Computational Model of a PEMFC with Liquid Water Transport. *J. Power Sources* 2004;128:173–184.
161. Amphlett J C, Mann R F, Peppley B A, Roberge P R, Rodrigues A. A Model Predicting Transient Responses of Proton Exchange Membrane Fuel Cells. *J. Power Sources.* 1996;61:183–188.
162. Kumar A, Reddy R G. Modeling of Polymer Electrolyte Membrane Fuel Cell with Metal Foam in the Flow-field of the Bipolar/End Plates. *J. Power Sources.* 2003;114:54–62.
163. Kumar A, Reddy R G. Effect of Channel Dimensions and Shape in the Flow-field Distributor on the Performance of Polymer Electrolyte Membrane Fuel Cells. *J. Power Sources.* 2003;113:11–18.
164. Armor J N. The Multiple Roles for Catalysis in the Production of H_2. *Appl. Catal. A.* 1999;176:159–176.
165. Minh N Q. Ceramic Fuel Cells. *J. Am. Ceram. Soc.* 1993;76(3):563–588.
166. Steele B C H. Survey of Materials Selection for Ceramic Fuel Cells II. Cathodes and Anodes. *Solid State Ionics.* 1996;86–88:1223–1234.
167. Badwal S P S, Foger K. Solid Oxide Electrolyte Fuel Cell Review. *Ceram. Int.* 1996;22:257–263.
168. Steele B C H. Materials for High-Temperature Fuel Cells. *Philos. Trans. R. Soc. London A* 1996;354:1695–1710.
169. Kawada T, Yokokawa H. Materials and Characterization of Solid Oxide Fuel Cell. *Key Eng. Mater.* 1997;125–126:187–248.
170. McEvoy A. Innovations for Anodes – Catalysis and Direct Oxidation. *Materialwissenschaft und Werstofftechnik* 2002;33:331–334.
171. Tietz F, Buchkremer H-P, Stöver D. Components Manufacturing for Solid Oxide Fuel Cells. *Solid State Ionics.* 2002;152–153:373–381.
172. Tietz F. Materials Selection for Solid Oxide Fuel Cells. *Mater. Sci. Forum.* 2003;426–432:4465–4470.
173. Ormerod R M. Solid Oxide Fuel Cells. *Chem. Soc. Rev.* 2003;32:17–28.
174. Quadakkers W J, Piron-Abellan J, Shemet V, Singheiser L. Metallic Interconnectors for Solid Oxide Fuel Cells – A Review. *Mater. High Temp.* 2003;20:115–127.
175. Brandon N P, Skinner S, Steele B C H. Recent Advances in Materials for Fuel Cells. *Annu. Rev. Mater. Sci.* 2003;33:183–232.
176. Metcalfe I S. Fuel Cells, Advanced Reactors and Smart Catalysis: The Exploitation of Ceramic Ion-Conducting Membranes. *Chem. Eng. Technol.* 2003;26:857–862.
177. S.C. Singhal. Advances in Solid Oxide Fuel Cell Technology. *Solid State Ionics.* 2000;135:305–313.
178. McIntosh S, Gorte R J. Direct Hydrocarbon Solid Oxide Fuel Cells. *Chem. Rev.* 2004;104:4845–4865.
179. Kharton V V, Marques F M B, Atkinson A. Transport Properties of Solid Oxide Electrolyte Ceramics: A Brief Review. *Solid State Ionics.* 2004;174:135–149.
180. Todd B, Young J B. Thermodynamic and Transport Properties of Gases for Use in Solid Oxide Fuel Cell Modelling. *J Power Sources.* 2002;110:186–200.
181. Aguiar P, Adjiman C S, Brandon N P. Anode-Supported Intermediate Temperature Direct Internal Reforming Solid Oxide Fuel Cell. I: Model-Based Steady-State Performance. *J. Power Sources.* 2004;138:120–136.
182. Haynes C, Wepfer W J. 'Design for Power' of a Commercial Grade Tubular Solid Oxide Fuel Cell. *Energy Conversion and Management* 2000;41:1123–1139.

183. Petruzzi L, Cocchi S, Fineschi F. A Global Thermo-Electrochemical Model for SOFC Systems Design and Engineering. *J. Power Sources.* 2003;118:96–107.
184. Chan S H, Khor K A, Xia Z T. A Complete Polarization Model of a Solid Oxide Fuel Cell and Its Sensitivity to the Change of Cell Component Thickness. *J. Power Sources.* 2001;93:130–140.
185. Virkar A V, Chen J, Tanner C W, Kim J. The Role of Electrode Microstructure on Activation and Concentration Polarizations in Solid Oxide Fuel Cells. *Solid State Ionics.* 2000;131:189–198.
186. Zhu H, Kee R J. A General Mathematical Model for Analyzing the Performance of Fuel-Cell Membrane-Electrode Assemblies. *J. Power Sources.* 2003;117:61–74.
187. Kim J, Virkar A V, Fung K, Mehta K, Singhal S C. Polarization Effects in Intermediate Temperature, Anode-supported Solid Oxide Fuel Cells. *J. Electrochem. Soc.* 1999;146: 69–78.
188. Horita T, Yamaji K, Sakai N, Yokokawa H, Weber A, Ivers-Tiffee E. Electrode Reaction of $La_{1-x}Sr_xCoO_{3-d}$ Cathodes on $La_{0.8}Sr_{0.2}Ga_{0.8}Mg_{0.2}O_{3-y}$ Electrolyte in Solid Oxide Fuel Cells. *J. Electrochem. Soc.* 2001;148:A456–A462.
189. Jiang S P, Badwal S P S. An Electrode Kinetics Study of H_2 Oxidation on Ni/Y_2O_3-ZrO_2 Cermet Electrode of the Solid Oxide Fuel Cell. *Solid State Ionics.* 1999;123:209–224.
190. Holtappels P, Vinke I C, de Haart L G J, Stimming U. Reaction of Hydrogen/Water Mixtures on Nickel-Zirconia Cermet Electrodes II. AC Polarization Characteristics. *J. Electrochem. Soc.* 1999;146(8):2976–2982.
191. Larrain D, Van herle J, Marechal F, Favrat D. Generalized Model of Planar SOFC Repeat Element for Design Optimization. *J. Power Sources.* 2004;131:304–312.
192. Aguiar P, Chadwick D, Kershenbaum L. Effect of Methane Slippage on an Indirect Internal Reforming Solid Oxide Fuel Cell. *Chemical Engineering Science.* 2004;59: 87–97.
193. Greene E S, Medeiros M, Chiu W K S. Application of an Anode Model to Investigation Physical Parameters in an Internal Reforming SOFC. Submitted to *ASME J. Fuel Cell Sci. Tech.* 2004.
194. Murthy S, Fedorov A G. Radiation Heat Transfer Analysis of the Monolith Type Solid Oxide Fuel Cell. *J Power Sources.* 2003;124:453–458.
195. Burt A C, Celik I B, Gemmen R S, Smirnov A V. A Numerical Study of Cell-to-Cell Variations in a SOFC Stack. *J Power Sources.* 2004;126:76–87.
196. Haberman B A, Young J B. Three-Dimensional Simulation of Chemically Reacting Gas Flows in the Porous Support Structure of an Integrated-Planar Solid Oxide Fuel Cell. *Int. J. Heat Mass Transfer.* 2004;47:3617–3629.
197. Lazzaretto A, Toffolo A, Zanon F. Parameter Setting for a Tubular SOFC Simulation Model. *J. Energy Resources Technology.* 2004;126:40–46.
198. Yakabe H, Hishinuma M, Uratani M, Matsuzaki Y, Yasuda I. Evaluation and Modeling of Performance of Anode-Supported Solid Oxide Fuel Cell. *J Power Sources* 2000;86: 423–431.
199. Li P W, Suzuki K. Numerical Modeling and Performance Study of a Tubular SOFC. *J. Electrochem. Soc.* 2004;151(4):A548–A557.
200. Prinkey M, Gemmen R S, Rogers W A. Application of a New CFD Analysis Tool for SOFC Technology. In Proceedings of IMECE 2001; ASME: New York, 2001; Vol. 369–4, p 291.
201. Yuan J, Rokni M, Sunden B. Three-Dimensional Computational Analysis of Gas and Heat Transport Phenomena in Ducts Relevant for Anode-Supported Solid Oxide Fuel Cells. *Int. J. Heat Mass Transfer.* 2003;46:809–821.

202. Bove R, Lunghi P, Sammes N M. SOFC Mathematic Model for Systems Simulations. Part One: From a Micro-Detailed to Macro-Black-Box Model. *Int. J. Hydrogen Energy*. 2005;30:181–187.
203. Bove R, Lunghi P, Sammes N M. SOFC Mathematic Model for Systems Simulations-Part 2: Definition of an Analytical Model. *Int. J. of Hydrogen Energy*. 2005;30: 189–200.
204. Yakabe H, Ogiwara T, Hishinuma M, Yasuda I. 3-D Model Calculation for Planar SOFC. *J. Power Sources*. 2001;102:144–154.
205. Hall D J, Colclaser R G. Transient Modeling and Simulation of a Tubular Solid Oxide Fuel Cell. *IEEE Transactions on Energy Conversion*. 1999;14(3):749–753.
206. Campanari S, Iora P. Definition and Sensitivity Analysis of a Finite Volume SOFC Model for a Tubular Cell Geometry. *J. Power Sources*. 2004;132:113–126.
207. Recknagle K P, Williford R E, Chick L A, Rector D R, Khaleel M A. Three-Dimensional Thermo-Fluid Electrochemical Modeling of Planar SOFC stacks. *J. Power Sources*. 2003;113:109–114.
208. Ferguson J R, Fiard J M, Herbin R. Three-Dimensional Numerical Simulation for Various Geometries of Solid Oxide Fuel Cells. *J. Power Sources*. 1996;58:109–122.
209. Pasaogullari U, Wang C Y. Computational Fluid Dynamics Modeling of Solid Oxide Fuel Cells. Proceedings of SOFC VIII, Eds. S. C. Singhal and M. Dokiya, pp 1403–1412, 2003.
210. Autissier N, Larrain D, Van herle J, Favrat D. CFD Simulation Tool for Solid Oxide Fuel Cells. *J. Power Sources*. 2004;131:313–319.
211. Modest M F. Radiative Heat Transfer, McGraw-Hill, New York, 1993.
212. Lehnert W, Meusinger J, Thorn F. Modelling of Gas Transport Phenomena in SOFC Anodes. *J. Power Sources*. 2000;87:57–63.
213. Drescher I, Lehnert W, Meusinger J. Structural Properties of SOFC Anodes and Reactivity. *Acta Electrochim*. 1998;43:3059–3068.
214. M. V. Twigg (Ed.), Catalyst Handbook, Wolfe Pub. Ltd., 1989.
215. Sunde S. Simulations of Composite Electrodes in Fuel Cells. *J. of Electroceramics*. 2000;5(2):153–182.
216. Fleig J. Solid Oxide Fuel Cell Cathodes: Polarization Mechanisms and Modeling of the Electrochemical Performance. *Annu. Rev. Mater. Res*. 2003;33:361–382.
217. Conway B E, Bockris JO'M, Yeager E, Khan S U M, eds. Comprehensive Treatise of Electrochemistry Vol. 7, New York: Plenum.
218. Fleig J. On the Width of the Electrochemically Active Region in Mixed Conducting Solid Oxide Fuel Cell Cathodes. *J. Power Sources*. 2002;105:228–238.
219. Mari C M, Dotelli G. A Random Resistor Model to Forecast the Electrical Properties of Crystalline Ionic Conductor Composites. *Solid State Ionics*. 2000;136:1315–1319.
220. Dotelli G, Mari C M. Modelling and Simulation of the Mechanical Properties of YSZ/Al_2O_3 Composites: A Preliminary Study. *Solid State Ionics*. 2002;148:527–531.
221. Hirschenhofer J H, Stauffer D B, Engleman R R, Klett M G. Fuel Cell Handbook, fourth edn., Department of Energy, Fossil Energy Technology Centre, 1998.
222. Rostrup-Nielsen J, Christiansen L J. Internal Steam Reforming in Fuel Cells and Alkali Poisoning. *Applied Catalysis A: General*. 1995;126:381–390.
223. Schultz T, Zhou S, Sundmacher Kai. Current Status of and Recent Developments in the Direct Methanol Fuel Cell. *Chem. Eng. Technol*. 2001;24:1223–1233.
224. Heizel A, Barragan. A Review of the State-of-the-Art of the Methanol Crossover in Direct Methanol Fuel Cells. *J. Power Sources*. 1999;84:70–74.
225. Baldauf M, Preidel W. Status of the Development of a Direct Methanol Fuel Cell. *J. Power Sources*. 1999;84:161–166.
226. Wasmus S, Küver. Methanol Oxidation and Direct Methnol Fuel Cells: A Selective Review. *J. Electroanalytical Chemistry*. 1999;461:14–31.

227. Ren X, Zelenay P, Thomas S, Davey J, Gottesfeld S. Recent advances in Direct Methanol Fuel Cells at Los Alamos National Laboratory. *J. Power Sources.* 2000;86: 111–116.
228. Argyropoulos P, Scott K, Taama W M. Carbon Dioxide Evolution Patterns in Direct Methanol Fuel Cells. *Electrochim. Acta.* 1999;44:3575–3584.
229. Sundmacher K, Scott K. Direct Methanol Polymer Electrolyte Fuel Cell: Analysis of Charge and Mass Transfer in the Vapor-Liquid-Solid System. *Chemical Engineering Science.* 1999;54:2927–2936.
230. Argyropoulos P, Scott K, Taama W M. Modelling Pressure Distribution and Anode/Cathode Streams Vapour-Liquid Equilibrium Composition in Liquid Feed Direct Methanol Fuel Cells. *Chemical Engineering Journal.* 2000;78:29–41.
231. Thomas S C, Ren X, Gottesfeld S. Influence of Ionomer Content in Catalyst Layers on Direct Methanol Fuel Cell Performance. *J. Electrochem. Soc.* 1999;146:4354–4359.
232. Liu L, Pu C, Viswanathan R, Fan Q, Liu R, Smotkin E S. Carbon supported and unsupported Pt-Ru anodes for liquid feed direct methanol fuel cells. *Electrochim. Acta* 1998;43:3657–3663.
233. Hayden E. The Promotion of CO Electro-oxidation on Platinum-bismuth as a Model for Surface Mediated Oxygen Transfer. *Catal. Today* 1997;38:473–481.
234. Page T, Johnson R, Hormes J, Noding S, Rambabu B. A Study of Methanol Electro-oxidation Reactions in Carbon Membrane Electrodes and Structural Properties of Pt Alloy Electro-Catalysts by EXAFS. *J. Electroanal. Chem.* 2000;485:34–41.
235. Narayanan S R, Frank H, Jeffries-Nakamura B, Smart M, Chun W, Halpert G, Kosek J, Cropley C. In Proton Conducting Membrane Fuel Cells I; Gottesfeld S, Halpert G, Landgrebe A, Eds.; The Electrochemical Society Proceedings Series: Pennington, NJ, 1995; PV 95–23, p 278.
236. Ren X, Zawodzinski T A, Uribe F, Dai H, Gottesfeld S. In Proton Conducting Membrane Fuel Cells I; Gottesfeld S, Halpert G, Landgrebe A, Eds.; The Electrochemical Society Proceedings Series: Pennington, NJ, 1995; PV 95–23, p 284.
237. Ren X, Springer T E, Zawodzinski T A, Gottesfeld S. Methanol Transport Through Nation Membranes: Electro-osmotic Drag Effects on Potential Step Measurements. *J. Electrochem. Soc.* 2000;147:466–474.
238. Valdez T I, Narayanan S R. In Proton Conducting Membrane Fuel Cells II; Gottesfeld S, Fuller T F, Halpert G, Eds.; The Electrochemical Society Proceedings Series: Pennington, NJ. 1998; PV 98–27, p 380.
239. Wang J T, Wasmus S, Savinell R F. Real-Time Mass Spectrometric Study of the Methanol Crossover in a Direct Methanol Fuel Cell. *J. Electrochem. Soc.* 1996; 143:1233–1239.
240. Argyropoulos P, Scott K, Shukla A K, Jackson C. A Semi-empirical Model of the Direct Methanol Fuel Cell Performance Part I. Model Development and Verification. *J. Power Sources.* 2003;123:190–199.
241. Scott K, Argyropoulos P, Sundmacher K. A Model for the Liquid Feed Direct Methanol Fuel Cell. *J. Electroanalytical Chemistry.* 1999;477:97–110.
242. Sandhu S S, Crowther R O, Krishnan S C, Fellner J P. Direct Methanol Polymer Electrolyte Fuel Cell Modeling: Reversible Open-Circuit Voltage and Species Flux Equations. *Electrochim. Acta.* 2003;48:2295–2303.
243. Dohle H, Wippermann K. Experimental Evaluation and Semi-Empirical Modeling of U/I Characteristics and Methanol Permeation of a Direct Methanol Fuel Cell. *J. Power Sources.* 2004;135:152–164.
244. Meyers J P, Newman J. Simulation of the Direct Methanol Fuel Cell I. Thermodynamic Framework for a Multicomponent Membrane. *J. Electrochem. Soc.* 2002;149: A710–A717.

245. Meyers J P, Newman J. Simulation of the Direct Methanol Fuel Cell II. Modeling and Data Analysis of Transport and Kinetic Phenomena. *J. Electrochem. Soc.* 2002;149: A718–A728.
246. Meyers J P, Newman J. Simulation of the Direct Methanol Fuel Cell III. Design and Optimization. *J. Electrochem. Soc.* 2002;149:A729–A735.
247. Divisek J, Fuhrmann J, Gärtner K, Jung R. Performance Modeling of a Direct Methanol Fuel Cell. *J. Electrochem. Soc.* 2003;150(6):A811–A825.
248. Baxter S F, Battaglia V S, White R E. Methanol Fuel Cell Model: Anode. *J. Electrochem. Soc.* 2000;146:437–447.
249. Kulikovsky A A, Divisek J, Kornyshev A A. Two-Dimensional Simulation of Direct Methanol Fuel Cell. A New (Embedded) Type of Current Collector. *J. Electrochem. Soc.* 2000;147:953–959.
250. Kulikovsky A A. Two-Dimensional Numerical Modelling of a Direct Methanol Fuel Cell. *J. Appl. Electrochem.* 2000;30:1005–1014.
251. Murgia G, Pisani L, Shukla A K, Scott K. A Numerical Model of a Liquid-Feed Solid Polymer Electrolyte DMFC and Its Experimental Validation. *J. Electrochem. Soc.* 2003;150:A1231–A1245.
252. Wang Z H, Wang C Y. Mathematical Modeling of Liquid-Feed Direct Methanol Fuel Cells. *J. Electrochem. Soc.* 2003;150(4):A508–A519.
253. Jeng K T, Chen C W. Modeling and Simulation of a Direct Methanol Fuel Cell Anode. *J Power Sources.* 2002;112:367–375.
254. Mawardi A, Yang F, Pitchumani R. Optimization of the Operating Parameters of a Proton Exchange Membrane Fuel Cell for Maximum Power Density. *J. Fuel Cell Sci. Tech.* 2005;2(2) In press.
255. Mawardi A, Pitchumani R. Effects of Parameter Uncertainty on the Performance Variability and Reliability of Proton Exchange Membrane (PEM) Fuel Cells. *J. Power Sources* 2005, In review.
256. Nelder J A, Mead R. A Simplex Method For Function Minimization. *Comput. J.* 1965;7:308–313.
257. Kirkpatrick S, Gelatt C D, Vecchi M P. Optimization by Simulated Annealing. *Science.* 1983;220(4598):671–680.
258. Press W H, Flannery B P, Teukolsky S A, Vettering W T. Numerical Recipes in FORTRAN. Cambridge University Press, New York, 1992.
259. Lagarias J C, Reeds J A, Wright M A, Wright P E. Convergence properties of the Nelder-Mead Simplex Method in Low Dimensions. *SIAM J. Optim.* 1998;9:112–147.
260. Iman R L, Shortencarier M J. A FORTRAN77 Program and User's Guide for Generation of Latin Hypercube and Random Samples for Use with Computer Models. *Technical Report, NUREG/CR-3624, SAND83-2365*, Sandia National Laboratories, 1984.

5

Fuels and Fuel Processing

Shaker Haji, Kinga A. Malinger, Steven L. Suib, Can Erkey

5.1 Introduction

H_2 is an ideal fuel for fuel cells (FCs) because of its high reactivity and zero-emission characteristics. Unfortunately, H_2 is not easily available, and neither its production nor distribution infrastructure are widely spread. Therefore, development of technologies for production of H_2 onboard and onsite from other sources such as natural gas, methanol, and gasoline is necessary. In the next section, different feedstocks that are suitable for H_2 production for fuel cell application are presented. Subsequent sections focus on reforming of hydrocarbons that are processed by a series of steps that include fuel desulfurization, reforming, water-gas shift reaction and carbon monoxide (CO) removal [1,2,3,4]. Figure 5.1 illustrates what is known as the fuel processing train with some options for the essential steps. This general fuel processing train is usually used for PEM fuel cells running on fossil fuels such as natural gas. The processing steps that the feedstock are subjected to depend on the type of the fuel and the fuel cell. For example, if methanol is the fuel for PAFC, the CO removal step might not be necessary. The last two steps, water-gas shift reaction and CO removal, are not necessary in the case of MCFC. If feedstocks heavier than methane are used as a feed to the SOFC, an additional process step known as pre-reforming might be used. Such variations are described in more detail in Section 5.3.

Currently, 48% of H_2 is produced from natural gas, 30% from petroleum (mostly consumed in refineries), 18% from coal, and the remaining 4% by water electrolysis worldwide [6]. At today's energy prices, it is considerably more expensive to produce hydrogen by water electrolysis than by reforming of fossil fuels. H_2 costs around $5.60 for every GJ when produced from natural gas, $10.30 per GJ from coal, and $20.10 per GJ from electrolysis of water [5]. Approximately half of the H_2 produced is consumed in the refineries to improve their product quality. Roughly, 40% of the H_2 produced is used in ammonia synthesis, and the rest is used in the manufacture of methanol and in other processes [6,7,8]. In general, the actual production of H_2 will have to be dramatically increased to fulfill the future energy requirements in fuel cell technology [9]. In the future a whole new infrastructure will be built for delivery and supply of H_2; however, reforming onboard and onsite will be the primary vehicle for generation of H_2 for fuel cells in the near future.

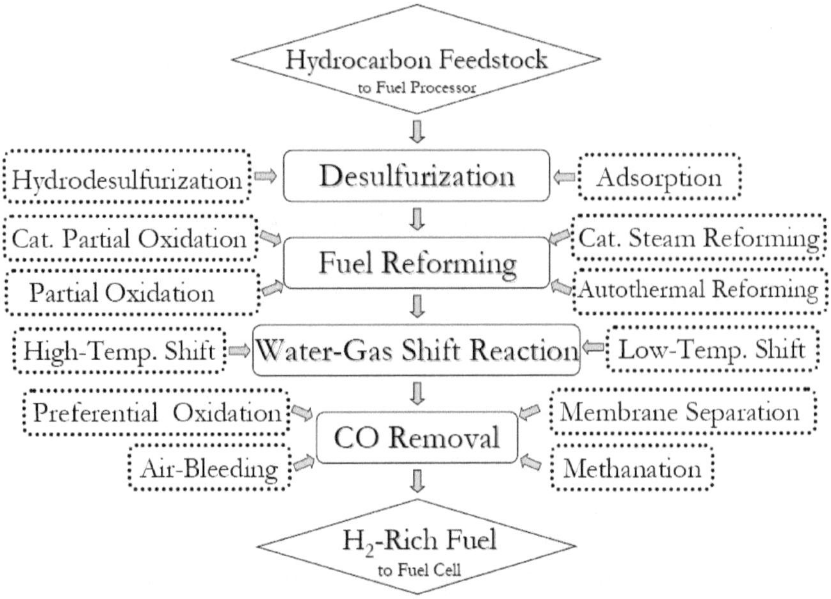

Figure 5.1. A general fuel processing train

5.2 Feedstocks for H_2 Production

Current feedstocks can be broadly categorized as: 1) renewable fuel sources such as biomass and ethanol, and 2) nonrenewable fuel sources such as natural gas. In the this section, we briefly describe (see Table 5.1) some common feedstocks (renewable and nonrenewable) that have been utilized in many studies [10,11,26] as potential candidates for H_2 manufacture – natural gas, gasoline, diesel, methanol, ethanol, ammonia, and biomass.

5.2.1 Natural Gas

Natural gas (NG) is a fossil fuel that occurs in porous rocks in the earth's crust. NG is often found in the same geologic formations as petroleum, but may also be found alone in separate reservoirs. Methane is usually the major constituent of natural gas, ranging between 75–96 vol.%. Nevertheless, other hydrocarbons (ethane, propane *etc.*), nitrogen, hydrogen sulfide, water, and carbon dioxide can be present in large amounts, which may pose particular processing difficulties [10]. Natural gas is relatively abundant around the world. In 2001, the annual world consumption of natural gas was 90.3×10^{12} ft^3 [12]. The well-established infrastructure together with proven reserves clearly indicates that natural gas is the most abundant primary fuel. Natural gas is an excellent means of energy storage and is the cleanest of all primary fossil fuels [11]. Currently, the cheapest source of industrial H_2 is the steam reforming of natural gas [11]. Natural gas is also the major source of H_2 production in the world. For safety reasons, natural gas is odorized by organosulfur compounds. The very distinct smell of sulfur compounds provides adequate warning of a gas leak.

5.2.2 Liquid Petroleum Gas

Propane, which is the main constituent of liquid petroleum gas (LPG) and HD-5 propane (95% propane), has been recently used as a fuel in both high and low-temperature fuel cells [13,14,16]. IdaTech (Oregon-based Company) has developed a PEM fuel cell (EtaGen™) with dual-fuel capability that can operate on either natural gas or commercial propane [14,15]. Recent interests in using propane arise from the fact that liquid propane has good energy density at low storage pressure (46.4 MJ/kg) and therefore it can be stored and transported economically. In addition, propane is available worldwide and is sometimes used to dilute methane to facilitate its storage and transport. Finally, propane is not a potential contributor to groundwater pollution like most of the liquid fuels and has relatively lower human toxicity characteristics compared to methanol or gasoline [16].

5.2.3 Liquid Hydrocarbon Fuels: Gasoline and Diesel

The existing production and distribution infrastructure of gasoline and diesel fuels makes them potential candidate fuels for H_2 production for fuel cell applications. It would be economical, especially for introduction into the market, if fuel cell vehicles could be fueled with fuels similar to those that are currently used in internal combustion engines. Another advantage is the high energy density of gasoline and diesel - the lower heating value of diesel is 42.5 MJ/kg while that of methanol is only 19.9 MJ/kg. However, reforming of diesel fuel is much more complicated and requires much higher temperatures [17]. In general, complex fuels like gasoline and diesel require reforming temperatures higher than 700°C for maximum H_2 production [107].

Gasoline and diesel duels, obtained by refining crude oil, consist of numerous chemical compounds consisting of paraffins (normal, branched, and cyclic), olefins, and aromatics. Gasoline contains C_4 to C_{10} hydrocarbons, while those in diesel fuel range between C_{10} to C_{22}. Fossil fuels are also contaminated with traces of sulfur, nitrogen or oxygen-containing compounds and organometallic compounds that lower their quality. The removal of organosulfur compounds is considered to be one of the challenges in utilizing gasoline and diesel fuel for fuel cell applications.

5.2.4 Alcohols: Methanol and Ethanol

Alcohols such as methanol and ethanol are among the feedstock candidates as they are convenient for storage, especially for portable applications. Methanol is commercially produced in large quantities via the reaction of CO and H_2 from synthesis gas [101]. The properties, grades, prices, and manufacture of methanol are reported elsewhere [1]. From a technological point of view, methanol has the following advantages: a) can be reformed at a relatively low temperature (250°C), b) high H_2-to-carbon ratio, c) no C-C bond, thus minimizing the risk of soot formation, d) high H_2 content in the product stream (up to 75%), and e) no sulfur in the fuel [18].

Methanol can be reformed by a wide variety of methods such as steam reforming, partial oxidation, and catalytic decomposition [1]. Each process has advantages and disadvantages. Currently, there is extensive research in the area of catalysis in order to develop highly efficient materials capable of transforming methanol selectively into H_2 and for such catalysts to remain active for long periods [18,19,20]. Methanol can also be directly fed into direct methanol fuel cells (DMFC) where this reactant, along with water, are electrochemically converted to carbon dioxide, protons, and electrons [21].

Ethanol, on the other hand, is commercially produced by hydrating ethylene [10] or by fermentation of starch or sugar [22]. Research suggests that ethanol may also be produced from lower-cost vegetation such as crop wastes [22]. Ethanol is readily and increasingly available in the United States because of the requirement for ethanol as an additive in gasoline fuels, and 2.8 billion gallons/year are now produced throughout the country by the fermentation of biomass at a cost of approximately $1 per gallon, which is competitive with petroleum fuels. H_2 generation from ethanol, using lanthanides, Ru, and Ni, via oxidation has been demonstrated [22] but ethanol-air mixtures are flammable over a wide composition range. Since the combustion reaction is highly exothermic, the resulting flames also form coke and soot, acetaldehyde, and ethylene, which are severe poisons to fuel cells [22].

5.2.5 Ammonia

Obtaining H_2 from a single-step process like ammonia decomposition can be an attractive alternative to hydrocarbon fuels for small-scale fuel cell applications. Ammonia is synthesized by catalytically reacting H_2 (from hydrocarbons) with nitrogen (from air) under high pressure. Ammonia can be conveniently stored in liquid form, has acceptable energy density (18.6 MJ/kg, LHV), and the safety issues concerning its storage and handling are well established. More importantly, since ammonia does not contain carbon atoms, coking is not a concern and the product stream (H_2/N_2) is carbon monoxide free [23]. Although ammonia decomposition has been utilized to power alkaline fuel cell prototypes, concerns about the effect of trace ammonia on the anode, and on the membranes of PEM fuel cells have largely limited applications for H_2 generation for PEM fuel cell [23]. For example, ammonia concentrations as low as 10–20 ppm can cause significant performance degradation to PEM fuel cells [24].

Table 5.1 Fuel comparison chart [26,25].

	Hydrogen	Natural Gas CNG/LCG[a]	Liquid Petroleum Gas (LPG)	Gasoline	No. 2 Diesel	Methanol (M85)[b]	Ethanol (E85)[b]	Ammonia
Main Fuel Source	*Natural Gas, Methanol, and other energy Sources*	*Underground reserves*	*A by-product of petroleum refining or natural gas processing*	*Crude Oil*	*Crude Oil*	*Natural gas, coal, or, woody biomass*	*Corn, Grains, or agricultural waste*	*Reaction of N_2 (air) with H_2 (Mainly natural gas also other hydrocarbons)*
Energy Content per Gallon	*~30,500 Btu[c] (113,000-134,000 Btu/gge)*	*CNG: 33-38 kBtu @ 3000 psi; 38- 44 kBtu @ 3600 psi. LNG: ~73.5 kBtu*	*~84,000 Btu*	*109,000 - 125,000 Btu*	*128,000 - 130,000 Btu*	*56,000 - 66,000 Btu*	*~ 80,000 Btu*	*~45,500 Btu[c]*
Average Cost at the pump[d]	*$15/gge[e]*	*CNG: $1.35/gge*	*$2.21/gge ($1.65/gal)*	*$1.48/gal*	*$1.48/gal*	*N/A*	*$1.70/gge ($1.36/gal)*	*N/A*
Fuel Availability	*There are only a small number of H_2 stations across USA. Most are available for private use only.*	*CNG: > 1,100 CNG stations in USA. Highest in CA. Home fueling available. LNG: limited stations (~35). Available via cryogenic liquids suppliers.*	*LPG is the most accessible alternative fuel in the U.S. There are more than 3,300 stations nationwide.*	*Available at all fueling stations.*	*Available at select fueling stations.*	*Methanol remains a qualified alternative fuel as defined by EPAct, but it is not commonly used.*	*Most of the E-85 fueling stations are located in the Midwest, but in all, ~150 stations are available in 23 states.*	*There are no NH_3 fueling stations. But available via cryogenic liquids suppliers.*
Safety Issues[f]	*Hydrogen has an excellent industrial safety record; codes and standards for consumer vehicle use are under development.*	*Pressurized tanks have been designed to withstand severe impact, high external temperatures, and automotive environmental exposure.*	*Ventilation is important for fueling due to LPG increased flammability. LPG tanks are 20 times more puncture resistant than gasoline tanks and can withstand high impact.*	*Gasoline is a relatively safe fuel since people have learned to use it safely. Gasoline is not biodegradable though, so a spill could pollute soil and water.*	*Diesel is a relatively safe fuel since people have learned to use it safely. Diesel is not biodegradable, so a spill could pollute soil and water.*	*Can form an explosive vapor in fuel tanks. In accidents; however, methanol is less dangerous than gasoline because its low evaporation speed keeps alcohol concentration in the air low and non explosive.*	*Can form an explosive vapor in fuel tanks. In accidents; however, ethanol is less dangerous than gasoline because its low evaporation speed keeps alcohol concentration in the air low and non explosive.*	*Safety issues concerning NH_3 storage and handling are well established. However, NH_3 is considered toxic and corrosive.*

a-CNG/LNG: Compressed or Liquefied Natural Gas. b- Blends of 85% methanol (M-85) or ethanol (E-85) with 15% gasoline. c-Based on liquid density at 1 atm and boiling point. d-The prices represent retail (at-the-pump sales prices) in Dec 2003, USA. Prices are per gallon or gasoline-equivalent gallon (gge). e-SunLine Transit Agency (hydrogen fueling station), Thousand Palms, CA. f-Without exception, all alternative fuel vehicles must meet today's OEM Safety Standards.

5.2.6 Biomass

H_2 could in principle be generated from renewable resources such as biomass. Biomass is a term used to describe organic matters that are available on a renewable basis. It includes forest and mill residues, agricultural crops and wastes, wood and wood wastes, animal wastes, livestock operation residues, aquatic plants, fast-growing trees and plants, and municipal and industrial wastes [26]. Biomass can be converted to biogas, which mainly consists of methane and carbon dioxide, via hydrogasification, anaerobic digestion, or pyrolysis. Biogas can also be generated in landfill sites. Methanol can be synthesized from the syngas obtained from biomass/biogas. Ethanol can be produced by biomass fermentation, and then used as a H_2 source, as discussed above. Biodiesel is another product of biomass that can be manufactured from vegetable oils, animal fats, or recycled restaurant greases. Biomass gasification with air produces a gas with low H_2 content, 8–14 vol %, accompanied by the production of several by-products [27]. The effectiveness of biomass as a H_2 source depends on the discovery of new generations of catalysts [28]. Biomass candidates for H_2 generation include sugar, starch, oils, and crop wastes. The production of H_2 from sugar by catalytic reactions has been demonstrated, but the process from glucose thus far has shown only 50% selectivity to H_2 and requires a long reaction time [22].

5.3 Fuel Processing for Fuel Cell Application

In fuel cell terminology, fuel processing is defined as the conversion of the raw primary fuel supplied to a fuel cell system into the fuel required by the stack [29]. The fuel reforming process, the main step in fuel processing, was developed for the first time in the 1920s to produce a H_2/N_2 mixture for ammonia synthesis [30, 31, 32, 33, 34, 35, 36, 37, 38] and was further developed in the 1930s when natural gas and other hydrocarbons became available on a large scale [39, 40]. Reforming processes have also been used for other applications such as conversion of higher hydrocarbons to a mixture of CH_4, CO, and H_2 called town gas [41] or to a CO/H_2 mixture called syngas [42, 43, 44, 45, 46] used for CH_3OH synthesis [47]. Current interests in fuel reforming are due to the use of H_2 in power production [48].

Methods to generate H_2 range from cracking of water by electrolysis, chemical conversion of fossil fuel, and biomass thermochemical and biological treatment [49, 50, 51, 52, 53, 54]. Today the most economical H_2 is produced by reforming natural gas and the technology is commonly used for stationary fuel cells (> 200 kW) [9,55]. In contrast, methanol is the fuel of choice for portable fuel cell applications such as mobile telephones (100 mW) and laptop computers (30 W) due to the ease of reloading. Fuel tank devices of the same size of present batteries can be reloaded with methanol just like a lighter or an ink pen, and each recharging can provide more than three times the power that current batteries can provide [9].

Table 5.1. The fuel specification required by the different types of fuel cells

Fuel Cell Type	Operating Temperature	H_2	CO	CH_4	CO_2 & H_2O	Sulfur (as H_2S and COS)
PEMFC	*60–100 °C*	*Fuel*	*Poison (>10ppm)*	*Diluent*	*Diluent*	*Poison (>0.1 ppm)*
AFC	*90–100 °C*	*Fuel*	*Poison*	*Diluent*	*Poison*[c]	*Unknown*
PAFC	*175–200 °C*	*Fuel*	*Poison (>0.5%)*	*Diluent*	*Diluent*	*Poison (>50 ppm)*
MCFC	*~650 °C*	*Fuel*	*Fuel*[a]	*Diluent*[a,b]	*Diluent*	*Poison (>0.5 ppm)*
SOFC	*600–1000 °C*	*Fuel*	*Fuel*[a]	*Diluent*[a,b]	*Diluent*	*Poison (>1 ppm)*

[a] In reality CO reacts with H_2O to produce H_2 and CO_2 by water gas shift reaction and CH_4 reacts with H_2O to form H_2 and CO faster than reacting as a fuel at the electrode. [b] A fuel in the internal reforming MCFC and SOFC. [c] CO_2 is a poison for AFC that more or less rules out its use with reformed fuels.

There are different types of fuel cells that operate at different conditions and thus require fuel feed in different specifications in terms of composition and purity [9]. Table 5.2 summarizes some of those requirements [29, 56, 57]. In general, the higher the operating temperature of the fuel cell, the more tolerant the fuel cell is to the impurities in the feed. For example, due to their high operating temperatures, SOFC and MCFC can utilize carbon monoxide via the water gas shift reaction, while a very small concentration of carbon monoxide can poison the PEM fuel cell. While PAFC is insensitive to CO_2, AFC is very sensitive to the acid gas that results in carbonate formation in the alkaline electrolyte. On the other hand, MCFC require the presence of CO_2 in the cathode to compensate for the carbonate transfer to the anode. Despite its low operating temperature, PAFC is found to be the most tolerant toward sulfur presence in the feed.

The following sections explain the basic steps in the fuel processing train and reviews some of the existing technologies and options utilized in fuel processing for fuel cell applications.

5.3.1 Desulfurization

An important consideration in processing hydrocarbon fuels is that they contain significant amounts of sulfur containing compounds such as thiophenes, benzothiophenes and dibenzothiophenes. In the case of natural gas, sulfur is present in the form of the odorants that are added for safety purposes. Typical sulfur-containing odorants are thiophenes (*e.g.* tetrahydrothiophene), mercaptans (*e.g.* ethyl mercaptans, tertiary butyl mercaptans), organic sulfides (*e.g.* diethyl sulfide), or mixtures of the above. Currently sulfur levels in diesel and gasoline are around 500 and 300 parts per million by weight (ppmw), respectively. The presence of such compounds at such concentrations has a detrimental effect on the performance of catalysts used in different sections of the fuel processor and the fuel cell stacks [39, 58]. For example, the presence of 100 ppm thiophene in a simulated gasoline mixture caused the H_2 content of the effluent from the fuel processing unit to drop from 60 to 40 vol.% over a 25 hour reaction time [3]. On a similar note, the presence of 50 ppm H_2S in the H_2 stream fed to a PEMFC (Pt-based electrode) operated at 70°C caused the current to decrease [58]. While the cell produced 12 A with pure H_2 feed, the current decreased 1.3 A in the first hour after the exposure to 50 ppm H_2S (voltage held at 0.69 V). The current reduced another 3.6 A in the second hour and the final value of 0.49 A was observed after 3.7 h of exposure to the H_2S-contaminated H_2 feed. In contrast to CO poisoning, adding Ru to Pt to the anode catalysts has no effect on increasing MEA tolerance toward H_2S poisoning [58]. Even the 15 and 30 ppmw S limits in gasoline and diesel that are imposed by EPA by the year 2006 are still too high for fuel cell applications. Palm *et al.* [59] studied the effect of sulfur in feed while assessing the feasibility of H_2 production from diesel fuel by autothermal reforming using precious metal catalyst. The H_2 production was compared for a feed of C_{13}-C_{15}- alkane mixture and that of the same mixture but doped with small amounts of 1-benzothiophene (11 and 30 ppmw S). The hydrocarbon conversion decreased significantly when the pure feed was replaced with the sulfur-contaminated feed: from ~93% down to 77% with 11 ppmw S and to 73% with 30 ppmw S. After the feed was switched back to the pure feed, the conversion did not increase back to its original level even after hours of operation. It is generally agreed that hydrocarbon fuels fed to PEMFC need to be desulfurized to concentrations less than 0.1 ppm sulfur that necessitates incorporation of desulfurization technologies into the fuel processor [29, 60, 108]. Thus, a desulfurization unit is required prior to the other fuel processing units. Two approaches for desulfurization, namely hydrodesulfurization (HDS) and sulfur removal by adsorption, are explained below; with the former being the most mature technology.

5.3.1.1 *Hydrodesulfurization (HDS)*

HDS is used in refineries to reduce the sulfur content in fossil fuels. In this process, the organosulfur compounds are converted to H_2S and sulfur-free organic compound(s) by reaction with H_2 in the presence of a catalyst as shown below:

(5.1)

HDS reactors are commonly operated at moderate temperatures (300–360°C) and at H_2 pressures of 3.0–5.0 MPa, usually with $CoMo/Al_2O_3$ or $NiMo/Al_2O_3$ catalysts [61, 62]. HDS has been adapted for PAFC systems, which operate with natural gas. HDS was also tested by Fuel Cell Energy (Danbury, CT) to be used in conjunction with direct carbonate fuel cells [192]. The process demonstrated the capability to reduce the sulfur content of NATO F-76 distillate fuel (3800 ppmv S) down to 0.1 ppmv S. While in refineries the resulting H_2S-from the HDS reaction- is eventually converted to elemental sulfur by a modified version of the Claus process [101], in fuel cell applications it is sufficient to eliminate the resulting H_2S from the feed by subsequently adsorbing H_2S onto a bed of zinc oxide to form zinc sulfide [63, 64, 65]:

$$H_2S + ZnO \rightleftharpoons ZnS + H_2O \tag{5.2}$$

In automotive fuel cell applications, preliminary calculations indicate that a fixed-bed adsorber containing about 1 kg of ZnO needs to be regenerated or changed every 10,000 miles (calculation is based on 10,000 mile/service, 45 mole H_2S adsorbed/mole ZnO [66], 80 mile/gal fuel consumption, diesel fuel with 430 ppmw S, and bed density (ZnO) of 5.606 g/cm^3). In an attempt to merge both of the processes together (the HDS and the H_2S adsorption) in one unit, Ni/ZnO catalyst was developed and evaluated for fuel cell application [66]. The Ni/ZnO was defined as "adsorptive HDS catalyst" where Ni catalyzed the hydrodesulfurization of kerosene, while ZnO was the support and H_2S the adsorbent. Adaptation of HDS for fuel cell applications is very challenging when the feedstock used to obtain H_2 is gasoline or diesel. Intermediate distillates, such as diesel, contain high boiling point organosulfur compounds such as benzothiophenes and dibenzothiophenes. The reactivity of such compounds toward HDS reaction is substantially lower than that of the low boiling point compounds such as sulfides and disulfides that are present in the light feedstocks or the mercaptans that are present in natural gas [61, 62]. The reactivity depends upon the local environment of the sulfur atom in the molecule and the overall shape of the molecule [61]. The reactivity of the organosulfur compounds in HDS is as follows [67]: thiophene > benzothiophene > benzonaphthothiophene > tetrahydrobenzonaphthothiophene > dibenzothiophene. In addition, the substitution of these compounds by ring alkylation further affects the reactivity. For example, the reactivity of naturally occurring dibenzothiophenes, with methyl substitution in different positions, decreases in the following order [68]: 2,8-dimethyldibenzothiophene (2,8-DMDBT) > dibenzothiophene (DBT) > 4-methyldibenthothiophene (4-MDBT) > 4,6-dimethyldibenzothiophene (4,6-DMDBT).

The exact mechanism of HDS of the refractory organosulfur compounds is still under investigation. However, HDS likely proceeds through the reaction network proposed by Houalla *et al.* [69, 70]. This is illustrated in Figure 5.2 for the HDS of DBT at 300°C and 102 atm [69]. The reaction of hydrogen with dibenzothiophene gave predominantly biphenyl (BiPh) as an organic product. In the parallel pathway mechanism, the primary reaction products formed directly from DBT were tetrahydrodibenzothiophene (THDBT) and/or hexahydrodibenzothiophene (HHDBT) from one path and biphenyl via another path. Both THDBT and HHDBT are very reactive intermediates that are difficult to isolate (for detection), and are further desulfurized to form the secondary product, which is cyclohexylbenzene (CHB). This pathway is referred to as the hydrogenation pathway (HYD) since the sulfur compound is hydrogenated prior to desulfurization. CHB is also produced via a finite contribution from the sequential hydrogenation of biphenyl (BiPh), which is formed via direct C-S bond hydrogenolysis of DBT. Desulfurization through this pathway is known as a hydrogenolysis pathway or direct desulfurization (DDS). The tertiary product is bicyclohexyl (BiCh) that is formed in traces via the slow hydrogenation reaction of CHB. The activity of $NiMo/Al_2O_3$ catalyst (per unit surface area) was about twice that of the $CoMo/Al_2O_3$ and the yield of CHB -at a given conversion- was about three times higher with the former catalysts than with the latter. This indicates that hydrogenation is a better route for increasing the desulfurization extent of the refractory compounds. In summary, the HDS of DBTs under industrial conditions occurs via two routes; the direct desulfurization of DBTs (DDS, hydrogenolysis) and the hydrogenation of the aromatic ring prior desulfurization (HYD: hydrogenation), the former reaction being faster than the latter. However, the hydrogenation rate becomes relatively fast as H_2S and/or H_2 concentration is increased in the reaction mixture, as there is/are methyl group(s) in the 4 or 4 and 6 positions, or as the catalyst is replaced with a more active one for hydrogenation ($NiMo/Al_2O_3$ vs. $CoMo/Al_2O_3$).

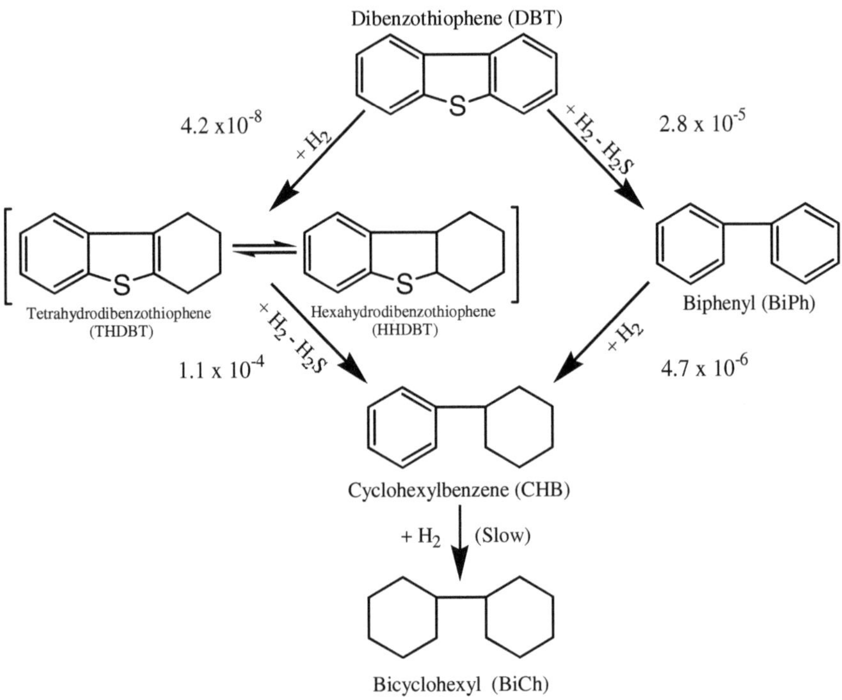

Figure 5.2. Proposed reaction network by Houalla *et al.* [69] for the conversion of DBT and H_2 in the presence of sulfided $CoMo/Al_2O_3$ at 300°C and 102 atm. The numbers next to the arrows are the pseudo first-order rate constants (m^3/kg cat.s) (reproduced by kind permission of Catal. Today)

An extensive number of studies have been reported in the literature on synthesis, characterization and activity of catalysts for HDS. Almost all the transition metals –in their sulfided form- have been tested for their activities toward HDS [71, 72]. When tested individually, the activity towards HDS varied over three orders of magnitude across the periodic table and the most active catalysts were found to be from the second and the third rows. The difference in activity was related to the electronic and structural properties. The synergetic effect in the bimetallic catalysts used in HDS was also studied [73, 74, 81, 87]. Examples of bimetallic catalysts are: the commercial catalysts ($CoMo/Al_2O_3$ and $NiMo/Al_2O_3$), NiW/Al_2O_3, CoW/Al_2O_3, and $PtPd/Al_2O_3$. Isoda *et al.* [75] performed a study using a blend of $CoMo/Al_2O_3$ and Ru/Al_2O_3 catalysts and compared its activity towards the HDS of 4,6-DMDBT in the presence of naphthalene to those of $CoMo/Al_2O_3$, $NiMo/Al_2O_3$, and Ru/Al_2O_3. The blend catalyst showed the highest rate of HDS of 4,6-DMDBT through its selective hydrogenation without excessive hydrogenation of naphthalene. Lecrenay *et al.* [76] studied the effect of different catalysts on the HDS reaction pathways of 4,6-DMDBT in decalin. It was found that the commercial $NiMo/Al_2O_3$ exhibited ~3 times higher activity than that of the commercial $CoMo/Al_2O_3$. This was ascribed to the higher hydrogenation activity of $NiMo/Al_2O_3$ catalyst vs. $CoMo/Al_2O_3$ deduced by comparison of the concentrations of methyl substituted-CHB and -BiPh in the product mixture (HYD/DDS = 12 for NiMo compared to 4 for CoMo). Hydrogenation of naphthalene in octane was used to compare the hydrogenation activity of both catalysts that showed that the Ni-based catalyst had ~3 times higher hydrogenation activity than the Co-based catalyst. Since the steric hindrance is considered to be the reason behind the low reactivity of 4,6-DMDBT [77], and hydrogenation of the aromatic ring leads to alleviation of this hindrance, catalysts with higher hydrogenation capability show higher catalytic activity towards the HDS of 4,6-DMDBT.

Supported platinum was also studied as a catalyst for HDS under standard industrial conditions. Vasudevan and coworkers [78] compared Pt-based catalysts with the commercial $CoMo/Al_2O_3$ in HDS of commercial diesel fuel. Both Pt/HY zeolite and Pt/ASA (amorphous silica-alumina) were more active than the $CoMo/Al_2O_3$ catalyst. Kabe *et al.* [79] also compared unsulfided Pt/Al_2O_3 (3 wt%), with the conventional $CoMo/Al_2O_3$ catalyst and showed that both had similar activity for HDS of DBT in decalin. 10% of DBT conversion occurred via the hydrogenation route with 2 wt% Pt/Al_2O_3 compared to 7% when conventional CoMo catalyst was used. Hypothesizing that hydrogenation followed by desulfurization will be the most likely route to desulfurize hindered dibenzothiophenes [61], van Veen and coworkers [80, 81] used catalysts with high hydrogenation activity to desulfurize 4-ethyl, 6-methyl dibenzothiophene (4E6MDBT). The HDS activity was found to be in the following order: Pt/ASA >> Pt/Al_2O_3 > NiW/Al_2O_3 >> $CoMo/Al_2O_3$ or $NiMo/Al_2O_3$. The superiority of Pt/ASA in HDS of 4E6MDBT was attributed to its superior hydrogenation activity. Both sulfided and unsulfided Pt/ASA had similar catalytic activities.

The catalyst preparation method also had an effect on the activity of catalysts for HDS. Venezia *et al.* [82] studied the influence of the preparation method of CoMo/silica catalysts on HDS of thiophene. The catalysts were prepared by either a total sol-gel route or incipient wetness impregnation/co-impregnation and the results were compared with catalysts supported on commercial silica. The catalyst supported on sol-gel silica with the two metals loaded by co-impregnation in the presence of nitrilotriacetic acid showed the highest activity. Kordulis *et al.* [83] found that in the case of $CoMo/Al_2O_3$, the catalyst prepared by depositing first the Mo precursor through equilibrium deposition filtration (EDF) and then the Co precursor via dry impregnation resulted in catalysts that had higher activity than the catalysts prepared by co-EDF or by a conventional impregnation technique. Recently, Haji *et al.* [86] evaluated Pt/Al_2O_3 catalysts prepared by a supercritical deposition method for HDS of DBT at atmospheric pressure.

Many different supports were also investigated in an attempt to enhance the HDS activity of the catalysts. Okamoto *et al.* [84] studied the effect of different supports on the activity of Co-Mo sulfide model catalysts. The supports used were Al_2O_3, TiO_2, ZrO_2 and SiO_2. The specific activity of the CoMoS phase supported on SiO_2 was found to be 1.7 times higher than those on the other supports mentioned above. Sakanishi *et al.* [85] studied the HDS of 4,6-DMDBT using Ru/C and NiMo/C. The carbons examined in the study were carbon blacks, granular activated carbons with moderate and large surface areas, and pitch-based activated fibers with large surface areas. The Ru/C showed significant activity for HDS of 4,6-DMDBT at 380°C, which was higher than that of the commercial NiMo/ Al_2O_3 catalyst. The NiMo/C catalysts exhibited higher activity for the HDS of 4,6-DMDBT at relatively higher temperatures of 340–380°C than a commercial NiMo/ Al_2O_3 catalyst regardless of the type of carbon support. The main route was the direct desulfurization (DDS) in this temperature range.

In the case of the commercial –natural gas fed– PAFC, sulfur removal is achieved by HDS process using the conventional catalyst at temperatures around 280 °C [29]. However, when heavier petroleum hydrocarbons are used, the conditions should be altered to achieve the desired sulfur removal level. The first approach is to use a higher metal loading. Other possibilities are to increase the operating temperature and H_2 flowrate, decrease the weight hourly space velocity (WHSV), and the use of ultra-low sulfur fuel. To demonstrate the effect of the above parameters on the conversation of sulfur compounds via HDS process, some examples are given below. These results are for the atmospheric hydrodesulfurization of surrogate sulfur contaminated diesel (dibenzothiophene (DBT) in *n*-hexadecane (*n*-HD)) at 310 °C [86]. The DBT conversion increased from 33% to 51% just by increasing the temperature from 310 to 330 °C. However, the temperature can only be increased to a point after which side reactions could take place reducing the active cites available for the reaction of interest, in addition to producing undesirable products. By increasing the H_2/n-HD mole ratio from 1.2 to 3.6, the DBT conversion (initially at 250 ppmw S) increased from 55% to 82%. In addition to consumption by the DBT conversion, H_2 was also consumed in hydrocracking *n*-HD. The H_2 consumed in hydrocracking increased with increasing H_2 flowrate, however, in fuel cell application this is not a concern as the H_2 consumed by hydrocracking should theoretically be recovered in reforming. Furthermore, lighter hydrocarbons are reformed easier. Increasing H_2 flowrate causes some other concerns, such as the increase in power consumed in recycling the unreacted H_2 to the HDS unit and the decrease in the residence time. As the sulfur compounds are considered self inhibitors, the lower concentration in the feed would lead to higher conversion. Decreasing the DBT concentration in the feed from 890 down to 110 ppmw S, increased the conversion from 43% up to 76%. Currently, the sulfur level in diesel is generally around 500 ppmw S, however, it ranges between 0.5 and 3000 ppmw depending on the fuel source. However, restricting the fuel cell feed with a certain specification is not practical. Fortunately, the sulfur content of diesel and gasoline are reducing even more by 2006. Another way of increasing the sulfur conversion in the HDS is to decrease the weight hourly space velocity either by increasing the catalyst amount in the reactor or decreasing the fuel flowrate. Decreasing the WHSV by a factor of 6 (from 28 to 4.5 h^{-1}) doubled the DBT conversion from 32% to 70%. Running at constant WHSV is possible with stationary applications where the power generated is constant. However, in automotive applications the rate at which the fuel is processed is dictated by the instantaneous required power. Therefore, the fact that the sulfur removal amount changes with the WHSV is problematic. Thus the HDS unit should be designed with the maximum WHSV in mind, or with variable H_2/feed ratio to meet the sulfur specification in the fuel.

The presence of aromatic compounds in the fuel inhibits the activity toward hydrodesulfurization. For example, the presence of 10 wt.% toluene in the *n*-hexadecane solution inhibited the hydrodesulfurization of DBT and the inhibition was more pronounced at low H_2/fuel ratios [86]. To maintain the DBT conversion of ~90% the H_2 flowrate had to be doubled when toluene was added to the solution. Inhibition due to the presence of aromatics in the feed was also noticed by others. Prins *et al.* [87] reported that traces of naphthalene had an inhibitive effect on the HDS of DBT and 4,6-DMDBT at 340°C and 5 MPa. Polar compounds such as nitrogen- and oxygen-containing compounds are also known to have strong inhibiting effects on HDS reactions [88,87]. H_2S, produced by the HDS reaction itself, inhibits the HDS reaction of hydrocarbon fuels. However, the inhibitive effect of H_2S is expected to be less pronounced in HDS reactions in fuel cell application than in a refinery because the feed to the fuel cell system has already been subjected to hydrodesulfurization (*i.e.* contains less organosulfur compounds). The inhibition to the HDS process is related to the competition between the inhibitors and the organosulfur compounds for adsorption on the catalyst active sites.

Low concentrations of H_2S or organosulfur compounds in the stream passing over the catalyst have advantages in terms of reducing inhibition. Sulfur species in the feed have a disadvantage for the activity of the sulfided catalysts. Sulfided catalysts ($CoMo/Al_2O_3$ and $NiMo/Al_2O_3$) maintain their activity as long as they are not reduced to their metal or oxide forms. The presence of sulfur compounds (mainly in the form of H_2S) helps maintaining the activity. Thus the catalysts activity can be compromised when the sulfur concentration is very low in the stream. Such problems can be overcome by using catalysts that are active without sulfiding. Haji and Erkey [86] and Kabe *et al.* [79] demonstrated that unlike Mo, Ni, or Ru-based catalysts, Pt-based catalysts are active without sulfiding the metal phase.

Another consideration of the HDS unit in the fuel cell application is their operation pressure. HDS reactors are currently operated under high H_2 pressure (3–5 MPa). High pressure operation is energy consuming. However, at atmospheric pressure and probably due to thermodynamic constraints, the HDS reaction proceeds via a hydrogenolysis route only [89,86]. Unfortunately, the HDS reaction, at atmospheric pressure, does not proceed through the hydrogenation route that is considered to be the effective route in desulfurizing refractory sulfur compounds such as 4,6-DMDBT. But the thermodynamic limits for the hydrogenation of aromatic rings at atmospheric pressure and at HDS temperatures (>300 °C) could be an advantage since the catalyst's active sites will not be occupied for the hydrogenation of aromatic compounds, although the inhibition in the presence of aromatics was noticed even when they are not hydrogenated. A study to determine the optimum operational pressure for the HDS unit for fuel cell application would be a valuable one.

5.3.1.2 Sulfur Removal by Adsorption

One of the approaches to reduce sulfur in fossil fuel is by selective adsorption. In this process, the fossil fuel is brought in contact with solid adsorbent that selectively adsorbs sulfur-containing compounds over paraffinic, olefinic, and aromatic hydrocarbons. Sulfur removal by adsorption was incorporated in the fuel processing train in fuel cell applications. Testings at Fuel Cell Energy (FCE) with natural gas fuel containing 2–12 ppmv S, have demonstrated the capability to achieve less than 0.1 ppmv S level by adsorption [108]. Engelhard (New Jersey-based catalysts specialist) has also developed adsorbents that are active at ambient temperature for removing organic and inorganic sulfur from natural gas and LPG used in fuel cell application [90]. One of the advantages of desulfurization of fuels by adsorption over HDS is operation at low temperature. Another advantage is that while H_2 is abundant in refineries for use in the HDS process, it is a precious reactant for generating power in fuel cell applications. In addition, recycling the H_2 exiting the reformer back to the pre-reformer hydrodesulfurization unit would add complexity to the fuel processing system. On another note, HDS cannot be easily applied to internal reforming fuel cell systems, such as molten carbonate (MCFC) and solid oxide (SOFC) fuel cells, since there is no H_2 rich stream to feed to the HDS reactor. Furthermore, the sulfur content of the fuel can be reduced to a very low level due to removal of the refractory sulfur compounds that cannot be removed easily by HDS. Desulfurization by adsorption faces two major challenges. The first is to develop easy regenerable adsorbents with a high adsorption capacity for sulfur compounds. The second challenge is to find adsorbents that selectively adsorb the sulfur compounds, which are mainly aromatic sulfur compounds that have not been removed in the HDS process in refinery, over the other aromatic and olefinic compounds present in the hydrocarbon fuel.

The suitability of certain adsorbent for sulfur removal can be tested by two different methods. The first method is static adsorption experiments that are run in a batch system where the adsorbent is in contact with the surrounding sulfur contaminated solution initially at a certain concenteration. The system is well mixed and the adsorption takes place for a sufficient time until the adsorbent and the surrounding fluid reach equilibrium. The second method is dynamic adsorption experiments that are run in a column packed with the adsorbent particles. The sulfur-contaminated solution is passed over the bed and the effluent concentration is monitored. The adsorption initially occurs in the inlet of the column and when saturated it proceeds toward the exit. The mass transfer of the adsorbate from the fluid phase to the adsorbent takes place in front of the saturated area and is hence called the "mass transfer zone." This zone keeps traveling through the column until the whole bed is saturated with the adsorbate. By following the adsorbate (sulfur) concentration in the effluent stream, the so-called "breakthrough curve" is obtained. An example of a breakthrough curve is shown in Figure 5.3 for total sulfur

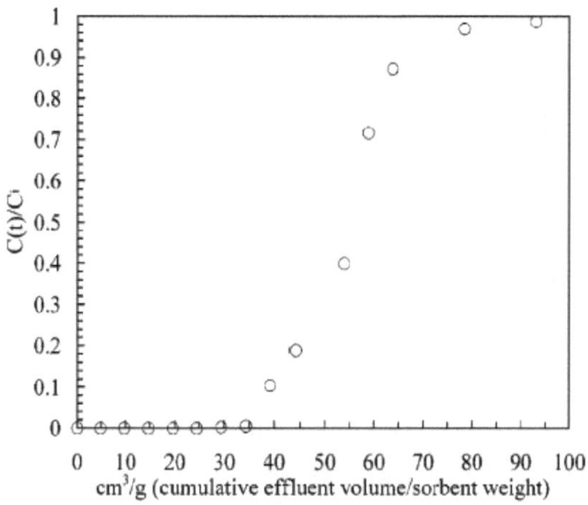

Figure 5.3. Breakthrough of total sulfur in a fixed-bed adsorber with AC/Cu(I)-Y adsorbent, with diesel fuel at room temperature. *Ci* is the total sulfur concentration of the feed. (reproduced by kind permission of Ind. Eng. Chem. Res.)

adsorbed (from commercial diesel fuel) by Cu(I)-Y zeolite combined with activated carbon (AC) guard bed [91].

If the batch system experiment is repeated using different initial sulfur concentrations or the dynamic method experiment using different sulfur concentration in the feed, adsorption isotherms can be obtained. The adsorption isotherm represents the amount of the adsorbate on the surface of the adsorbent in equilibrium with the adsorbate concentration in the solution at a given temperature. Figure 5.4 is an example of an adsorption isotherm curve for the adsorption of dibenzothiophene (DBT) onto two types of carbon aerogels (CA) with different average pore sizes (22 and 4 nm) prepared by using different formulations [92]. The adsorption isotherm along with the transport coefficients such as external mass transfer coefficient and effective diffusion coefficient are necessary to design an adsorption column and to predict its dynamic capacity without conducting extensive experimentation.

Several studies have been conducted on the desulfurization of hydrocarbon fuels by adsorption. Examples of the adsorbents studied are: zeolites, activated carbon, alumina, zirconia and silica gel. Using ZSM-5 zeolite as a sorbent, Weitkamp *et al.* [93] found that thiophene is adsorbed more selectively than benzene when passed (with nitrogen as carrier gas) over a fixed-bed adsorber. On the basis of the π-complexation principle, which is stronger than the van der Waals interactions between the sorbate and the sorbent, Yang et al [94, 95]developed new sorbents for desulfurization process. Comparing the vapor phase adsorption isotherm for the different sorbents they studied, CuY and AgY zeolites had the best thiophene adsorption capacity. The adsorption capacities followed the order Cu-Y and Ag-Y >> Na-ZSM-5 > activated carbon > Na-Y > modified alumina and H-USY. Using FT-IR, Busca et al [96] studied the adsorption of benzothiophene (BT), dibenzothiophenes (DBT) and 4,6-dibenzothiophene (4,6-DMDBT) on alumina, zirconia and magnesia from the vapor phase. The experiments were carried out over a range of

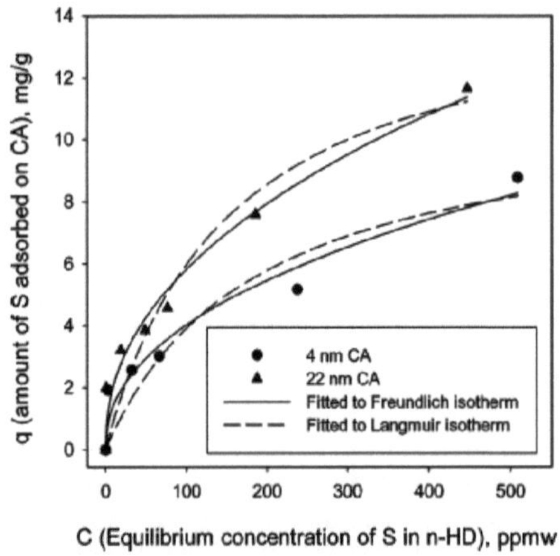

Figure 5.4. Adsorption isotherms for adsorption of DBT onto CA at room temperature. The experimental data are fitted to the equations of Freundlich and Langmuir isotherms. (reproduced by kind permission of Ind. Eng. Chem. Res.)

temperatures from 298 to 723 K. Adsorption on alumina was the strongest and that 4,6-DMDBT adsorption was limited, likely due to a steric effect.

Lee *et al.* [97] studied sulfur removal from diesel fuel contaminated with methanol. In their study, ten different activated carbons (AC) obtained from different sources were screened. Coconut shell-based carbons activated by high-temperature steam were more effective at sulfur removal than coal-based or wood-based carbons. In an attempt to find an adsorbent in which the metal interacts with the sulfur atom and not the C=C double bond of the thiophenic compounds, Xiaoliang Ma *et al.* [98] explored various transition metal-based adsorbents. Further fixed-bed breakthrough experiments were conducted with the most effective transition metal compound. The proprietary metal compound was supported on porous silica gel with 5% loading. The adsorbent was tested for removal of DBT and 4,6-DMDBT from model diesel. The adsorbent had significant selectivity toward the sulfur compounds over naphthalene (NA) and 2-methylnaphthalene (2-MNA). By comparing the sulfur uptake of the DBT and 4,6-DMDBT, the methyl groups at the 4- and 6-positions of the DBT inhibit somewhat the interaction between sulfur compounds and adsorption sites. In another study, Haji and Erkey [92] investigated the adsorption of DBT from *n*-hexadecane by carbon aerogels (CA) of pore sizes 4 and 22 nm and CA with the bigger pore size having a higher sulfur uptake and adsorption capacity. For example the predicted adsorption capacity of the 22 CA was 15.1 mg S/g CA vs. 11.2 mg S/g CA for the 4 nm CA. With low sulfur concentration (43.5 ppmw S = 1.04 mM), the amount of sulfur adsorbed was reduced by around 6% when equal moles of naphthalene (1.04 nm) was added to the solution. From the last two studies, similarities can be seen between the inhibitors of both desulfurization processes, HDS and sulfur adsorption, namely the presence of aromatics in the feed and steric effect of the organosulfur compounds. One of the highest capacity adsorbents in the literature is the combination of Cu(I)-Y zeolite with activated carbon prepared by Yang *et al.* [91] with total sulfur uptake (saturation loading) of 18.9 mg S/g sorbent for commercial diesel fuel (430 ppmw S) fed to the column at room temperature. The breakthrough loading was calculated to be 10.9 mg S/g sorbent. Therefore, for automotive fuel cell application, around 16 kg adsorbent is required in the sulfur removal bed if it is to be changed or regenerated every 10,000 miles (calculation is based on 10,000 mile/service, 80 mile/gal fuel consumption, diesel fuel with 430 ppmw S). That would constitute a substantial fraction of the volume of the vehicle. Thus, there is great need to improve adsorbents for fuel cell applications.

5.3.2 Fuel Reforming

Fuel reforming involves the conversion of hydrocarbons into H_2 and carbon monoxide or carbon dioxide. Fuel reforming is the main step in the fuel processing train. There are three common methods of reforming: steam reforming, partial oxidation and autothermal reforming. They are explained individually below.

5.3.2.1 Steam Reforming

Steam reforming is widely used in industry to produce H_2 and syngas. Steam reforming is an endothermic reaction of steam with the fuel in the presence of a catalyst to produce H_2 and carbon monoxide. In some fuel cell systems, the required thermal energy is provided by oxidation of the unreacted H_2 exiting from the anode [29, 99]. In other fuel cell systems, the thermal energy is provided by burning the raffinate stream from the hydrogen-permeable membrane through which the reformate is passed [15, 189]. While H_2 permeates through the membrane to be used as a fuel in the stack, the raffinate (residual gases), which contains water, carbon oxides, low concentration of organic compounds, and some hydrogen, are combusted in a burner to provide the heat of reforming. If necessary, the anode gas off or the raffinate can be supplemented by the fresh fuel to provide the required heat especially during the startup [100]. A typical steam-reforming reactor consists of a catalyst bed packed in a vertical tubular design surrounded by a source of heat (furnace). Steam reforming of natural gas is usually catalyzed by supported nickel at temperatures above 750°C. For example, nickel particles dispersed on alumina are typically used. Other transition metals are also active for steam reforming, but are not as attractive as nickel for various reasons. Iron and cobalt tend to oxidize under high pressure conditions, while noble metals such as platinum, palladium, ruthenium and rhodium are prohibitively expensive [40]. H_2 can be produced by steam reforming of natural gas (methane), LPG or HD-5 (mainly propane), and liquid hydrocarbons fuels (gasoline or diesel). The following chemical equation illustrates the steam reforming of natural gas:

$$CH_4 + H_2O \rightleftharpoons CO + 3H_2 \qquad \Delta H^{o}_{298} = 206\ \text{kJ/mol} \tag{5.3}$$

One of the advantages of steam reforming is that, besides to the two moles of H_2 obtained from methane, an extra mole of H_2 is obtained from the added water. As can be concluded from Le Chatelier's principle, increasing the temperature and keeping the pressure low shifts the equilibrium toward H_2 production. Excess steam is used to prevent carbon formation on the catalyst (coking) and to force the reaction to completion. An associated reaction to the reforming reaction is the water-gas shift reaction, which also produces H_2 as a product:

$$CO + H_2O \rightleftharpoons CO_2 + H_2 \quad \Delta H^o_{298} = -41\ \text{kJ/mol} \tag{5.4}$$

Unlike the steam reforming reaction, which is endothermic, the water-gas shift reaction is exothermic. So according to Le Chatelier's principle, at such elevated temperatures the equilibrium of the water-gas shift reaction is shifted away from the products. Therefore, CO cannot be totally converted to CO_2 in the reformer. Since CO is a poison for low-temperature fuel cells, further treatment is required to dramatically decrease its concentration, as will be discussed in Section 5.3.3 and 5.3.4. Hence a typical steam reformer product stream contains approximately 75% H2, 8% CO, and 15% CO_2 on a dry basis [101] besides the unreacted methane and steam. As mentioned earlier, commercial PEMFC and PAFC (*e.g.* Ballard 250 kW PEMFC) for stationary power generation utilizes steam reforming of natural gas to supply the stack with H_2.

In practice, reaction (5.3) is only one of a whole series of reactions that occur in the steam reformer (Scheme 5.1). Reactions (5.2) and (5.5) in the scheme can occur during the reforming process and lead to carbon deposition, which can deactivate the catalyst. Appropriate steam to carbon ratios can be chosen to minimize this carbon formation as shown in reaction (5.3). Steam to carbon ratio of ~2.5 is found to be sufficient to avoid coking [102]. In addition to the CO formed via the main reforming reaction (5.1), coke also can be formed by the carbon gasification reaction (5.3). As explained above, CO can be oxidized to CO_2 by the water-gas shift reaction (5.4), which uses the reductive properties of CO to reduce additional H_2O to H_2.

Scheme 5.1 Reactions involved in the steam reforming of methane [40]

$$CH_4 + H_2O \leftrightarrow CO + 3H_2 \tag{1}$$

$$CH_4 \leftrightarrow C + 2H_2 \tag{2}$$

$$C + H_2O \leftrightarrow CO + H_2 \tag{3}$$

$$CO + H_2O \leftrightarrow CO_2 + H_2 \tag{4}$$

$$2CO \leftrightarrow C + CO_2 \tag{5}$$

Methane can be activated in a direct dissociative adsorption step. The limiting step in the steam reforming process is the dissociative adsorption due to the extremely low sticking coefficients for methane adsorption, which can be enhanced by vibrational excitation of the molecule [103]. Complete studies of the steam reforming on well-defined single crystals are not available yet but the reverse reaction has been studied extensively [104]. Scheme 5.2 shows the overall reaction from computational chemistry as described in the literature [40]:

Scheme 5.2 Reaction mechanism derived from computational chemistry [40]. The asterisk refers to active sites

$$CH_4 + 2^* \rightarrow CH_3{}^* + H^* \qquad (1)$$

$$CH_3{}^* + {}^* \leftrightarrow CH_2{}^* + H^* \qquad (2)$$

$$CH_2{}^* + {}^* \leftrightarrow CH^* + H^* \qquad (3)$$

$$CH^* + {}^* \leftrightarrow C^* + H^* \qquad (4)$$

$$H_2O + 2^* \leftrightarrow HO^* + H^* \qquad (5)$$

$$HO^* + {}^* \leftrightarrow O^* + H^* \qquad (6)$$

$$C^* + O^* \rightarrow CO^* + {}^* \qquad (7)$$

$$CO^* \leftrightarrow CO + {}^* \qquad (8)$$

$$2H^* \leftrightarrow H_2 + 2^* \qquad (9)$$

Steam reforming can also be applied to higher hydrocarbons. The general steam reforming of hydrocarbons is described by the following chemical reaction equation:

$$C_nH_m + n\,H_2O \rightleftharpoons nCO + (n + m/2)\,H_2 \qquad \Delta H^o_{298} > 0 \qquad (5.5)$$

Steam reforming of synthetic diesel and gasoline was investigated for portable fuel cell applications at Pacific Northwest National Laboratory [105]. When iso-octane was reformed (GHSV of 144,000 h^{-1}, S:C = 3) complete conversion was achieved at around 640°C. Synthetic diesel fuel (sulfur-free) was tested at 530°C, GHSV of 144,000 h^{-1}, and S:C = 3 and conversion as high as 95% was achieved with essentially no deactivation. However, when JP-8 (3000 pm) was reformed as received, sulfur in the fuel caused significant catalyst deactivation within 3 h. InnovaTek Inc. [106] has developed a proprietary catalyst formulation for the steam reforming of various hydrocarbons, such as *iso*-octane, surrogate gasoline and diesel. The catalyst has shown very stable performance for the steam reforming of *iso*-octane for over 300 h at 800°C with S/C ratio of 3.6. The catalyst was also tested with retail gasoline (sulfur contaminated) at 800°C, S/C = 6.6, and a gasoline feed rate of 0.12 g/min. During the entire testing period (50 h), the catalyst maintained its high selectivity and activity. The product H_2 concentration was above 70% (dry basis). The low CO concentration for the test was a result of a high steam/C ratio (=6.6). The equilibrium composition for steam reforming of different hydrocarbons can be determined thermodynamically using Gibbs free energy minimization. For example, the equilibrium product composition for steam reforming of hexadecane (surrogate diesel) using a steam-to-carbon ratio of 2 is shown in Figure 5.5 [107].

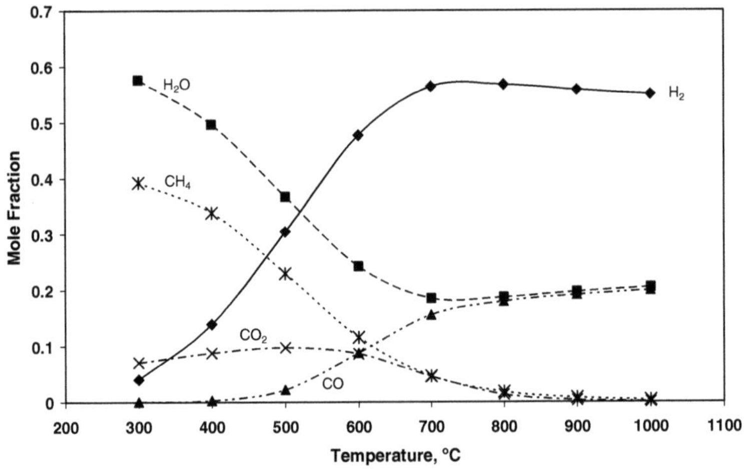

Figure 5.5. Equilibrium steam reforming product composition of hexadecane using a steam-to-carbon ratio of 2 and at atmospheric pressure [107]

An additional process step to steam reforming, termed pre-reforming, can sometimes be used to convert heavier hydrocarbons to a mixture of smaller molecules such as CH_4, H_2 and carbon oxides at low temperatures [15, 108, 109]. This methane-rich gas mixture [110, 111, 112, 113, 114, 115, 116] can then be used directly as feedstock to internally produce H_2 in a special type of fuel cell called a direct carbonate fuel cell [117, 118, 119]. The use of pre-reforming processes for higher hydrocarbons helps to reduce the steam to carbon ratio that would be required to produce H_2 without pre-reforming. Rostrup-Nielsen *et al.* [120] also proposed converting higher hydrocarbons into methane and carbon oxides at conditions where carbon formation does not occur to avoid coking problems. Propane has been recently utilized as a feedstock for a direct carbonate fuel cell [108, 117], low-temperature SOFC [16], and PEMFC [15] where propane first gets processed in a pre-reformer. The processing of propane will be discussed in more detail in Section 5.3.6.

H_2 can also be obtained from steam reforming of alcohols, such as methanol and ethanol. Where the steam reforming of fossil fuel mainly produces H_2 and carbon monoxide, steam reforming of alcohols mainly produces H_2 and carbon dioxide as illustrated by the steam reforming of methanol and ethanol:

$$CH_3OH + H_2O \rightleftharpoons CO_2 + 3\,H_2 \qquad \Delta H^o_{298} = 49\ \text{kJ/mol} \tag{5.6}$$

$$C_2H_5OH + 3\,H_2O \rightleftharpoons 2\,CO_2 + 6\,H_2 \qquad \Delta H^o_{298} = 173.8\ \text{kJ/mol} \tag{5.7}$$

This makes alcohols good candidates for the CO-poisoned fuel cells. However, CO in small quantities is produced as the water-gas shift reaction of (5.4) is reversible. So, further CO removal from the reformate is needed before use in the PEMFC, although the CO concentration could be tolerated by the PAFC. Another advantage of the steam reforming of alcohols over fossil fuels (mainly alkanes) can be deduced by comparing the heat requirement, for instance, of steam reforming of methane vs. methanol (reactions 5.3 and 5.5). At room temperature, steam reforming of methanol requires about four times less heat than natural gas. This is translated to the moderate temperature (200–300°C) at which the reaction is carried out. This feature, along with the storage convenience and low CO concentration in the reformate, makes alcohol suitable for PEM fuel cell vehicles. In the years 1997–1998 Toyota, DaimlerChrysler, and GM demonstrated prototype vehicles that are fueled by steam reforming of methanol [121].

Regarding the steam reforming of methanol, the catalysts commonly used are copper based mixed oxides. The active component is copper, which is combined with other metal oxides as promoters to enhance the activity, prevent sintering or work as a structural support. Examples of catalysts used are Cu/ZnO and $Cu/ZnO/Al_2O_3$. There are many kinds of preparation methods for the copper based catalysts [40]. Equation (5.6) is not the only reaction occurring in the steam reforming of methanol as shown by Scheme (5.3) [40]. Reactions (5.4) and (5.5), in Sceme (5.3), can take place during reforming and lead to coke formation, which can deactivate the catalyst. Appropriate steam to carbon ratios can be chosen to minimize carbon formation as shown in reaction (5.5). On the other hand, the decomposition of methanol (5.2) forms CO, a byproduct that poisons the catalyst used in fuel cells. The water-gas shift reaction (5.3) can help in removing CO or can act as a source of CO production, depending on the parameters that affect the thermodynamic equilibrium.

Scheme 5.3 Chemical reactions involved in the steam reforming of methanol [40]

$$CH_3OH + H_2O \leftrightarrow CO_2 + 3H_2 \quad (1)$$

$$CH_3OH \leftrightarrow CO + 2H_2 \quad (2)$$

$$CO + H_2O \leftrightarrow CO_2 + H_2 \quad (3)$$

$$2CO \leftrightarrow C + CO_2 \quad (4)$$

$$C + H_2O \leftrightarrow CO + H_2 \quad (5)$$

Steam reforming of ethanol (equation 5.7) at atmospheric pressure over noble and non-noble catalysts was also investigated for H_2 production for fuel cell applications [122, 123, 124]. Steam reforming of ethanol is carried out at temperatures higher than that of methanol [125]. Batista *et al.* [124] studied the steam reforming of ethanol (3:1 H_2O:Ethanol molar ratio) at 400°C on Co/Al_2O_3 and Co/SiO_2 catalysts with a cobalt content of 8 and 18% (w/w). A reformate with CO concentration as low as 800 ppm was obtained. The lower CO level occured over the $Co(18)/Al_2O_3$ catalyst with almost 100% of ethanol conversion and H_2 molar concentration in the gaseous product of around 60%, but with the production of non-negligible quantities of methane (5–10%). Ethylene formation occurred only on the Co/Al_2O_3 catalyst with small Co contents ($\leq$8%). CO-free ethanol reformate using Co based catalyst was also reported by Llorca *et al.* [126]. The performance of Co/ZnO catalysts for the steam reforming of ethanol was found to be dependent on the cobalt-precursor used. Co(CO)/ZnO prepared from $Co_2(CO)_8$ showed high stability selectivity for the production of CO-free H_2 at reaction temperature as low as 623 K. Goula *et al.* [123] investigated H_2 production by ethanol steam reforming over a commercial Pd/γ-Al_2O_3 catalyst with water to ethanol molar ratios ranging from 3 to 15 and temperature range of 220–700°C. Ethanol was totally converted even at very low temperatures (300–350°C). For the examined H_2O/ethanol molar ratios, CO concentration exhibited a minimum at a temperature close to 450°C. Furthermore, carbon formation was found to be negligible even for a H_2O/ethanol molar ratio equal to the stoichiometric one. Moreover, the most significant effect upon increasing water content was reported to be the increase in the H_2/CO molar ratio (*i.e.* the increase of the H_2 and the decrease of the CO production). However, the CO concentration in the reformate necessitates further CO removal if H_2 is to be used in PEMFC as mentioned by the authors. In general the selectivity toward H_2 production increases by increasing the reformer temperature [126, 127, 128], following the trend given by thermodynamic equilibrium [127, 129].

5.3.2.2 Partial Oxidation

The second option for fuel reforming is partial oxidation (POX), which is an exothermic reaction and occurs when the feed reacts directly with air or pure oxygen at carefully balanced oxygen to fuel ratios. POX can be described as fuel oxidation with less than the stoichiometric amount of oxygen needed for complete fuel combustion to water and carbon dioxide. Partial oxidation can be applied to natural gas, propane and alcohols, but its chief advantage is that POX can reform liquid hydrocarbons such as gasoline, diesel fuel and heavy fuel oil [101] even without the use of a catalyst. Partial oxidation of gasoline has been used to produce H_2 by several car manufacturers in the US and Europe [130, 131].

All the commercial partial oxidation processes (Texaco, Shell, Montecatini) employ non-catalytic partial combustion of the hydrocarbon feed with oxygen in the presence of steam in a combustion chamber at flame temperatures between 1300 and 1500°C [101]. Since this is a non-catalytic process and takes place at high temperatures, a hydrodesulfurization unit is not needed prior to the reformer, although the produced H_2S has to be removed before the reformate enters other units. The reaction rates are higher for partial oxidation than for steam reforming. In addition to its exothermic nature, these two reasons make the partial oxidation suitable for applications that require rapid startups. Below is an illustration of partial oxidation of methane (natural gas) and a generalized equation:

$$CH_4 + 1/2\,O_2 \rightleftharpoons CO + 2\,H_2 \qquad \Delta H^o_{298} = -247\ \text{kJ/mol} \tag{5.8}$$

$$C_nH_m + n/2\,O_2 \rightleftharpoons n\,CO + m/2\,H_2 \qquad \Delta H^o_{298} < 0 \tag{5.9}$$

Partial oxidation of methane, reaction (5.7), is the net reaction of a group of reactions described in Scheme 5.4 [101]:

Scheme 5.4 Reactions involved in partial oxidation of methane

$$CH_4 + 2O_2 \leftrightarrow CO_2 + 2H_2O \qquad (1)$$

$$CH_4 + CO_2 \leftrightarrow 2CO + 2H_2 \qquad (2)$$

$$2\,CH_4 + 2H_2O \leftrightarrow 2CO + 6H_2 \qquad (3)$$

Another advantage arises from the fact that water is not a reactant in this process; therefore steam generators are not required, leading to a simpler system. However, this can also be a disadvantage since water is a source of H_2 too. Comparison of equations 5.3 and 5.7 shows that less H_2 is produced per mole of methane in partial oxidation than in steam reforming which results in lower H_2 partial pressure in the reformate stream. Another factor that lowers the H_2 partial pressure is the introduction of nitrogen into the stream, when air is used as the oxygen source in partial oxidation. Lower H_2 partial pressure in the anode causes lower efficiencies according to the Nernst equation. Although the non-catalytic partial oxidation is commercialized on a large scale (Texaco, Shell, Montecatini), this process does not scale down well and the control of the reaction is problematic [29]. Thus, due to the reactor size, POX is not suitable for fuel cell mobile applications, which prefer compact fuel processors [56].

The use of catalysts in partial oxidation can reduce the temperature required for hydrocarbon reforming by hundreds of degrees [132, 133], enabling the use of inexpensive materials of construction and reducing the CO concentration in the reformate. The reaction is initiated catalytically (flameless) and the process is then known as Catalytic Partial Oxidation (CPO). COP catalysts may contain either noble metals (Pt, Pd, and Rh) or non-noble transition metals (Ni, Cu, Fe, and Co) supported on zirconia, ceria, or alumina and they can be active at temperatures as low as 450–850°C depending on the feedstock [65, 132, 134]. CPO experiences some major operating problems such as over-heating or hot spots [56] due to the exothermic nature of the reactions. Furthermore, coking is a serious concern, but fortunately a catalysts' resistance to carbon formation can be improved [135, 136, 137]. Shell developed a Rh-based catalyst for the partial oxidation of light hydrocarbons (C_1-C_5) containing upto 100 ppm sulfur without the need for prior sulfur removal [138]. The reaction, however, occurs at elevated temperatures (950–1300°C). In 2003, HydrogenSource (CT-based, dissolved in summer 2004) developed an onboard gasoline-to-hydrogen fuel processor for fuel cell vehicles (50 kWe PEMFC) [139]. The system utilized a patented CPO technology and was reported to take less than 4 min from room temperature until production of fuel cell quality hydrogen.

5.3.2.3 Autothermal Reforming

Autothermal reforming is a combination of partial oxidation and steam reforming. The fuel is mixed with steam and substoichiometric amounts of oxygen or air where the ratios of oxygen to carbon (O:C) and steam to carbon (S:C) are properly adjusted so that the partial combustion (equation 5.9) supplies the necessary heat for endothermic steam reforming (equation 5.3). The autothermal reformer consists of two zones; the thermal zone or the partial oxidation zone where partial combustion occurs and the heat generated is supplied to the subsequent endothermic steam reforming occurring in the catalytic zone. The reactions can either be run in a single reactor or in separated reactors that are in good thermal contact. The autothermal reforming of methane is presented below:

$$CH_4 + 3/2\,O_2 \rightleftharpoons CO + 2\,H_2O \qquad \Delta H^{o}_{298} = -519\ \text{kJ/mol} \tag{5.10}$$

$$CH_4 + H_2O \rightleftharpoons CO + 3H_2 \qquad \Delta H^{o}_{298} = 206\ \text{kJ/mol} \tag{5.11}$$

and yields between 2 to 3 moles H_2 per mole of methane as can be concluded by comparing equations (5.3) (steam reforming) and (5.7) (partial oxidation).

A chief participant in autothermal reforming of liquid hydrocarbons such as gasoline and diesel fuel is Argonne National Laboratory (ANL) with their work on liquid fuel reformer development [102, 140, 141]. For example, Pt supported on doped ceria showed excellent activity toward reforming commercial-grade gasoline, yielding 60% H_2 (dry, N_2-free basis) in the product stream at 750°C [102]. The Pt-based catalysts were also reported to have an excellent sulfur resistance, where sulfur-contaminated isooctane (0–1300 ppm) was reformed at 700°C [102]. ANL also studied other transition metals supported on doped ceria [102]. The second- and third-row transition metals (Ru,Pd,Pt) exhibited higher H_2 selectivity (>60%) than the first-row transition metals (Fe,Co,Ni,Cu) at temperatures greater than 650°C and at temperatures greater than 600°C, 95% conversion was achieved by all investigated metals (Ag being an exception in both cases). Autothermal reforming of diesel fuels was also studied in ANL, where hexadecane was reported to yield products containing 60% H_2 (dry, N_2-free basis), while the maximum H_2 product yields for the diesel fuels (certified low-sulfur grade 1 and standard grade 2 diesel) was reported to be near 50% (water:fuel ratio = 2, oxygen:carbon ratio = 1, T = 850°C) [141]. Johnson Matthey HotSpot™ is an example of an autothermal reformer that can be integrated into fuel cell vehicles for onboard H_2 production [142]. This autothermal reformer uses methanol as a fuel and produces 750 liters of H_2/hour from a 245 cm^3 reactor. The HotSpot™ demonstrates very quick start-up; 100% output is achieved in 50 s, which is attained by initially running the reactor at a high rate of partial oxidation. The concentration of CO is negligible in the few seconds after the cold-start up, and as the reactor reaches its near autothermal steady state the [CO] increases to a maximum of 2–3% (dry basis). In conjunction with the U.S. Department of Energy, Nuvera Fuel Cells, a company formed through the merger of Epyx Corp. (MA, USA, a subsidiary of Arthur D. Little, Inc) and De Nora Fuel Cells (Milan Italy), developed an autothermal-based fuel processor specifically for the automotive market [143]. The fuel processor, named Star™ (Substrate Transportation Autothermal Reformer), utilizes gasoline as a fuel and meets the performance requirements for system size, power, efficiency and gas purity needed for PEMFC-powered vehicles.

5.3.2.4 Common Challenges in Reforming Processes

Carbon deposition is one of the challenges in reforming. Carbon is mainly formed by CO and methane decomposition, but it can also form via other gaseous reactions (*e.g.* reaction of H_2 with CO_2). Solid carbon forms on the catalysts if the reaction is run under conditions in which carbon formation is favored thermodynamically or if its formation rate is faster than its removal rate. Carbon formation can cause blockage of catalyst pores and catalyst deterioration, leading to premature reactor shutdown. Thus, avoiding conditions leading to coke formation is important. In steam reformers, carbon deposition can be minimized in practice by use of a sufficiently high steam to carbon ratio. However, the amount of steam required may be considerably above that required by the thermodynamic equilibrium. For economical consideration, this ratio should be reduced to the minimum consistent with adequate catalyst life. Liquid petroleum feedstocks are more susceptible to carbon formation than natural gas due to the presence of aromatics and polyaromatic sulfur compounds. Carbon deposition can also be minimized by addition of promoters, which can result in secondary non-desirable effects that compromise catalytic performance or stability. Examples of promoters used for this purpose are: CaO, La_2O_3, and CeO_2 [144]. The addition of molybdenum to Ni/Al_2O_3 (steam reforming catalyst) was also examined in order to increase the resistance to coking in the steam reforming of hydrocarbons via increasing the number of oxygen on the catalyst surface [145].

Apart from coke formation, another factor that can cause the decay of the catalyst activity with time on stream is sintering of the metal particles. Sintering results in a decrease of the surface area of the active site. There are many factors that affect the catalysts sintering rate [146, 147, 148]. The reaction environment is one of those factors, whether it is an inert, reducing, oxidizing atmosphere, gas or liquid phase. The particle growth rate is also dependent on the metal species. In an inert or reducing atmosphere particle growth is predicted to be related inversely to the strength of the cohesive forces in a metal crystal [146]. In studies of alumina supported catalysts, Fiedorow *et al.* [149] reported that in oxygen atmospheres the sequence of thermal stability was found to be Rh > Pt > Ir, while in H_2 atmospheres the sequence was Ir > Rh > Pt. The sintering of nickel during steam-reforming was also studied [150, 151, 152]. For instance, above temperatures of ~600°C at 40 bar and ~700°C at 1 bar total pressure, the rate and the activation energy of sintering of supported nickel catalysts were reported to increase considerably [151].

Many studies have been performed for improving the resistance to sintering and coking of the catalysts used in reforming by promoting with different metals and using various preparation methods [153, 154, 155]. The performance of MgO supported Pd, Rh, Ni and Co catalysts in the steam reforming of simulated bio-ethanol at MCFC operative conditions (T = 650°C) has been investigated by Frusteri *et al.* [156]. In spite of its lower selectivity towards H_2 production, Rh/MgO showed the best performance both in terms of activity and stability. Ni, Co and Pd catalysts were found to deactivate mainly due to metal sintering. A very low coke formation rate was observed on Rh/MgO catalyst, however, even on Ni/MgO catalyst coke formation occurs at a modest rate.

As mentioned previously, in the low-temperature fuel cell, CO is a non-desirable byproduct from reforming processes. Designing novel preparation methods that can translate into catalysts with enhanced activity combined with minimal CO formation would be ideal. Meanwhile, the removal of CO from the reformate to adequate concentrations is necessary for PEMFC , AFC, and PAFC, which is the topic of the following two subsections.

5.3.3 Water-Gas Shift Reaction

The CO concentration in the reformate varies with the reforming type and feedstock used. While high temperature fuel cells (MCFC and SOFC) are capable of processing CH_4 and CO in the anode by internal reforming, a substantial amount of CO is considered to be poisonous in the low-temperature fuel cells (PEMFC and PAFC), as indicated in Table 5.2. At low temperatures and high concentrations, the Pt in the anode-catalyst in fuel cells preferentially adsorbs CO, consequently blocking access of H_2 to the catalytic sites and resulting in significant decreased performance of the fuel cell stack. For instance, an introduction of 100 ppm CO to a pure H_2 stream (500 cm^3/min, no-humidity) feeding a PEMFC (0.3 mg Pt/cm^2 electrode) operating at 0.6 V, 1 bar and 70°C, reduced the current density from 164 to 45 mA/cm^2 [157]. A water-gas shift reaction (WGSR), in a separate stage(s), is the primary means for reducing the CO concentration in the fuel gas. Through WGS, the CO formed from reforming reactions further reacts with excess steam to form carbon dioxide and H_2. Thus, the H_2 yield is increased, and poisonous CO is decreased. As mentioned in the reforming section, the water-gas shift reaction (WGSR):

$$CO + H_2O \rightleftharpoons CO_2 + H_2 \qquad \Delta H^o_{298} = -41\ kJ/mol \quad (5.12)$$

occurs simultaneously with the steam reforming reaction in the reformers. WGSR is a reversible, exothermic reaction. Thus, whereas the H_2 yield from reforming is increased at higher temperatures, WGS conversion is limited at high temperatures by thermodynamic equilibrium. Therefore, the WSGR should be run at temperatures much lower than that of the reformers. After exiting the reformer, the reformate temperature is reduced by a heat exchanger to around 315–400°C before being introduced, along with extra steam, to the first stage water-gas shift reactor (high-temperature shift reactor, HTS). Since the temperature in the latter reactor is much lower than that in the reformer, the equilibrium of the WGSR is shifted toward CO conversion to CO_2 and 80 to 95% conversion is obtained. The reaction mixture's temperature increases due to the exothermic nature of the reaction leading to enhancement in the reaction rate, however, adversely affecting the equilibrium. To obtain a satisfactory CO conversion, the reaction mixture is cooled again via an interstage cooler and introduced to a second stage water-gas shift reactor (low-temperature shift, LTS) where the temperature is around 200–250°C. Methane could form in the CO converters, which is considered to be a disadvantage. Iron or iron-chromium oxides are used as catalysts in the HTS reactor. Co/Mo catalysts were found to be more active and sulfur-resistant than the industrial Fe/Cr catalysts [158]. Mixtures of Cu/Zn oxides are used as catalysts in the LTS reactor. Thompson *et al.* [159] recently found that Mo_2C catalyst was more active than the commercial catalyst under atmospheric pressure and did not catalyze the methanation reaction. Mo_2C should also be more sulfur-tolerant as ZnO (in the commercial catalyst) converts to ZnS if H_2S was present in the feed to the CO converter (see equation 5.2). Precious metal catalysts such as Ru, Pt, Pd, and Rh supported on zirconia or ceria are also being investigated for WGS reactions [160, 161, 162]. For instance, Pt/ceria catalysts (developed by NexTech Materials) exhibited superior performance compared to commercial copper-based WGS catalysts at temperatures above 250°C [160]. HydrogenSource (CT-based, dissolved in summer 2004) has evaluated a proprietary noble metal that demonstrated high performance for both HTS and LTS reactors, and sustained performance over many thousands of hours, even in the presence of sulfur [163]. These advanced noble metal catalysts have been utilized in HydrogenSource's products such as the 5kWe Vega 5™ and the 150 kWe stationary platforms, which also employ the HydrogenSource's proprietary Catalytic Partial Oxidation (CPO) technology mentioned above.

Two WGS reactors bring the CO concentration down to around 0.5%, which is tolerable for the PAFC. When methanol is the fuel, the reformate may contain as little as 0.1% CO, depending on the temperature, pressure and water/methanol ratio. Therefore, WGS reactors may not be required which in turn reduces the system complexity and fuel processor volume. However, 0.5% or even 0.1% is still 500–100 times more than the tolerable CO concentration in the feed for PEMFC. Thus, further CO cleanup is required prior the introduction of such feed to PEMFC, and this will be explained in the next section.

5.3.4 Carbon Monoxide Removal

PEMFC anode catalyst is comprised of Pt and its alloys and operates under low temperature. Especially at concentrations greater than 10 ppm, CO strongly adsorbs onto Pt preventing the H_2 adsoption, therefore deactivating the catalyst and reducing the fuel cell efficiency. It is very difficult for the water-gas shift reactors to reach such low CO concentration. Further CO removal is achieved by one of the following three methods:

5.3.4.1 Preferential Oxidation

The anode catalyst is deactivated because CO is bonded very strongly to Pt surfaces at low/moderate temperature preventing H_2 oxidation reaction. Utilizing the same mechanism, CO is oxidized to CO_2 in the preferential oxidation reactor by bleeding air (~ 2%) in to the fuel stream entering the reactor in the presence of noble metals such as Pt, Ru, Rh supported on alumina. The term preferential oxidation (PROX) refers to selective oxidation of CO (equation 5.11) over H_2 (equation 5.12) in spite of the the high concentration of H_2 in the stream:

$$CO + 1/2\,O_2 \rightleftharpoons CO_2 \qquad \Delta H^{o}_{298} = -283\ kJ/mol \qquad (5.13)$$

$$H_2 + 1/2\,O_2 \rightleftharpoons H_2O \qquad \Delta H^{o}_{298} = -242\ kJ/mol \qquad (5.14)$$

The air has to be added in careful measures to avoid oxidizing the H_2 and/or producing an explosive mixture. CO was reduced to ~100 ppm on a Pt/Al_2O_3 catalyst with the addition of excess air corresponding to [O2]/[CO] = 2 [164]. Identical to the WGS, multi-stage PROX reactors might be needed to achieve the required 10 ppm CO concentration. Yoon *et al.* [165] reported that CO was reduced to less than 10 ppm by a two-stage reactor with a total additive air corresponding to [O2]/[CO] = 1.5. However, increasing the number of stages in any process would increase the system complexity. Echigo *et al.* [166] found that a Ru-based catalyst, which is reduced by H_2/N_2 after aqueous reduction, had extremely high practical CO removal performance even in a single-stage reactor. In their experiments, CO was reduced from 0.5 vol.% to < 1 ppm between 100 and 120°C at $[O_2]/[CO] = 1.5$ for a reactant gas mixture containing CO_2 and H_2O. Plzak *et al.* [167] reported high oxidation rates when Fe_2O_3-Au catalysts were used for CO PROX at 80°C with 5% air bleed. They claimed the unit to be very suitable for integration into PEMFC. Fe was also found to have a promotional effect of Pt-based catalysts, where the iron addition increased the overall CO oxidation rate [168]. Kahlich *et al.* [169] calculated, using kinetic data measured in simulated methanol reformate (1% CO, 65% H_2, 10% H_2O, rest CO_2), the minimum amount of noble metal required for the complete removal of CO from the feed gas (≤10 ppm) and found that this concentration is lower by two orders of magnitude for Rh/MgO operated at 250°C than for Pt/Al_2O_3 and Ru/Al_2O_3 catalysts at 200 and 150°C, respectively, at a comparable excess O_2 level.

Another approach similar to PROX is sometimes referred to as oxygen-bleeding or air-bleeding [170]. In this approach, the oxygen (~2%) or air is bled into the fuel stream just before entering the fuel cell allowing the CO oxidation to occur in the anode itself instead of the use of separate PROX reactor. Gottesfeld et. al [170] reported that injection of small amounts of oxygen (2–5%) into a 100 ppm CO-contaminated H_2 stream, fed to a PEMFC operating at 80°C, can completely correct for the deleterious effect of the CO. However, due to the low lower explosive limit of O_2/H_2 mixtures (5% O_2 in H_2), ~100 ppm is the maximum level of CO that can be treated effectively by the oxygen-/air-bleeding method at 80°C [171].

5.3.4.2 Methanation

The methanation reaction is the reverse of the steam reforming of methane:

$$CO + 3\,H_2 \rightleftharpoons CH_4 + H_2O \qquad \Delta H^{o}_{298} = -206\ kJ/mol \qquad (5.15)$$

Obviously, this method is very convenient in removing CO as no extra reactant has to be added to the fuel stream, hence reducing the system complexity. In addition there is no concern for formation of an explosive mixture, as is the case in PROX. The major disadvantage of the methanation process is H_2 consumption, especially if the CO concentration is as high as a few percent. However, this is not a big concern if the CO concentration is as small as in the case of methanol reforming. Methane in the fuel stream is not poisonous, but it acts as a diluent. A competitive reaction of the CO methanation is CO_2 methanation as CO_2 is present in higher concentrations:

$$CO_2 + 4\,H_2 \rightleftharpoons CH_4 + 2\,H_2O \qquad \Delta H^{o}_{298} = -165\ \text{kJ/mol} \tag{5.16}$$

For example, Ni and Ni/Al_2O_3 catalysts showed higher methanation rate for CO_2 over CO though both were present in the feed at same concentrations ($CO{=}CO_2{=}0.1$ atm, rest H_2) [172]. For PEMFC applications, Echigo *et al.* [173] investigated the CO methanation in simulated natural gas steam reformate gas (0.54% CO, 21% CO_2, balance H_2 (dry basis)) over Ru/Al_2O_3 and found that, due to the presence of CO_2 in the stream, CO concentration could not be reduced to <10 ppm in a single-stage methanation reactor, however, it was possible after three stages, after which the amount of CO_2 converted to CH_4 was 0.27 of that of CO. In a subsequent study to the one mentioned in the PROX section, Echigo *et al.* [166] suggested that the high activity and the high selectivity of the activated Ru-based catalyst are achieved by combination of both PROX and methanation reactions. They suggested that PROX of CO reduced its concentration from 0.5 vol.% to ~100 ppm and followed by methanation of the residual CO from ~100 ppm to less than 1 ppm at low temperatures where the influence of CO_2 methanation can be ignored [174].

5.3.4.3 Membrane Separation

A mature technology for H_2 purification is permeation through metal membranes, namely palladium. The membrane is only permeable to H_2 allowing separation of H_2 from the other fuel constituents. This method not only removes the poisonous CO but also eliminates other inert gases such CO_2, N_2, and other trace impurities causing an increase of the H_2 partial pressure in the anode feed. To enhance the H_2 permeability and reduce the cost, dense metal membranes are avoided and instead composite membranes consisting of a thin metallic film formed on a porous inorganic substrate are used. Such composites are manufactured by different techniques such as chemical vapor deposition [181, 175], electroless-plating [176], electroplating [177], sputter-deposition [178], spray pyrolysis [179], and rolling a thin metallic film over ceramic porous tubes [180]. Platinum is another candidate [181] which was supported on porous alumina. Others synthesized and/or investigated bimetallic membranes that are palladium-silver alloys supported on alumina [182, 183]. Los Alamos scientists [184] developed a membrane where a very thin film of palladium (1 μm) was deposited, via chemical vapor deposition, on each side of a tantalum sheet (13 μm).

Due to its high adsorptivity to noble metals, CO was found to retard the H_2 permeation through the membrane. Chabot *et al.* [185] reported that the poisoning effect of CO is a concern for the H_2, purification process using a Pd/Ag alloy membrane while the presence of CO_2 and CH_4 appeared to have no practical effect on the permeation rate. They showed that the H_2 permeation is reduced in the presence of CO at temperatures below 300°C. Emonts *et al.* [186] studied a similar membrane (commercial 100 μm thick Pd/23% Ag) for purifying methanol reformate consisting of 66.8% H_2, 2.6% CO, 20.5% CO_2, and 0.2% methanol. They suggested a feed-side pressure of 21 bar and a minimum operation temperature of 350°C in the presence of CO. Other disadvantages of this approach are the cost of the noble-metal membrane, its durability, and the required high differential pressure across the membrane.

The membrane separation method was put into great use when it was utilized to develop a compact and integrated purification and production reactor to directly produce high purity H_2 with a supported palladium membrane reactor [187, 188]. Lin *et al.* [189] introduced an interesting reformer/H_2 purification/CO oxidation integrated reactor. The double-jacketed membrane reactor is a methanol steam reformer that produces a pure H_2 stream (free of CO and CO_2) without extra purification facilities, clean emission containing only CO_2 and H_2O (CO-free), and is energy self-balanced. The reactor is a concentric module that includes a supported palladium membrane tube located at the center position and two stainless steel tubes separately assembled as double outer jackets. A copper-based catalyst is used for the steam reforming reaction and is loaded adjacent to the outer surface of the membrane tube. At the outer surface of the first jacket, a Pd/Al_2O_3 catalyst is used for oxidation of gases rejected from the membrane tube. The mixture of methanol and water is fed into the reforming catalyst bed in order to produce H_2, CO and CO_2. Under the driving force of the transmembrane pressure difference, the instantaneously produced H_2 penetrates through the Pd-membrane tube and is directly removed as high purity H_2. The rejected gases, H_2, CO, CO_2 and/or trace amount of unreacted methanol, then enter the oxidation catalyst bed and are converted into CO_2 and H_2O as a result of total oxidation reaction with air. The exothermal oxidation reaction provides the heat required for the reforming reaction.

Among the three approaches, the preferential oxidation appears to be the preferred solution. As mentioned above, high CO removal by one of the above methods is only required in conjunction with PEMFC. The CO tolerance can be improved by developing new anode catalysts and electrodes. A well established method to increase the anode CO tolerance is by introducing Ru to the Pt/C at 1:1 atomic ratio. Even then the tolerance is not increased above 30 ppm CO. Venkatraman *et al.* [190] reported that sulfided platinum showed enhanced CO tolerance (up to 104 ppm) and only a little adsorbed CO was observed in CO stripping studies.

If PEMFC could be operated at higher temperatures (>100°C), the anode could be more tolerant to higher concentration of CO. However, higher temperature operation (under low humidity) leads to deterioration in the conductivity of the Nafion®-based membrane. If the membrane is modified to be able to run at higher temperatures, this would increase the low feed CO concentration limit, and would decrease the system complexity due to elimination of the last purification step. For instance, membrane electrode assemblies (MEAs) made from polyetheretherketone (PEEK) showed good fuel cell performance and thermal stability in the presence of substantial CO at elevated temperatures (120°C) in proton exchange membrane fuel cells (PEMFCs) [191]. For example, the current from a MEA made from PEEK membrane at 0.6 V and 120°C was 0.50 A/cm^2 when run on pure H_2 and 0.45 A/cm^2 when run on reformate (50% H_2, 1300 ppm CO, and balance N_2). However, the ionic conductivity of the PEEK membrane at 120°C was approximately three times lower than Nafion®.

5.3.5 HD-5 Propane Processing for Direct Carbonate Fuel Cell (DFC™) Applications

In this section a detailed discussion is presented on one of the fuel processing studies conducted in the laboratories of Connecticut Global Fuel Cell Center (CGFCC). High-temperature fuel cells, namely solid oxide (SOFC) and molten carbonate fuel cells (MCFC) have advantages over the low temperature ones in that they are capable of internal fuel reforming. This system is based on the idea that the heat generated in the fuel cell stack by the electrochemical reaction can be used for the endothermic steam reforming of low molecular weight hydrocarbons (*e.g.* natural gas). The fuels can either be reformed directly or indirectly. In direct internal reforming (DIR), the reforming reactions take place within the anode compartment of the fuel cell stack. Whereas in the indirect internal reforming (IIR), the reforming reactions are carried out in a separate chamber that is in good thermal contact with the fuel cell stacks. There are many advantages of internal reforming. Internal reforming reduces the complexity of the system, its cost, the amount of air needed to cool the system, and the amount of steam required, since SOFC and MCFC produce steam on the anode. In addition, since produced H_2 is readily consumed, especially in the case of DIR, the reforming reaction is shifted toward more H_2 production.

The internal reforming fuel cells can be coupled with the pre-reforming system, described earlier, enabling utilization of higher hydrocarbons fuels. In the pre-reformers heavier hydrocarbons are converted to a mixture of smaller molecules such as CH_4, H_2 and carbon oxides at low temperatures [192, 193]. This methane-rich gas mixture [194-200] can then be used directly as fuel to produce H_2 internally in the internal reforming fuel cells. The use of pre-reforming processes for higher hydrocarbons helps to reduce the steam to carbon ratio that would be required to produce H_2 without pre-reforming. One of the commercialized internal reforming fuel cells is called a direct carbonate fuel cell (DFC™) [195, 196], which has been developed by Fuel Cell Energy and is a type of molten carbonate fuel cell. In terms of fuel reforming, DFC is a hybrid system, which incorporates both direct internal reforming (DIR) and indirect internal reforming (IIR). Figure 5.6 shows the hybrid arrangement in which the reforming catalyst is placed in the anode compartment of each fuel cell and the reforming unit is placed in-between fuel cell groups. DFC has increased efficiency, high performance, longer lifetime, better thermal management and reduced costs. Theoretically DFC can operate on a variety of hydrocarbon fuels, however fuel supplied to the fuel cell has to undergo a

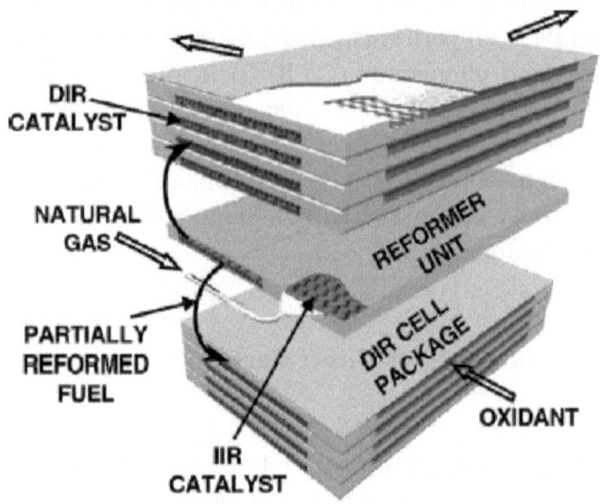

Figure 5.6. The fuel cell stack in the Direct Carbonate Fuel Cell (DFC): IIR-DIR stack concept (reproduced by kind permission of Catal. Today) [197]

two-step refinement process involving desulfurization and fuel pre-reforming to remove higher hydrocarbons that would cause coking on internal fuel cell catalysts. This process can also both, provide H_2 atmosphere and serve as a sulfur guard for internal reforming catalysts. Methane-rich fuel from the pre-reformer is then reformed internally within the fuel cell to produce the needed H_2 [197].

The Gas Processors Association (GPA) defines propane HD-5 as the best commercial grade of propane and fuel suitable for internal combustion engines [198]. HD-5 is commercially available in the USA and in many other countries. HD-5 grade propane is mainly composed of propane (preferably ≥ 95%), 2–5% of propylene, and smaller amounts of other C_2-C_5 hydrocarbons. Ethyl mercaptan or thiophane (tetrahydrothiophene) are typically added to HD-5 as odorants. In industry, HD-5 is initially desulfurized by passing it over an activated carbon bed before reforming. HD-5 propane can be used as a fuel for DFC applications upon proper pre-reforming. In this section we present part of the work done in the CGFCC regarding methane production by HD-5 steam reforming for DFC applications. There are three major reactions taking part in steam reforming of propane:

The endothermic reforming reaction:

$$C_3H_8 + 3\,H_2O \rightleftharpoons 3CO + 6H_2 \qquad \Delta H^\circ_{298} = 498 \text{ kJ/mol} \tag{5.17}$$

The exothermic water-gas shift reaction:

$$CO + H_2O \rightleftharpoons CO_2 + H_2 \qquad \Delta H^\circ_{298} = -41 \text{ kJ/mol} \tag{5.18}$$

And finally, the exothermic methanation reaction:

$$CO + 3\,H_2 \rightleftharpoons CH_4 + H_2O \qquad \Delta H^\circ_{298} = -206 \text{ kJ/mol} \tag{5.19}$$

The overall pre-reforming reaction is endothermic, which favors the exothermic shift and methanation reactions.

A mixture of 5 vol.% of propylene in propane was prepared using ultra high purity grade gases from Air-Gas Company. This mixture resembled HD-5 propane composition. Separate gas tanks of helium, H_2, and air were built-in in the setup for the catalyst pretreatment procedures. Steam pre-reforming experiments were carried out in a continuous flow fixed-bed tubular glass reactor at pressure close to one atmosphere. 0.050 g-samples of the catalyst were used for all the experiments. An ice-cooled trap was used to condense and collect the water. The reactor setup is shown in Figure 5.7.

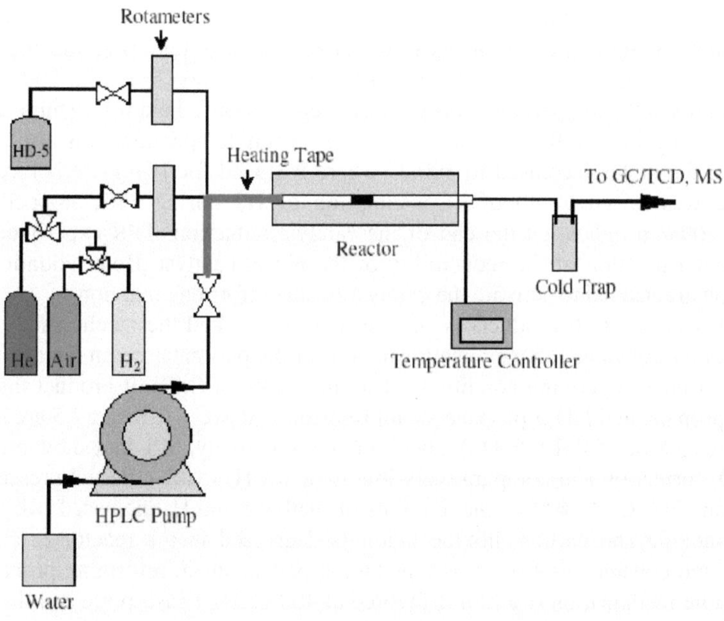

Figure 5.7. HD-5 propane steam pre-forming experimental setup

A commercial Ni-based catalyst from Sud-Chemie was used for the steam reforming of HD-5 propane. The 0.050 g catalyst samples were activated before each experiment by oxidation in air at 300°C, followed by reduction in H_2 at 400°C. Temperature Programmed Reduction (TPR) experiments were performed to study the pretreatment process. 0.500 g of the catalyst was first degassed with He. Subsequently, a reducing gas (H_2) was passed continuously over the sample at a flowrate of 1 ml/min. A furnace equipped with a temperature controller and in-situ sample temperature sensing was used to heat the catalyst and control temperature ramping. The reaction between steam and the reducing gas was monitored by a MKS-UTI PPT quadruple mass spectrometer. The temperature program included ramping from 25°C to 400°C with a ramping rate of 5 deg/min followed by heating the catalyst sample at 400°C for 4 hours.

The steam reforming experiments were done isothermally at three different temperatures: 200°C, 300°C, 400°C, at approximately atmospheric pressure and at two different steam to carbon ratios (S/C): 1.8, and 2.5. Gas hourly space velocity (GHSV) was set to be around 4000/h. The reaction products were identified with an online MKS-UTI PPT quadruple mass spectrometer. Product concentrations and reactant conversions were quantified by a Hewlett-Packard Model 5890 Series II chromatograph equipped with a thermal conductivity detector (TCD). The instrument was equipped with a 5 ft × 1/8 In SS Carboxen 1000 column from Supelco for separation of H_2, CO, CH_4. Gas-Pro GSC 30 m × 0.32 mm capillary column was used to separate propane. The Raman spectra were taken at room temperature in the spectral range 500 - 3500 cm-1 using a Renishaw 2000 microscope system that includes an optical microscope and a CCD camera for multichannel detection.

Product distribution experiments were carried out with pure propane and HD-5 propane for comparison. For HD-5 SR at S/C =1.8, reaction started at 200oC, and after 8 minutes the temperature was increased up to 300oC, held for 19 minutes and then raised again to 400oC. At S/C = 2.5, the starting temperature was also 200oC and after 8 minutes increased to 300oC where it stayed for 18 minute before temperature was raised to 400oC. Consumption of H2 during TPR stopped after around 200 min indicating the end of the catalyst reduction. TPR experiments allowed determination of the reducibility of the nickel catalyst. Both oxidation and reduction are needed to activate the catalyst for the reforming reaction.

Identification of all products is necessary to understand the mechanism of propane steam reforming and methane formation for the particular commercial catalyst employed under operating conditions. The mass spectra (MS) of product distribution in propane and HD-5 propane steam reforming at S/C = 1.8 and 2.5 are shown in Figures 5.8 (a), 5.8 (b), 5.9 (a), and 5.9 (b), respectively. MS data show high CO and CO_2 formation and comparatively low H_2 and CH_4 relative partial pressures at 200°C and 300°C. At 400°C, the amounts of methane and H_2 increased, while carbon monoxide and carbon dioxide amounts decreased in the reactor exit gases. Taking into consideration the reactions taking part in steam reforming process, an exothermic methanation reaction is favored at 400°C. H_2O adsorption and dissociation on the catalyst may be accelerated at high temperatures. This can result in more

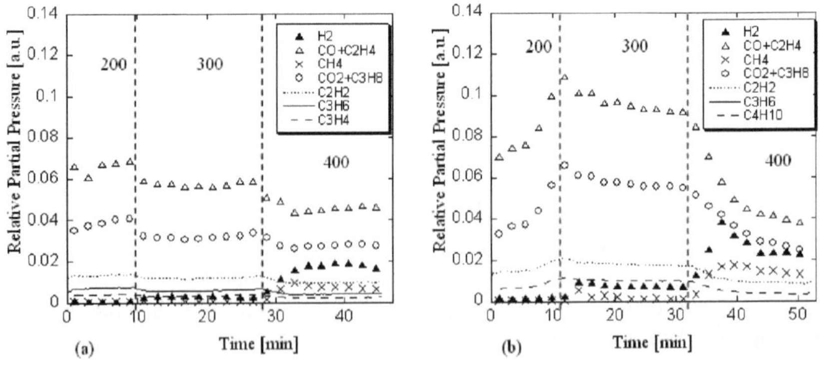

Figure 5.8. Steam reforming of propane at (a) S/C = 1.8 and (b) S/C = 2.5

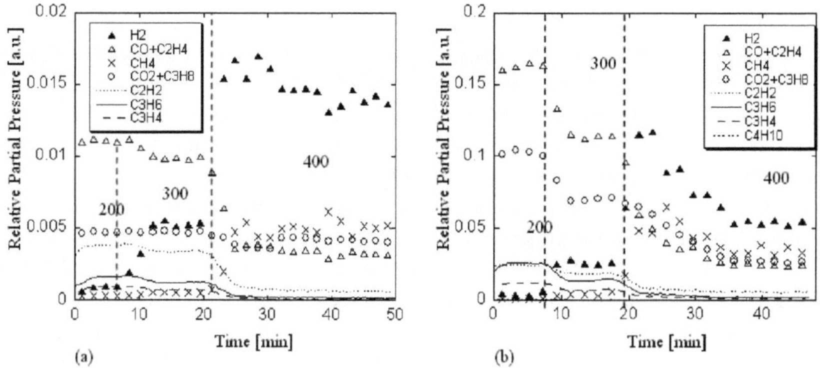

Figure 5.9. Steam reforming of HD-5 Propane at (a) S/C = 1.8 and (b) S/C = 2.5

oxygen and H_2 species being available for shift and methanation reactions. Oxygen species are necessary to prevent coke formation and to provide good stability and activity of the catalyst. Therefore more methane formation and better catalyst activity at a higher S/C ratio are expected. MS data displayed in Figures 5.8 and 5.9 show that more H_2 and methane were formed than carbon dioxide and carbon monoxide in HD-5 propane SR compared to pure propane SR. This observation may indicate that the presence of unsaturated hydrocarbon helps in formation of methane. In SR of HD-5 higher S/C ratio also leads to higher methane concentration in the products.

From gas chromatography data of propane SR products, fuel conversion and selectivity to methane were calculated. Figures 5.10 (a) and (b) illustrate conversion of propane at different S/C ratios. Figure 5.10 (a) shows propane conversion at different temperatures and at S/C = 1.8. A considerable increase in conversion after changing the temperature from 200°C to 300°C was observed. At 400°C, a slightly increased propane conversion as compared to its conversion at 300°C was observed, however, with time the conversion went down. The drop in catalyst activity over a longer reaction period may be due to a decrease in available area because of coke formation or another deactivation factor. Steam is generally used to avoid coke formation and thus steam to carbon ratio (S/C) was changed from 1.8 to 2.5.

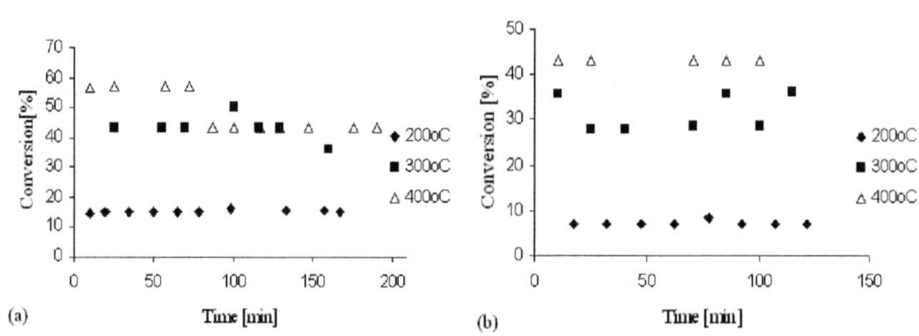

Figure 5.10. Conversion of propane in steam reforming at (a) S/C=1.8 and (b) S/C = 2.5

We observed a constant reaction rate (catalyst activity) over the entire period of reaction as shown in Figure 5.10 (b). This shows that the presence of steam prevents coke depositing on the catalyst surface. However, conversion rates are slightly on the lower side as compared to S/C of 1.8. This also suggests that some parallel reaction other than propane conversion is taking place on the catalyst surface thereby reducing the area available for the reaction of interest. The side reaction could be the oxidation of carbon to CO/CO_2.

Figure 5.11 (a), (b), and (c) shows the selectivity of propane to methane conversion reaction at different temperatures and S/C ratios. For an S/C ratio of 1.8, the selectivity increased from 28% to 72% with a temperature increase from 200°C to 300°C but decreased to 51% with further increase in temperature to 400°C. The same trend was noticed with S/C of 2.5 where the selectivity changed from 37% up to 64% then down to 40% when the temperatures were 200, 300, and 400°C, respectively. While at higher temperatures (300 and 400°C) the selectivity was higher with lower S/C ratio, at lower temperature (200°C) the opposite was observed. Since the propane conversion was higher at higher temperatures, operation at low S/C might be suggested. However, higher steam to carbon ratios are recommended as such ratios would aid in reducing coke formation thereby restoring or retaining catalyst activity. For operating at elevated temperatures where higher

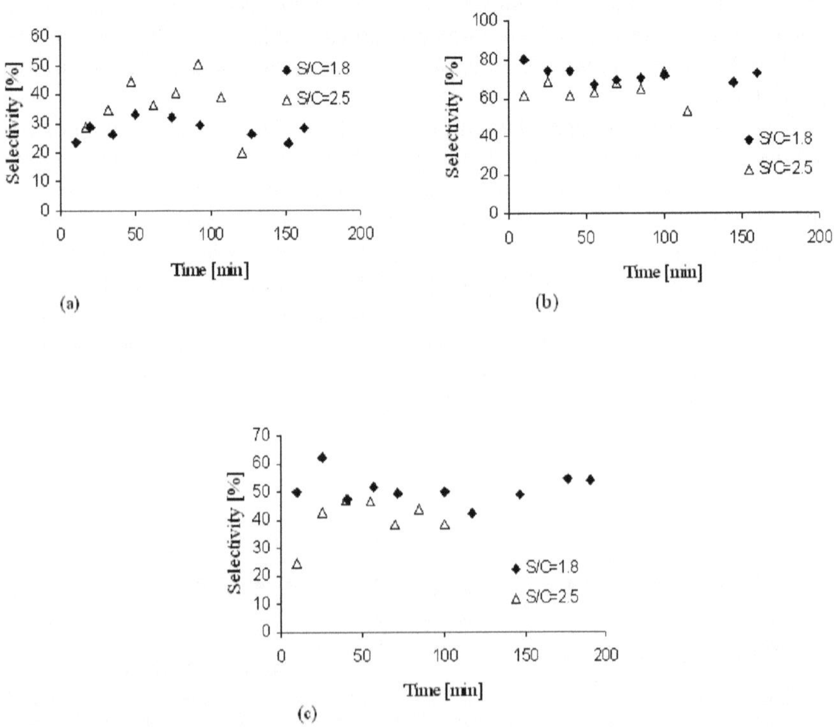

Figure 5.11. Steam reforming selectivity of propane to methane at (a) 200, (b) 300, and (c) 400°C

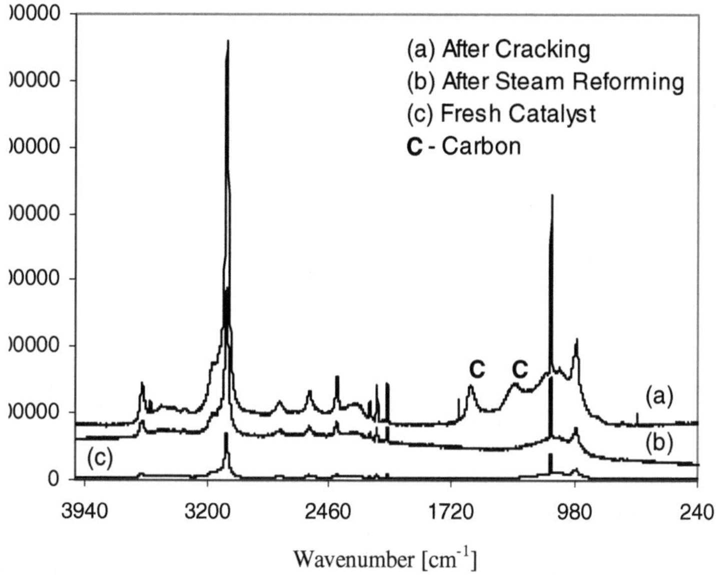

Figure 5.12. Raman spectra of the catalyst

conversions were obtained, an optimum S/C ratio that is intermediate between 1.8 and 2.5 is recommended. Such an optimum ratio will likely result in stable catalyst activity and acceptable selectivity, both of which are desired.

Figure 5.12 shows the Raman spectra of the fresh catalyst, the catalyst after cracking, and the catalyst after steam reforming reaction at S/C =1.8. The catalyst after SR does not show peaks attributed to the presence of amorphous carbon and highly disordered graphite in contrast to the catalyst examined after cracking of propane. Therefore the presence of steam during the reforming process prevents coke formation on the catalyst's surface.

5.4 Conclusions and Directions for Future Research

The development of coke-resistant catalysts makes it possible to process hydrodesulfurized heavy hydrocarbon oil. This process is not only promising in industrial H_2 production but also attractive in producing H_2 for fuel cells on merchant ships, which require much longer startup time than automobiles. By this process, H_2 can be produced directly from hydrodesulfurized heavy hydrocarbon oil without first catalytically cracking it to diesel and gasoline.

Vehicles, trains, and naval ships require short startup time but merchant ships and submarines can tolerate extended startup time. Based on start-up time requirements, partial oxidation reformers are more suitable for short startup applications than steam reformers since the POX reformers have low weight, rapid starting, are compact, and dynamically responsive devices. However, steam reformers have higher energy efficiency due to the H_2 generation from water molecules, which makes them more attractive for long startup applications.

Most steam-reforming catalysts are also good catalysts for partial oxidation reforming. But most commercial reforming catalysts are conventional Ni-based or Cu-based catalysts designed for stationary processes and not for onboard application. The latter has space limitation so the catalysts for self-moving equipment should be stable at higher space velocities compared to the stationary processes. The activity of catalysts depends on the nickel or copper surface area [199]. Commercial Ni-based reforming catalysts are usually active at temperatures above 450°C. Metallic nickel particles are usually large (200–1000 Å) due to strong thermal sintering [200]. This also means that the nickel area is relatively small, on the order of a few square meters per gram catalyst. Copper-based catalysts are more susceptible to thermal sintering than nickel catalysts [201]. Therefore, copper-based catalysts have to be operated at relatively low temperature, usually not higher than 300°C.

Research on advanced catalyst preparation technology is extremely important for onboard reforming/fuel cell systems. New promoter-support series with high metal dispersion and strong resistance to thermal sintering are attractive. This kind of research is not only important for reforming catalysts, but also useful for the development of prereforming, water-gas shift, and preferential oxidation catalysts that are normally composed of metals on oxide supports.

Modifying the acid-base properties of catalysts and enhancing the coke-resistant capability of reforming catalysts is important. Nonmetallic catalysts were introduced [199] to avoid sulfur poisoning but the activities of these catalysts are still under development. Efforts should be directed at long-term testing of catalysts.

5.5 References

1. Brown L F. Int. J. Hydrogen Energy 26 (2001) 381.
2. Joensen F, Rostrup-Nielsen J R,.J. Power Sources 1 (2001) 4566.
3. Moon D J, Sreekumar K, Lee S D, Lee B G, H.S. Kim, Appl. Catal. A 1 (2001) 215.
4. Yanhui W, Diyong W. Int.J. Hydrogen Energy 26 (2001) 795.
5. Audus H, Kaarstad O, Kowal M. Decarbonisation of Fossil Fuels: Hydrogen as an Energy Carrier, CO2 Conference, Boston/Cambridge 1997, published in Energy Conversion Management, Vol. 38, Suppl., pp. 431–436.
6. Scholz W H. Processes for industrial production of hydrogen and associated environmental effects. Gas Sep. Purif. 1993;7:131–9.
7. Armor J N. The multiple roles for catalysis in the production of H_2. Applied Catalysis A: General 176 (1999) 159–176.
8. Blomen L J M J, Mugerwa M N. Fuel Cell System. New York: Plenum Press, 1993.
9. Avadikyan A, Cohendet P, Heraud J A. The Economic Dynamics of Fuel Cell technologies. Springer-Verlag: Berlin, 2003, pp. 23–42.
10. Lee F. Brown, International Journal of Hydrogen Energy 26 (2001) 381–397.
11. Dicks A L, Journal of Power Sources 61 (1996) I 1.3–124.
12. U.S. Energy Information Administration. Available at http://www.eia.doe.gov. Accessed August 26, 2004.

13. Zhan Z, Liu J, Barnett S A. Operation of anode-supported solid oxide fuel cells on propane–air fuel mixtures. Applied Catalysis A: General, Volume 262, Issue 2, 20 May 2004, Pages 255–25.
14. European certification for IdaTech's EtaGen. Fuel Cells Bulletin, Volume 2004, Issue 3, March 2004, Page 8.
15. Loffler D G, Taylor K, Mason D. A light hydrocarbon fuel processor producing high-purity hydrogen. Journal of Power Sources 117 (2003) 84–91.
16. Chen F, Zha S, Dong J, Liu M. Pre-reforming of propane for low-temperature SOFCs. Solid State Ionics, Volume 166, Issues 3–4, 30 January 2004, Pages 269–273.
17. Amphlett J C, Mann R F, Peppley B A, Robnerge P R, Rodrigues A, Salvador J P, J. Power Sources 1998, 71:179–184
18. Lindstrom B, Peterson L J, Govind Menon P. Applied Catalysis A: General 234 (2002) 111–125.
19. Fisher I A, Bell A T. Journal of Catalysis 184 (1999) 357–376.
20. Peppley B A, Amphlett J C, Thurgood C P, Mann R F. Topics in Catalysis, 22 3–4 (2003) 253–259.
21. Wasmus S, Kuver A. Methanol oxidation and direct methanol fuel cells: a selective review. Journal of Electroanalytical Chemistry 461 (1999) 14–31.
22. Deluga G A, Salge J R, Schmidt L D, Verykios X E. SCIENCE vol. 303 (2004) 13.
23. Chellappal A S, Fischer C M, Thom W J. Applied Catalysis A: General 227 (2002) 231–240.
24. Uribe F A, Gottesfeld S, Zawodzinski Jr T A. Effect of ammonia as potential fuel impurity on proton exchange membrane fuel cell performance, J. Electrochem.l Soc. 149 (3) (2002) A293–A296.
25. Air liquide. Available at: airliquide.com. Accessed Sep 8, 2004.
26. US Dept. of Energy, Energy Efficiency and Renewable Energy. Available at: http://www.eere.energy.gov. Accessed August 2, 2004.
27. Lv P, Chang J, Wang T, Fu Y, Chen Y. Energy & Fuels 18 (2004) 228–233.
28. Huber G W, Shabaker J W, Dumesic J A. SCIENCE vol. 300 (2003) 27.
29. Larminie J, Dicks A. Fuel Cell Systems Explained. Chichester: John Wiley & Sons, 2002.
30. Wu S, Chen J X, Zheng X F, Zeng H S, Zheng C M, Guan N J. Novel preparation of nanocrystalline magnesia-supported caesium-promoted ruthenium catalyst with high activity for ammonia synthesis. Chemical Communications, 2003, 19, 2488–2489.
31. Wang R, Wei K, Lin J X, Yu X J, Mao S L. Preparation of Ru/AC ammonia synthesis catalyst reduced with hydrazine hydrate. Chinese Journal of Catalysis, 2003, 24, 929–932.
32. Trukhan S N, Ivanov V P, Kochubey D I, Tsyru'nikov P G, Dobrynkin N M, Noskov A S. Ammonia formation on (Ru plus Cs)/C catalysts by the interaction of adsorbed nitrogen with hydrogen that diffused from the bulk of cesium-ruthenium particles. Kinetics and Catalysis, 2003, 44, 648–651.
33. Pernicone N, Ferrero E, Rossetti I, Forni L, Canton P, Riello P, Fagherazzi G, Signoretto M, Pinna F. Wustite as a new precursor of industrial ammonia synthesis catalysts. Applied Catalysis A: General, 2003, 251, 121–129.
34. Lytken O, Waltenburg H N, Chorkendorff I. Ammonia synthesis on Au modified Fe(111) and Ag and Cu modified Fe(100) surfaces. Surface Science, 2003, 543, 207–218.
35. Logadottir A, Norskov J K. Ammonia synthesis over a Ru(0001) surface studied by density functional calculations. Journal of Catalysis, 2003, 220, 273–279.
36. Kowalczyk Z, Krukowski M, Rarog-Pilecka W, Szmigiel D, Zielinski J. Carbon-based ruthenium catalyst for ammonia synthesis - Role of the barium and caesium promoters and carbon support. Applied Catalysis A: General, 2003, 248, 67–73.

37. Volpi A, Clary D C. Theoretical investigation of the surface reaction N-(ads)+H-(ads)-> NH(ads) on Ru(0001) using density functional calculations, variational transition-state theory, and semiclassical tunneling method. Journal of Physical Chemistry B, 2004, 108, 336–345.
38. Wang R, Wei K, Lin J X, Yu X J, Mao S L. Preparation of Ru/AC ammonia synthesis catalyst reduced with hydrazine hydrate. Chinese Journal of Catalysis, 2003, 24, 929–932.
39. Satterfield C N. Heterogeneous Catalysis in Practice. McGraw Hill: New York, 1980.
40. Chorkendorff I, Niemantsverdriet J W. Concepts in modern Catalysis and Kinetics. WILEY-VCH: Weinheim, 2003, pp. 301–348.
41. Ishiguro T. Make Town Gas from Naphtha. Hydrocarbon Processing, 1968, 47, 87–96.
42. Wilhelm D J, Simbeck D R, Karp A D, Dickenson R L. Syngas production for gas-to-liquids applications: technologies, issues and outlook. Fuel Processing Technology, 2001, 71, 139–148.
43. Warnatz J. Hydrocarbon oxidation high-temperature chemistry. Pure and Applied Chemistry, 2000, 72, 2101–2110.
44. Pena M A, Gomez J P, Fierro J L G. New catalytic routes for syngas and hydrogen production. Applied Catalysis A: General, 1996, 144, 7–57.
45. Jenner G. Ruthenium Catalyzed Hydroformylation Reactions Using in Situ Generation of Synthesis Gas from Aqueous Methyl Formate. Applied Catalysis, 1991, 75, 289–298.
46. Schmidt L D, Klein E J, Leclerc C A, Krummenacher J J, West K N. Syngas in millisecond reactors: higher alkanes and fast lightoff. Chemical Engineering Science, 2003, 58, 1037–1041.
47. Vosloo A C. Fischer-Tropsch: a futuristic view. Fuel Processing Technology, 2001, 71, 149–155.
48. Young S. Tomorrow's Energy: Hydrogen, Fuel Cells and the Prospects for a Cleaner Planet by Peter Hoffmann. Nature, 2001, 414, 487–488.
49. Wang Z F, Wang W P, Lu G X. Studies on the active species and on dispersion of Cu in CU/SiO_2 and $Cu/Zn/SiO_2$ for hydrogen production via methanol partial oxidation. International Journal of Hydrogen Energy, 2003, 28, 151–158.
50. Wang W P, Lu G X. Advances in catalytic generation of hydrogen from ethanol. Progress in Chemistry, 2003, 15, 74–78.
51. Segal S R, Carrado K A, Marshall C L, Anderson K B. Catalytic decomposition of alcohols, including ethanol, for in situ H_2 generation in a fuel stream using a layered double hydroxide-derived catalyst. Applied Catalysis A: General, 2003, 248, 33–45.
52. Pfeifer P, Schubert K, Liauw M A, Emig G. Electrically heated microreactors for methanol steam reforming. Chemical Engineering Research & Design, 2003, 81, 711–720.
53. Larsen H, Feidenhans'l R, Petersen L S. Hydrogen system energy technologies: Hydrogen system energy technologies in global, European and Danish perspective. Risø Energy Report 3. Risø National Laboratory.
54. McHugh K. Hydrogen Production Method. Prepared for MPR Associates, Inc. Febuary 2005.
55. Xu S, Wang X L, Zhao R. Study on the production of hydrogen from methane. Progress in Chemistry, 2003, 15, 141–150.
56. Song C. Fuel processing for low-temperature and high-temperature fuel cells. Challenges, and opportunities for sustainable development in the 21st century. Catal. Today 2002; 77:17–49.
57. Thomas S, Zalbowitz M. Fuel Cells-Green Power, Los Alamos National Laboratory.

58. Mohtadi R, Lee W-K, Cowan S, Van Zee J W, Murthy M. Effects of Hydrogen Sulfide on the Performance of a PEMFC. Electrochemical and Solid-State Letters, 6 (12) A272–A274 (2003).
59. Palm C, Cremer P, Peters R, Stolten D. J Power Sources 2002, 106:231–237.
60. Nielsen B. Appl.Catal. 11 (1984) 5.
61. Shafi R, Hutchings G J. Catal. Today 59 (2000) 423.
62. Whitehurst D D, Isoda I, Mochida I. Adv. Catal. 42 (1998) 345.
63. Novochinskii II, Song C, Ma X, Liu X, Shore L, Lampert J, Farrauto R J. Low-Temperature H_2S Removal from Steam-Containing Gas Mixtures with ZnO for Fuel Cell Application. 1. ZnO Particles and Extrudates. Energy & Fuels (2004), 18(2), 576–583.
64. Fan H, Li Y, Li C, Guo H, Xie K. The apparent kinetics of H_2S removal by zinc oxide in the presence of hydrogen. Fuel 2002, 81(1), 91–96.
65. Ahmed S, Kumar R, Krumpelt M. Fuel processing for fuel cell power systems. Fuel Cell Bull 1999;12:4–7.
66. Tawara K, Nishimura T, Iwanami H, Nishimoto T, Hasuike T. New Hydrodesulfurization Catalyst for petroleum-Fed Fuel Cell Vehicle and Cogenerations. Ind. Eng. Chem. Res. 2001; 40:2367–70.
67. Nag N K, Sapre A V, Broderick D H, Gates B C. J. Catal. 57 (1979) 509.
68. Kilanowski D R, Teeuwen H, Gates B C, Beer V H J D, Schuit G C A, Kwart H. J. Catal. 55 (1978) 129.
69. Houalla M, Nag N K, Sapre A V, Broderick D H, Gates B C. AIChE J. 24 (1978) 1015.
70. Houalla M, Broderick D H, Sapre A V, Nag N K, Beer V H J D. J. Catal. 61 (1980) 523.
71. Pecoraro T A, Chianelli R R. Hydrodesulfurization catalysis by transition metal sulfides. J. Catal. 1981, 67, 430.
72. Ledoux M J, Michaux O, Agostini G, Panissod P. The influence of sulfide structures on the hydrodesulfurization activity of carbon-supported catalysts. J. Catal. 1986, 102, 275.
73. Grange P, Vanhaeren X. Hydrotreating catalysts, an old story with new challenges. Catal. Today 1997, 36, 375.
74. Qian W, Yoda Y, Hirai Y, Ishihara A, Kabe T. Hydrodesulfurization of dibenzothiophene and hydrogenation of phenanthrene on alumina-supported Pt and Pd catalysts. Appl. Catal., A 1999, 184, 81.
75. Isoda T, Nagao S, Ma X, Korai Y, Mochida I. Hydrodesulfurization of Refractory Sulfur Species. 1. Selective Hydrodesulfurization of 4,6–Dimethyldibenzothiophene in the Major Presence of Naphthalene over $CoMo/Al_2O_3$ and Ru/Al_2O_3 Blend Catalysts. Energy Fuels 1996, 10, 482.
76. Lecrenay E, Sakanishi K, Mochida I. Catalytic hydrodesulfurization of gas oil and model sulfur compounds over commercial and laboratory-made CoMo and NiMo catalysts: Activity and reaction scheme. Catal. Today 1997, 39, 13.
77. Bej S K, Maity S K, Turaga U T. Search for an Efficient 4,6–DMDBT Hydrodesulfurization Catalyst: A Review of Recent Studies. Energy Fuels 2004, 18(5), 1227.
78. Navarro R, Pawelec B, Fierro J L G, Vasudevan P T, Cambra J F, Arias P L. Appl. Catal. A 137 (1996) 269.
79. Qian W, Yoda Y, Hirai Y, Ishihara A, Kabe T. Appl. Catal. A 184 (1991) 81.
80. Robinson W R A M, van Veen J A R, de Beer V H J, van Santen R A. Fuel Processing Technology, 61 (1991) 89.
81. Robinson W R A M, van Veen J A R, de Beer V H J, van Santen R A. Fuel Processing Technology, 61 (1991) 103.
82. Venezia A M, Parola V L, Deganello G, Cauzzi D, Leonardi G, Predieri G. Appl. Catal. 229 (2002) 261.

83. Papadopoulou C, Vakros J, Matralis H K, Kordulis C, Lycourghiotis A. J. Colloid Interface Sci. 261 (2003) 146.
84. Okamoto Y, Ochiai K, Kawano M, Kobayashi K, Kubota T. Appl. Catal. A 226 (2002) 115.
85. Sakanishi K, Nagamatsu T, Mochida I, Whitehurst D D. J. Mol. Catal. A 155 (2000) 101.
86. Haji S, Zhang Y, Kang D, Aindow M, Erkey C. Hydrodesulfurization of Model Diesel using Pt/Al_2O_3 Catalysts Prepared by Supercritical Deposition. Catal. Today Volume 99, s 3–4, 30 (2005) 365–373
87. Egorova M, Prins R. J. Catal. 224 (2004) 278.
88. Nagai M, Kabe T. J. Catal 1983; 81:440.
89. Bartsch R, TaniellianC. J. Catal. 35 (1974) 353.
90. Engelhard develops fuel sulfur removal. Fuel Cells Bulletin, Volume 2003, Issue 11, November 2003, Page 9.
91. Hernandez-Maldonado A J, Yang R T. Desulfurization of Commercial Liquid Fuels by Selective Adsorption via π-Complexation with Cu(I)-Y Zeolite. Ind. Eng. Chem. Res. 2003; 42: 3103–10.
92. Haji S, Erkey C. Removal of Dibenzothiophene from Model Diesel by Adsorption on Carbon Aerogels for Fuel Cell Applications. Ind. Eng. Chem. Res. 2003; 42: 6933–37.
93. Weitkamp J, Schwark M and Ernest S. J. Chem. Soc., Chem. Commun. (1991) 1133.
94. Yang R T, Takahashi A, Yang F H. Ind. Eng. Chem. Res. 40 (2001) 6236.
95. Takahashi A, Yang F H, Yang R T. Ind. Eng. Chem. Res. 42 (2002) 2487.
96. Larrubia M A, Gutierrez-Alejandre A, Ramirez J, Busca G A. Appl. Catal. A 224 (2002) 167.
97. Lee S H D, Kumar R, Krumpelt M. Sep. Purif. Technol. 26 (2002) 247.
98. Ma X, Sun L, Song C. Catal. Today 77 (2002) 107.
99. Hoogers G. Fuel Cell Technology Handbook. Boca Raton: CRC Press; 2003.
100. Hohlein B, Bee M, Bogild-Hansen J, Brockerhoff P, Colsman G, Emonts B, Menzer R, Riedel E. Hydrogen from methanol for fuel cells in mobile systems: development of a compact reformer. Journal of Power Sources 61 (1996) 143–147.
101. Austin G T. Shreve's Chemical Process Industries. New York: McGraw-Hill; 1984.
102. Krumpelt M, Krause T R, Carter J D, Kopasz J P, Ahmed S. Fuel Processing for Fuel Cell Systems in Transportation and Portable Power Applications. Catal. Today 2002;77:3–16.
103. Holmblad P M, Wambach J, Chorkendorff I. J. Chem. Phys., 1995, 102, 8255.
104. Goodman D W. Acc. Chem. Res., 1984, 17, 194.
105. Hu J, Wang Y, VanderWiel D, Chin C, Palo D, Rozmiarek R, Dagle R, Cao J, Holladay J, Baker E. Chemical Engineering Journal 93 (2003) 55–60.
106. Ming Q, Healey T, Lloyd Allen, Irving P. Steam reforming of hydrocarbon fuels. Catalysis Today 77 (2002) 51–64.
107. Ziemlewski J I. Catalytic Steam Reforming of Diesel for Portable Fuel Cell Applications [dissertation]. University of Connecticut; 2003.
108. Katikaneni S, Yuh C, Abens S, Farooque M. The direct carbonate fuel cell technology: advances in multi-fuel processing and internal reforming, Catalysis Today, 2002, 77, 99–106.
109. Katikaneni S, Yuh C, Lee T, Farooque M. Operation of direct carbonate fuel cell with propane Preprints of Symposia - American Chemical Society, Division of Fuel Chemistry, 2003, 48, 854–x.
110. Bery R N. SNG from Naphtha Chemical Engineering, 1973, 80, 90–91.
111. Suzuki T, Iwamoto O, Kitahara T. Development of supported ruthenium catalyst for SNG process, 2001, 44, 183–187.

112. Ninomiya A, Sato M, Nishiguchi H, Ishihara T, Takita Y. Atmospheric production process for substitute natural gas from saturated hydrocarbons 1. Utilization of catalytic reaction for atmospheric methane formation, Kagaku Kogaku Ronbunshu, 1999, 25, 316–321.
113. Nagase S, Takami S, Hirayama A, Hirai Y. Development of a high efficiency substitute natural gas production process. Catalysis Today, 1998, 45, 393–397.
114. Ishigaki Y, Uba M, Nishida S, Inui T. Application of Co-Mn_2O_3_Ru Catalyst to the Process for Producing High-Calorie Substitute Natural-Gas from Coke-Oven Gas. Applied Catalysis, 1989, 47, 197–208.
115. Lorant F, Behar F, Vandenbroucke M, McKinney D E, Tang Y C. Methane generation from methylated aromatics: Kinetic study and carbon isotope modeling. Energy & Fuels, 2000, 14, 1143–1155.
116. Behar F, Budzinski H, Vandenbroucke M, Tang Y. Methane generation from oil cracking: Kinetics of 9–methylphenanthrene cracking and comparison with other pure compounds and oil fractions. Energy & Fuels, 1999, 13, 471–481.
117. Carrette L, Friedrich K A, Stimming U. Fuel cells – fundamentals and applications. Fuel Cells, 2001, 1, 5–9.
118. Wojtowicz M A, Katikaneni S, Privette R. The 2001 symposium on recent advances in fuel cells. Fuel, 2002, 81, 2147–2149.
119. Katikaneni S, Farooque M. The direct carbonate fuel cell technology - an overview. American Institute of Chemical Engineers, Spring National Meeting, 2003, 1689–1696.
120. Rostrup-Nielsen J, Christensen T S, Dybkjaer I. Studies Surf Sci Catalysis 1998, 113: 81–95.
121. Advances in Automotive Reforming Technology. Fuel Cell Catalyst. Vol. 2, No. 2, Winter 2002.
122. Sun J, Qiu X, Wu F, Zhu W, Wang W, Hao S. Hydrogen from steam reforming of ethanol in low and middle temperature range for fuel cell application. International Journal of Hydrogen Energy 2004;29(10):1075–81.
123. Goula M, Kontou S, Tsiakaras P. Hydrogen production by ethanol steam reforming over a commercial Pd/γ-Al2O3 catalyst. Applied Catalysis B: Environmental 2004; 49(2):135–44.
124. Batista M, Santos R, Assaf E, Assaf J, Ticianelli E. High efficiency steam reforming of ethanol by cobalt-based catalysts. Journal of Power Sources, Volume 134, Issue 1, 12 July 2004, Pages 27–32.
125. Brown L F. A comparative study of fuels for on-board hydrogen production for fuel-cell-powered automobiles. International Journal of Hydrogen Energy 26 (2001) 381–397.
126. Llorca J, Ramııez de la Piscina P, Dalmon J, Sales J, Homs N. CO-free hydrogen from steam-reforming of bioethanol over ZnO-supported cobalt catalysts Effect of the metallic precursor. Applied Catalysis B: Environmental 43 (2003) 355–369.
127. Liguras D, Kondarides D, Verykios X. Production of hydrogen for fuel cells by steam reforming of ethanol over supported noble metal catalysts. Applied Catalysis B: Environmental 43 (2003) 345–354.
128. Comas J, Mariño F, Laborde M, Amadeo N. Bio-ethanol steam reforming on Ni/Al_2O_3 catalyst. Chemical Engineering Journal 98 (2004) 61–68.
129. Ioannides. Thermodynamic analysis of ethanol processors for fuel cell applications. Journal of Power Sources 92 (2001) 17–25.
130. Ogden J, Steinbugler M, Kreutz T, Journal of Power Sources 79–2 (1999) 143–168.
131. Docter A, Lamm A, Journal of Power Sources 84 (1999) 194–200.
132. Newson E, Truong T B. Low-temperature catalytic partial oxidation of hydrocarbons (C1–C10) for hydrogen production. International Journal of Hydrogen Energy 28 issue 12 (2003) 1379–1386.

133. Ahmed S, Krumpelt M, Kumar R, Lee SHD, Carter JD, Wilkenhoener R, Marshall C. Catalytic Partial Oxidation Reforming of Hydrocarbon Fuels. Fuel Cell Seminar. Palm Springs, California. Nov 16–19, 1998.
134. Naidja A, Krishna C R, Butcher T, Mahajan D. Cool flame partial oxidation and its role in combustion and reforming of fuels for fuel cell systems. Progress in Energy and Combustion Science 29 (2003) 155–191.
135. Galuszka J, Pandey R N, Ahmed S. Methane conversion to syngas in a palladium membrane reactor. Catalysis Today 46 (1998) 83–89.
136. Freni S, Calogero G, Cavallaro S. Hydrogen production from methane through catalytic partial oxidation reactions. Journal of Power Sources 87 2000 28–38.
137. Xu S, Zhao R, Wang X. Highly coking resistant and stable Ni/Al_2O_3 catalysts prepared by W/O microemulsion for partial oxidation of methane. Fuel Processing Technology (2004), in press.
138. De Jong K P, Schoonebeek R J, Vonkeman K A. Process for the Catalytic Partial Oxidation of Hydrocarbons. US Patent 5,720,901, 1998.
139. HydrogenSource demonstrates gasoline processor. Fuel Cell Bull. Volume 2003, Issue 6, Page 7 (June 2003).
140. Pereira C, Wilkenhoener R, Ahmed S, Krumpelt M. DOE Hydrogen Program Annual Review, July 2, 1999.
141. Pereira C, Bae J, Ahmed S, Krumpelt M. DOE Hydrogen Program Annual Review, May 9, 2000.
142. Edwards N, Ellis S R, Frost J C, Golunski S E, van Keulen A N J, Lindewald N G, Reinkingh J G. On-bored Hydrogen Generation for Transport Applications: The Hot-Spot Methanol Processor. J. Power Sources, 1998:7(1–2), 123–128.
143. Nuvera gasoline reformer achieves DOE targets. Fuel Cells Bulletin, Volume 2004, Issue 4, April 2004, Page 9.
144. Wang S, Lu G Q. J Chem Technol Biotechnol , 2000, 75, 589–595.
145. Borowiecki T, Denis A, Gac W, Dziembaj R, Piwowarska Z, Drozdek M. Oxidation–reduction of Ni/Al2O3 steam reforming catalysts promoted with Mo. Applied Catal. A. 2004:article in press.
146. Satterfield C N. Heterogeneous Catalysis in Practice. New York: Mc-Graw-Hill, 1980; 136–143.
147. Wanke S E, Flynn P C. Sintering of supported metal catalysts. Catalysis Reviews - Science and Engineering (1975), 12(1), 93–135.
148. Ruckenstein E, Sushumna I. Hydrogen effects in sintering of supported metal catalysts. Chemical Industries (Dekker) (1988), 31(Hydrogen Eff. Catal.), 259–91.
149. Fiedorow R M J, Chahar B S, Wanke S E. The sintering of supported metal catalysts: II. Comparison of sintering rates of supported Pt, Ir, and Rh catalysts in hydrogen and oxygen. Journal of Catalysis 1978;51(2): 193–202.
150. Sehested J. Sintering of nickel steam-reforming catalysts. Journal of Catalysis 217 (2003) 417–426.
151. Sehested J, Gelten J A P, Remediakis I N, Bengaard H, Nørskov J K. Sintering of nickel steam-reforming catalysts: effects of temperature and steam and hydrogen pressures. Journal of Catalysis 223 (2004) 432–443.
152. Rasmussen F B, Sehested J, Teunissen H T, Molenbroek A M, Clausen B S. Sintering of Ni/Al_2O_3 catalysts studied by anomalous small angle X-ray scattering. Applied Catalysis A: General 267 (2004) 165–173.
153. Chen H-W, Wang C-Y, Yu C-H, Tseng L-T, Liao P-H. Carbon dioxide reforming of methane reaction catalyzed by stable nickel copper catalysts. Catal. Today 2004: in press.
154. Kameoka S, Tanabe T, Tsai AP. Al–Cu–Fe quasicrystals for steam reforming of methanol: a new form of copper catalysts. Catal. Today 2004: in press.

155. Frusteri F, Freni S, Chiodo V, Spadaro L, Bonura G, Cavallaro S. Potassium improved stability of Ni/MgO in the steam reforming of ethanol for the production of hydrogen for MCFC. Journal of Power Sources 132 (2004) 139–144.
156. Frusteri F, Freni S, Spadaroa L, Chiodo V, Bonura G, Donato S, Cavallarob S. H_2 production for MC fuel cell by steam reforming of ethanol over MgO supported Pd, Rh, Ni and Co catalysts. Catalysis Communications 2004;article in press.
157. Narusawa K, Hayashida M, Kamiya Y, Roppongi H, Kurashima D, Wakabayashi K. Deterioration in fuel cell performance resultingfrom hydrogen fuel containing impurities: poisoning effects by CO, CH_4, HCHO, and HCOOH. JSAE Review 24 (2003) 41–46.
158. Urban P M, Funke A, Müller J T, Himmen M, Docter A. Catalytic processes in solid polymer electrolyte fuel cell systems. Applied Catal A 221 (2001) 459–470.
159. Patt J, Moon D J, Phillips C, Thompson L, Molybdenum Carbide Catalysts for Water-Gas Shift. Catal. Lett. 2000; 65(4): 193–5.
160. Bunluesin T, Gorte R J, Graham G W. Studies of the water-gas-shift reaction on ceria-supported Pt, Pd, and Rh: Implications for oxygen-storage properties. Applied Catalysis B: Environmental 15 (1998) 107–l 14.
161. Swartz S L, Seahaugh M M, Holt C T, Dawson W. Fuel Processing Catalysts based on Nanoscale Ceria. Fuel Cells Bulletin, Volume 4, Issue 30, March 2001, Pages 7–10.
162. Goerke O, Pfeifer P, Schubert K. Water gas shift reaction and selective oxidation of CO in microreactors. Applied Catalysis A: General 263 (2004) 11–18.
163. HydrogenSource demos noble WGS catalysts. Fuel Cells Bulletin, Volume 2004, Issue 3, March 2004, pages 8–9.
164. Gottesfeld S, Pafford J. J. Electrochem. Soc. 135 (1988) 2651.
165. Yoon W, Kim C, Han M, Park J, Jeong J. Abstracts of 2002 Fuel Cell Seminar, Palm Springs, CA, 18–21 November 2002, p. 639.
166. Echigo M, Tabata T. Appl. Catal. A 251 (2003) 157.
167. Rohland B, Plzak V. The PEMFC-integrated CO Oxidation-A Novel method of Simplifying the Fuel Cell Plant. J. Power Sources 1999; 84(2):183–6.
168. Sirijaruphan A, Goodwin J G, Rice R W. Effect of Fe promotion on the surface reaction parameters of Pt/γ -Al_2O_3 for the selective oxidation of CO. J. of Catal. 2004; 224:304–13.
169. Han Y-F, Kahlich M J, Kinne M, Behm R J. CO removal from realistic methanol reformate via preferential oxidation – performance of a Rh/MgO catalyst and comparison to Ru/Al_2O_3, and Pt/Al_2O_3. Appl. Catal. B 2004; 50: 209–18.
170. Gottesfeld S, Pafford J. J. Electrochem. Soc. A new approach to the problem of carbon monoxide poisoning in fuel cells operating at low temperatures 1988;135(10):2651–2652.
171. Alkire R C, Gerischer H, Kolb D M, Tobias C W. Polymer Electrolyte Fuel Cells in Advances in Electrochemical Science and Engineering. Volume 5. Wiley-VCH, New York, 1997; 219–225.
172. Fujita S-I, Takezawa N. Difference in the Selectivity of CO and CO_2 Methanation Reactions. Chem. Eng. J. 1997; 68:63–68.
173. Echigo M, Tabata T. CO Removal from Reformed Gas by Catalytic Methanation for Polymer Electroyte Fuel Cell Applications. J. Chemical Engineering of Japan 2004; 37(1):75–81.
174. Echigo M, Tabata T. Development of novel Ru catalyst of preferential CO oxidation for residential polymer electrolyte fuel cell systems. Catal. Today 2004; 90: 269–75.
175. Xomeritakis G, Lin Y. S. Fabrication of a thin palladium membrane supported in a porous ceramic substrate by chemical vapor deposition. Journal of Membrane Science, Volume 120, Issue 2, 13 November 1996, Pages 261–272.

176. Cheng Y S, Yeung K L. Palladium-silver composite membranes by electroless plating technique. Journal of Membrane Science 158 (1999) 127–141.
177. Quicker P, Höllein V, Dittmeyer R. Catalytic dehydrogenation of hydrocarbons in palladium composite membrane reactors. Catal. Today 2000; 56(1–3):21–34.
178. Jayaraman V, Lin Y S, Pakala M, Lin R Y. Fabrication of ultrathin metallic membranes on ceramic supports by sputter deposition. Journal of Membrane Science 1995;99(1): 89–100.
179. Li Z Y, Maeda H, Kusakabe K, Morooka S, Anzai H, Akiyama S. Preparation of palladium-silver alloy membranes for hydrogen separation by the spray pyrolysis method. Journal of Membrane Science 1993;78(3):247–254.
180. Tosti S, Bettinali L, Castelli S, Sarto F, Scaglione S, Violante V. Sputtered, electroless, and rolled palladium–ceramic membranes. Journal of Membrane Science 2002; 196(2):241–249.
181. Kajiwara M, Uemiya S, Kojima T, Kikuchi E. Hydrogen permeation properties through composite membranes of platinum supported on porous alumina. Catalysis Today 2000; 56:65–73
182. Jayaraman V, Lin Y S. Synthesis and hydrogen permeation properties of ultrathin palladium-silver alloy membranes. Journal of Membrane Science 104 (1995) 251–262.
183. Uemiya S, Matsuda T, Kikuchi E. Hydrogen permeable palladium-silver alloy membrane supported on porous ceramics. Journal of Membrane Science 1991; 56(3):315–325.
184. Peachey N M, Snow R C, Dye R C. Composite Pd/Ta metal membranes for hydrogen separation. Journal of Membrane Science 1996; 111(1):123–133.
185. Chabot J, Lecomte J, Grumet C., Sannier J. Fuel clean-up system: poisoning of palladium-silver membranes by gaseous impurities. Fusion Technology 1988;14:614–18.
186. Emonts B, Bogild Hansen J, Loegsgaard Jorgensen S, Hohlein B, Peters R. Compact methanol reformer test for fuel-cell powered light-duty vehicles. J. Power Sources 1998; 71(1,2):288–293.
187. Lin Y M, Rei M H. An integrated purification and production of hydrogen with a palladium membrane-catalytic reactor. Catal. 1998; 44:343–349.
188. Kuroda K, Kobayashi K, O'Uchida N, Ohta Y, Shirasaki Y. Study on Performance of Hydrogen Production from city gas equipped with Palladium membranes. Mistubishi Juko Giho 1996; 33(5):346–349.
189. Lin Y M, Rei M H. Process development for generating high purity hydrogen by using supported palladium membrane reactor as steam reformer. International Journal of Hydrogen Energy 2000;25:211–219.
190. Venkataraman R, Kunz H R, Fenton J M. CO-Tolerant, Sulfided Platinum Catalysts for PEMFCs. Journal of the Electrochemical Society 2004;151(5):A710–15
191. Lakshmanan B, Huang W, Olmeijer D, Weidner JW. Polyetheretherketone Membranes for Elevated Temperature PEMFCs. Electrochemical and Solid-State Letters 2003;6(12):A282–5.
192. Katikaneni S, Yuh C, Abens S, Farooque M. The direct carbonate fuel cell technology: advances in multi-fuel processing and internal reforming. Catalysis Today, 2002, 77, 99–106.
193. Katikaneni S P, Yuh C Y, Lee T W, Farooque M. Operation of direct carbonate fuel cell with propane. Preprints of Symposia - American Chemical Society, Division of Fuel Chemistry 2003;48: 854.
194. Bery R N. SNG from Naphtha. Chemical Engineering, 1973, 80, 90–91.
195. Wojtowicz M A, Katikaneni S, Privette R. The 2001 symposium on recent advances in fuel cells. Fuel 2002;81:2147–2149.
196. Katikaneni S, Farooque M. The direct carbonate fuel cell technology - an overview. American Institute of Chemical Engineers, Spring National Meeting 2003; 1689–1696.

197. Katikaneni S, Yuh C, Abens S, Farooque M. The direct carbonate fuel cell technology: advances in multi-fuel processing and internal reforming. Catal. Today 2002;77:99–106.
198. LPG Premium Propane Newsletter, V.2, Issue 2, Fall/Winter (1998)
199. Rostrup-Nielsen J R, Rostrup-Nielsen T. Cattech, 2002;6(4):150.
200. Larsen J H, Chorkendorff I. Surface Science Reports, 1999;35:163.
201. Twigg M V, Spencer M S. Deactivation of copper metal catalysts for methanol decomposition, methanol steam reforming and methanol synthesis. Topics in Catalysis 2003; 22(3–4):191.

6

System-level Modeling of PEM Fuel Cells

Xingjian Xue and Jiong Tang

6.1 Introduction

A proton exchange membrane fuel cell (PEMFC) is an electrochemical device that converts the chemical energy of hydrogen and oxygen, with the aid of electrocatalysts, directly into electrical energy. After four decades of research and development, this device has reached the test and demonstration phase [1]. Typically, the analysis and design of PEMFCs are centered around the membrane-electrode assembly (MEA), which involves the characterization of the physical environment of the electrochemical reaction, the transport phenomena of gas (hydrogen, oxygen, water vapor, *etc.*), liquid water, proton and current, and the relationships between the fuel cell voltage, current, temperature, material (electrode, catalyst and membrane) properties and transport parameters.

Over the years, PEMFCs have been modeled at various levels of complexity with different focuses. Early work was mainly on the modeling of MEA using mechanistic methods [2–5]. Attempts have been made to investigate the multi-species diffusion through the substrate and the diffusion layer of the electrode, the reaction kinetics in the catalyst layers, and proton and water transport through the membrane. The resultant models are generally governed by a set of complex partial differential equations. While the mechanistic models are built upon the rigorous mathematical descriptions of fundamental physics, they normally have very complicated expressions and often certain key physical parameters are difficult to quantify, which may in turn reduce the modeling effectiveness. Empirical modeling, by mapping the fuel cell voltage as a function of various contributing variables [6–8], on the other hand, has certain advantages in some applications. Although this approach does not examine in depth the underlying electrochemical process, it utilizes the phenomenological description based on benchmark experimental studies to reduce the model complexity as well as computational time. Typically, such empirical MEA sub-models are used in large fuel cell system-level models [9–12]. It should be mentioned that water plays a critical role in PEMFC operations. In order to efficiently conduct the hydrogen protons and prevent localized hot spots, the Nafion membrane and the Nafion within the catalyst layers need to be appropriately hydrated [3]. To achieve sufficient hydration, water is introduced into the cell by a variety of methods, among which the humidification of reactants by passing them through humidifiers before entering the cell is most commonly used [4, 13]. Usually, the water liquid-vapor two-phase change occurs during PEMFC operations, which is strongly affected by the fuel cell temperature. When the fuel cell temperature goes down, condensation takes place and, correspondingly, the vapor partial pressure decreases. When the fuel cell temperature goes up, evaporation will occur if liquid water exists, and the vapor partial pressure increases. The variation of the vapor partial pressure will inversely affect the partial pressures of other species and thus the species flow and fuel cell power will all be affected. The change of these parameters will affect the fuel cell operation/performance and, in turn, affect the temperature of the fuel cell. Wang *et al.* [14] explored the detailed two-phase flow, transport of reactants and products in the air cathode, and their effects on the polarization curve. Their model was further improved to include the catalyst layer of the cathode side by You and Liu [15]. In these studies, multi-phase mixture flow modeling approach was employed, while a constant fuel cell temperature was assumed. Nguyen and White [13] developed a two-dimensional water and heat management model for PEMFC, where a constant phase change rate was employed to account for the condensation and evaporation of water reaction.

Fuel cells are very complicated *dynamic* systems. The aforementioned studies, albeit extensive, essentially only concern the steady-state characteristics of PEMFCs. In order to carry out system-level optimal design and controls, there is a need to develop PEMFC dynamic models to capture the non-steady state and transient behaviors (due to, *e.g.*, sudden load change, *etc.*). Amphlett *et al.* [9] proposed a PEMFC dynamic model taking into account the fuel cell temperature variations where, however, the steady-state species flow was used and water vapor was treated as a saturated one. Pukrushpan *et al.* [10] developed a fuel cell system-level model that includes several components such as compressor, supply and return manifolds *etc.*, but the fuel cell temperature was kept as a constant. Also, the water phase was simply treated such that vapor condenses into the liquid form once the relative humidity of the gas exceeds 100%. Similar work was performed by Iqbal [11] who studied a hybrid energy system with an embedded fuel cell. Yerramalla *et al.* [16] considered the humidifier and stack pressure in the system level fuel cell modeling. Building upon these dynamics studies, Xue *et al.* [12] developed a system-level PEMFC dynamic model that considers both the fuel cell temperature variation and the species flow and their effects on cell voltage output. Nevertheless, the two-phase change effect was not incorporated in their model.

It is the goal of this study to develop a system-level dynamic model of PEMFC that incorporates the complicated temperature, gas flow, membrane hydration and inlet gas humidification effects, *etc.* The focus is on the dynamic and transient properties of the system, and the two-phase change characterization of water liquid-vapor is explicitly embedded in the model. We will use the control volume approach [17] to develop a set of complete dynamic equations that govern the system dynamics. The fuel cell system is divided into four control volumes, the anode channel, the cathode channel, the fuel cell body, and the cooling water micro channels. For each control volume, the establishment of a lumped-parameter dynamic model is realized using a combination of intrinsic mechanistic relations and empirical modeling. The system-level model is simulated under a MATLAB® environment. Numerical studies are correlated to existent benchmark results for model verification. A series of analyses are carried out to investigate the complicated dynamic interaction between fuel cell state variables. This model is capable of characterizing the transient dynamic properties of a fuel cell system, and can be readily employed in the optimization and real-time control of PEMFCs installed in practical automotive or stationary applications.

6.2 PEMFC System-level Dynamic Modeling

In this research, we develop a dynamic model of PEMFC with phase change effect by using the control volume approach with lumped parameters. As shown in Figure 6.1, hydrogen is humidified with water in the humidifier and enters the anode channel. Part of the hydrogen and water vapor diffuse into the MEA, where the hydrogen dissociates into protons and electrons at the anode catalyst layer, while the water vapor diffuses through the anode electrode into the membrane and further diffuses into the cathode side through the membrane with the aid of protons. At the cathode side, the oxygen is humidified in the cathode side humidifier and enters the cathode channel. Part of the oxygen and vapor diffuse into the MEA where the oxygen dissociates and combines with the protons and electrons to form water at the cathode catalyst layer, and the remaining water diffuses either into the cathode channel or back into the anode channel.

During this process, the vapor condenses into liquid water if oversaturated vapor exists. Meanwhile, the liquid water evaporates into vapor if the vapor is not saturated. The heat generated by the electrochemical reaction and the latent heat associated with

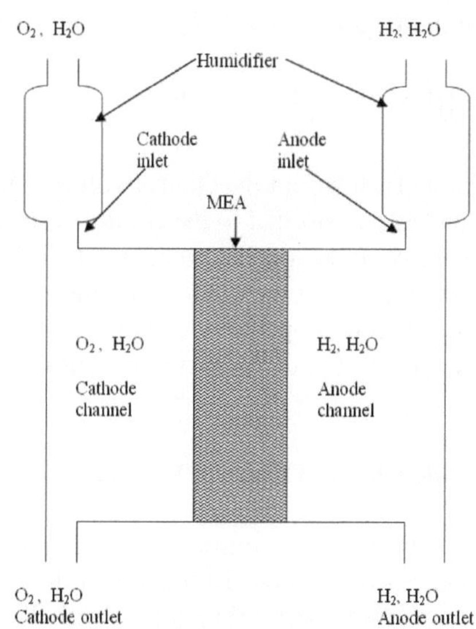

Figure 6.1. Schematic diagram of fuel cell model

condensation and evaporation will affect the whole process. In order to quantify the complicated mass and energy interactions as well as the phase change within a fuel cell, we use the control volume approach and define the following four control volumes: the anode channel, the cathode channel, the fuel cell body, which includes MEA as part of the control volume [12], and the cooling water micro channels. Throughout this chapter, we make the following assumptions:

1) The physical properties within each control volume are uniform;
2) Heat transfer by conduction in the gas phase is negligible;
3) The electrochemical reaction takes place in the MEA, and generates heat to the fuel cell body control volume. The generated water is in the form of vapor [13];
4) Water leaves the channels and enters the MEA in the form of vapor only [13];
5) The gas mixture is ideal;
6) Liquid water formed by condensation is in the form of small droplets and their volume is negligible [13]. Part of the liquid water droplets flow out of the channels with the aid of the outlet gases;
7) The electro-osmotic drag coefficient and the diffusion coefficient of the water in the membrane are estimated using the activity of the water vapor in the anode channel [4, 13].

The dynamic modeling is based upon the mass and energy conservations [12, 17, 18], which are given as follows. The detailed description of the mathematical model for each control volume will be given in the subsequent subsections.

For an arbitrary control volume, the continuity equation in integral form is given as [17],

$$\frac{\partial}{\partial t}\iiint_{\text{C.V.}}\rho dV + \oiint_{\text{C.S.}}\rho(\vec{V}\cdot\vec{n})dA = 0 \tag{6.1}$$

and the energy equation in integral form is [17]

$$\frac{\partial}{\partial t}\iiint_{\text{C.V.}}\rho e_t dV + \oiint_{\text{C.S.}}\rho h(\vec{V}\cdot\vec{n})dA = \frac{dQ}{dt} + \frac{d\omega}{dt} \tag{6.2}$$

where ρ is the material density within the control volume, $\vec{V}$ is the velocity vector of flow fluid, $\vec{n}$ is unit vector normal to the control surface toward the outside, e_t is specific internal energy, h is specific enthalpy, Q is heat added to the control volume, and ω is work done on the control volume. Note that the typical gas flow velocity is low under fuel cell operating conditions. The kinetic and potential energy of the gases is thus neglected in the energy equation.

6.2.1 Cathode and Anode Channel Control Volumes

For the control volumes corresponding to the anode and cathode channels, the first term in Equation (6.1) becomes the mass change rate of the respective species. The second term in Equation (6.1) is the mass flowrate through the surface of the control volumes. For an individual channel, three kinds of mass flowrate are involved, which correspond to the gas entering the flow channel from the inlet, the gas leaving the channel from the outlet, and the gas diffusion through the anode/cathode diffusion layers, respectively, as described in Figure 6.1,

$$\oiint_{\text{C.S.}} \rho(\vec{V}\cdot\vec{n})dA = \iint_{\text{inlet}} \rho(\vec{V}\cdot\vec{n})dA + \iint_{\text{outlet}} \rho(\vec{V}\cdot\vec{n})dA + \iint_{\text{electrode}} \rho(\vec{V}\cdot\vec{n})dA \tag{6.3}$$

The velocity profiles are required for the calculation of Equation (6.3). In general, the instant velocity profile can be obtained by solving the momentum equation simultaneously. This chapter deals with the macroscopic level dynamic behavior of the fuel cell, and the mass flow through the inlet/outlet control surfaces will be simplified as the traditional valve flowrate equation [10],

$$\iint \rho(\vec{V}\cdot\vec{n})dA = k(\Delta p) \tag{6.4}$$

where Δp denotes the pressure difference, and k is the mass flowrate coefficient that in turn depends upon temperature and the pressure difference.

The third term on the right hand side of Equation (6.3) involves a complicated process. Hydrogen diffuses through the porous gas-diffusion anode to the anode catalyst, where the electrochemical reaction takes place to form a proton and an electron; oxygen diffuses through the porous gas-diffusion cathode to the cathode catalyst, where the reduction reaction takes place to form water and heat. During the electrochemical/reduction reaction, current is generated. Although such processes at the catalyst layer are complicated and still subjected to intensive studies, they comply with the macroscopic level mass and species conservation law, *i.e.*, the consumed hydrogen and oxygen have a definitive relationship with the current generated (10),

$$\iint_{\text{electrode}} \rho(\vec{V}\cdot\vec{n})dA = N\frac{i}{nF}M \tag{6.5}$$

where i denotes the current, N is the number of cells in a stack, F , M and n are Faraday constant, gas mole mass and constant corresponding to different gas species, respectively.

The general energy equation (Equation (6.2)) can be applied to the anode and cathode channels in a similar manner. The first term on the left hand side of Equation (6.2) represents the internal energy change rate within the control volume. The second term in equation (6.2) is the heat transfer rate due to the mass transport through the control surface. Such mass transport involves the hydrogen in the anode channel that diffuses through the porous anode diffusion layer to the MEA (part of the cell body) and the oxygen in the cathode channel that diffuses through the porous cathode diffusion layer to the cell body. On the right hand side of Equation (6.2), the first term $\frac{dQ}{dt}$ represents the rate of heat added to the control volume, which is generally caused by convection heat transfer or possibly radiation between the cell body and the channels and between the cell body and the surrounding ambience. As the temperature of a PEMFC body is relatively moderate, the radiation effect can be neglected. In this study the viscous normal stresses at the inlet, outlet and electrode interface of mass diffusion are neglected, and the work done on the control volume ($\frac{d\omega}{dt}$) disappears.

6.2.1.1 Cathode Channel

The model of the cathode control volume describes the cathode gas flow, temperature, and phase change behaviors. Using the aforementioned mass and energy conservation principles, we can obtain the following governing equations.

Mass conservation

We first employ the mass conservation principle on the cathode control volume, which yields,

$$\frac{dm_{O_2}}{dt} = W_{O_2,in} - W_{O_2,out} - M_{O_2}\frac{NI}{4F} \tag{6.6}$$

$$\frac{dm_{ca,v}}{dt} = W_{ca,v,in} - W_{ca,v,out} + M_v\frac{NI}{2F} + W_{v,mem} + \{\mathrm{sgn}\}_p W_{ca,v,phase} \tag{6.7}$$

$$\frac{dm_{ca,w}}{dt} = -\{\mathrm{sgn}\}_p W_{ca,v,phase} - W_{ca,w,out} \tag{6.8}$$

where $W_{O_2,in}$, $W_{O_2,out}$, $W_{ca,v,in}$, $W_{ca,v,out}$, $W_{v,mem}$, and $W_{ca,w,out}$ are the inlet and outlet species flowrates and the water transport rates across the membrane. $W_{ca,v,phase}$ refers to the water phase change rate, and will be explained later in detail. The sign function $\{\mathrm{sgn}\}_p$ is used in characterizing the direction of phase change, (+1) indicating evaporation and (-1) indicating condensation. Equation (6.6) shows the inlet, outlet, and consumed oxygen (due to the electrochemical reaction) with respect to the oxygen mass conservation within the cathode control volume. Equation (6.7) indicates that the change of vapor mass in the cathode channel is caused by the following factors: inflow vapor through the inlet, outflow vapor through the outlet, vapor produced by the electrochemical reaction, vapor transported across the MEA that includes the combined effects of electro-osmotic drag force and back diffusion, and phase change contribution. The liquid water balance, as shown in Equation (6.8), involves the phase change rate and the rate of outflow liquid (dragged by the gas phase species). The species mass conservation equations are actually temperature dependent, which is reflected in the energy conservation equations that follow.

Energy conservation

Using the energy conservation principle on the cathode channel control volume, we obtain,

$$\begin{aligned}\frac{d}{dt}[(\sum_i m_i c_i)T_{ca}] &= \sum_j W_{j,in}T_{s,ca} - \sum_j W_{j,out}T_{ca} - M_{O_2}\frac{NI}{4F}c_{p,O_2}T_{ca} + M_v\frac{NI}{2F}c_{p,v}T_{body} \\ &+ (hA)_{ca}(T_{body} - T_{ca}) + W_{v,mem}c_{p,v}[\{\mathrm{sgn}\}_d T_{body} + (1-\{\mathrm{sgn}\}_d)T_{ca}] \\ &+ \{\mathrm{sgn}\}_p W_{ca,v,phase}(-h_{latent})\end{aligned} \tag{6.9}$$

Here, the summation $\sum_i$ involves the oxygen, vapor and liquid water. The summation $\sum_j$ generally involves the oxygen and vapor only, but for the outlet, it may also involve liquid water. In Equation (6.9), the sign function $\{\text{sgn}\}_d$ is used to indicate the vapor transport direction in MEA, $\{\text{sgn}\}_d = +1$ if vapor is transported from the anode channel to the cathode channel, and $\{\text{sgn}\}_d = 0$ otherwise. $(hA)_{ca}$ represents the convective heat transfer coefficient between the cathode channel control volume and the fuel cell body. The gas flow in the cathode channel could be either pure oxygen or air. In this study we assume pure oxygen is used. However, the case of air can be similarly derived by additionally considering the mass conservation of nitrogen and its contribution to the energy balance.

In typical PEMFC operations, before entering the cathode channel the oxygen is humidified in the cathode side humidifier. Assume that the temperature of the cathode side humidifier is T_s^{ca}, and the oxygen partial pressure is P_s^{ca}. Given the relative humidity in the humidifier as ϕ_{ca}, we may obtain the vapor partial pressure in the humidifier as [10, 18],

$$P_{hum,ca,v} = \phi_{ca} P_{hum,ca,sat}^{T_s^{ca}} \tag{6.10}$$

where $P_{hum,ca,sat}^{T_s^{ca}}$ is the water vapor saturation pressure at temperature T_s^{ca}. The humidity ratio of the oxygen water vapor mixture in the cathode side humidifier can then be written as [10, 18],

$$\beta_{ca,hum} = \frac{M_v}{M_{o_2}} \frac{\phi_{ca} P_{hum,ca,sat}^{T_s^{ca}}}{P_s^{ca}} \tag{6.11}$$

Similarly, the humidity ratio of oxygen water vapor mixture in the cathode channel is [10, 18]

$$\beta_{ca,channel} = \frac{M_v}{M_{o_2}} \frac{P_{ca,v}}{P_s^{ca}} \tag{6.12}$$

where $P_{ca,v}$ is the water vapor partial pressure in the cathode channel.

The inlet and outlet gas flowrates between the adjacent control volumes will be calculated using the valve flowrate equation [6.10]. Since the pressure difference in our analysis is generally small, a linear flowrate equation is assumed. The total inlet and outlet flowrates of the cathode channel can be written as, respectively,

$$W_{ca,in} = K_{ca,in}(P_s^{ca} + P_{hum,ca,v} - P_{ca,o_2} - P_{ca,v}) \tag{6.13}$$

$$W_{ca,out} = K_{ca,out}(P_{ca,o_2} + P_{ca,v} - P_{atm}) \tag{6.14}$$

where $K_{ca,in}$ and $K_{ca,out}$ are the inlet and outlet flowrate coefficients, and P_{atm} is the ambient pressure. Here we assume the gas emission of the cathode channel directly flows into the surrounding ambient.

By virtue of the humidity ratios given in Equations (6.11) and (6.12) and the total flowrates shown in Equations (6.13) and (6.14), we may obtain the respective species flowrates involved in Equations (6.6)-(6.9) as follows,

$$W_{O_2,in} = \frac{1}{1+\beta_{ca,hum}} W_{ca,in} \tag{6.15}$$

$$W_{O_2,out} = \frac{1}{1+\beta_{ca,channel}} W_{ca,out} \tag{6.16}$$

$$W_{ca,v,in} = \frac{\beta_{ca,hum}}{1+\beta_{ca,hum}} W_{ca,in} \tag{6.17}$$

$$W_{ca,v,out} = \frac{\beta_{ca,channel}}{1+\beta_{ca,channel}} W_{ca,out} \tag{6.18}$$

The volume of liquid water in the cathode channel is assumed to be negligible. To account for the flowrate of outlet liquid water dragged by the outlet gases, we define the following ratio,

$$\beta_{ca,w} = \frac{m_{ca,w}}{m_{ca,v} + m_{O_2}} \tag{6.19}$$

The definition of (6.19) is similar to that of the humidity ratio and is called the pseudo-humidity ratio in this chapter. Based on Equation (6.19), the liquid water flowrate at the cathode outlet is proportional to the mass ratio between liquid water and total species in cathode channel, which is a reasonable approximation. The liquid water outlet flowrate $W_{ca,w,out}$ can then be expressed as,

$$W_{ca,w,out} = \frac{\beta_{ca,w}}{1+\beta_{ca,w}} W_{ca,out} \tag{6.20}$$

Dynamic two-phase change

The water liquid-vapor two-phase change is a salient feature of PEMFC. In principle, if the vapor partial pressure is larger than the in-situ saturation vapor pressure, liquid water will be formed through condensation. On the other hand, if the vapor partial pressure is less than the corresponding in-situ saturation pressure, the existent liquid water will evaporate until the vapor partial pressure reaches the saturation pressure. In microscopic level modeling, a multiple-phase mixture based two-phase flow model is generally employed, where the phases are assumed to be distinct and separable components with non-zero interfacial areas, and the phase change is characterized by smoothly varying phase compositions in their mixture [14, 15]. Nguyen and White [13] defined a steady-state condensation/evaporation rate using the homogeneous rate constant. In the system-level dynamic modeling of PEMFCs, however, the phase-change effect has only been modeled under simplified assumptions. Amphlett *et al.* [9] provided the phase-change calculation based on standard thermodynamic principles under steady-state flow and saturation vapor assumptions. Pukrushpan *et al.* [10] developed a constant temperature model in which it was assumed that the vapor condenses into the liquid form once the relative humidity of the gas exceeds 100%. Condensation/evaporation is a very complex process, and the rate of phase change is dependent on several factors including temperature, vapor partial pressure, and total liquid water surface exposed to the cathode channel, *etc.* One fundamental principle governing this process is that the vapor partial pressure within the channel is prone to becoming saturated in a limit period of time regardless of the saturation level, provided that enough liquid water exists. We now assume that the vapor partial pressure is governed by a first order dynamic equation and the time parameter is $\tau(T_{ca}, P_{ca,v}, \alpha_{ca})$. The time parameter τ might depend on the cathode channel temperature T_{ca}, vapor partial pressure $P_{ca,v}$, and a surface configuration related parameter α_{ca}. We then obtain,

$$\tau(T_{ca}, P_{ca,v}, \alpha_{ca})\frac{dP_{ca,v}}{dt} = P^{T_{ca}}_{ca,v,sat} - P_{ca,v} \tag{6.21}$$

where $P^{T_{ca}}_{ca,v,sat}$ is the saturation vapor pressure at the cathode channel at temperature T_{ca}. The expression for the saturation vapor pressure as a function of temperature (^{0}C) is provided in [4, 13]. Here we give a different curve-fitting result for temperature in K,

$$P^{T}_{v,sat} = 8.09 \times 10^{-4} \exp(5.06 \times 10^{-2} T) \tag{6.22}$$

Taking the time derivative of the ideal gas state equation of vapor, $P_{ca,v}V_{ca} = m_{ca,v,phase}R_vT_{ca}$, we obtain,

$$\frac{dP_{ca,v}}{dt}V_{ca} = \frac{dm_{ca,v,phase}}{dt}R_vT_{ca} + m_{ca,v,phase}R_v\frac{dT_{ca}}{dt} \tag{6.23}$$

One can see from Equation (6.23) that two factors contribute to the vapor partial pressure changes within the cathode channel control volume: the phase change and the temperature change. Since analyses have shown that the temperature change rate is generally small, in what follows we neglect the temperature effect in Equation (6.23) and obtain

$$\frac{dm_{ca,v,phase}}{dt} = \frac{V_{ca}}{R_vT_{ca}}\frac{dP_{ca}}{dt} \tag{6.24}$$

Combining Equations (6.21) and (6.24), we have the phase change rate,

$$W_{ca,v,phase} = \frac{dm_{ca,v,phase}}{dt} = \frac{V_{ca}}{R_vT_{ca}}\frac{1}{\tau(T_{ca}, P_{ca,v}, \alpha_{ca})}(P^{T_{ca}}_{ca,v,sat} - P_{ca,v}) \tag{6.25}$$

Equation (6.25) indicates that the water (liquid or vapor) in the cathode channel either condenses when the vapor is over-saturated or evaporates when the vapor is not saturated. In the latter case, we assume enough water liquid exists.

6.2.1.2 Anode Channel

Similar to those obtained for the cathode channel control volume, the governing equations of the anode channel control volume can be derived by using mass/energy conservation principles and phase-change relations. Here we omit the detailed derivation, and only present the relevant equations.

Mass conservation

$$\frac{dm_{H_2}}{dt} = W_{H_2,in} - W_{H_2,out} - M_{H_2}\frac{NI}{2F} \tag{6.26}$$

$$\frac{dm_{an,v}}{dt} = W_{an,v,in} - W_{an,v,out} - W_{v,mem} + \{\mathrm{sgn}\}_p W_{an,v,phase} \tag{6.27}$$

$$\frac{dm_{an,w}}{dt} = -\{sign\}_p W_{an,v,phase} - W_{an,w,out} \tag{6.28}$$

Energy conservation

$$\begin{aligned}\frac{d}{dt}[(\sum_i m_i c_i)T_{an}] = {} & \sum_j W_{j,in}T_{s,an} - \sum_j W_{j,out}T_{an} - M_{H_2}\frac{NI}{2F}c_{p,H_2}T_{an} + \{\mathrm{sgn}\}_p W_{an,v,phase}(-h_{latent}) \\ & + (hA)_{an}(T_{body} - T_{an}) - W_{v,mem}c_{p,v}[\{\mathrm{sgn}\}_d T_{an} + (\{\mathrm{sgn}\}_d - 1)T_{body}]\end{aligned} \tag{6.29}$$

$$P_{hum,an,v} = \phi_{an} P^{T_s^{an}}_{hum,an,sat} \tag{6.30}$$

$$\beta_{an,hum} = \frac{M_v}{M_{H_2}}\frac{\phi_{an} P^{T_s^{an}}_{hum,an,sat}}{P_s^{an}} \tag{6.31}$$

$$\beta_{an,channel} = \frac{M_v}{M_{H_2}}\frac{P_{an,v}}{P_s^{an}} \tag{6.32}$$

$$W_{an,in} = K_{an,in}(P_s^{an} + P_{hum,an,v} - P_{an,H_2} - P_{an,v}) \tag{6.33}$$

$$W_{an,out} = K_{an,out}(P_{an,H_2} + P_{an,v} - P_{atm}) \tag{6.34}$$

$$W_{H_2,in} = \frac{1}{1+\beta_{an,hum}}W_{an,in} \tag{6.35}$$

$$W_{H_2,out} = \frac{1}{1+\beta_{an,channel}}W_{an,out} \tag{6.36}$$

$$W_{an,v,in} = \frac{\beta_{an,hum}}{1+\beta_{an,hum}}W_{an,in} \tag{6.37}$$

$$W_{an,v,out} = \frac{\beta_{an,channel}}{1+\beta_{an,channel}}W_{an,out} \tag{6.38}$$

$$\beta_{an,w} = \frac{m_{an,w}}{m_{an,v} + m_{H_2}} \tag{6.39}$$

$$W_{an,w,out} = \frac{\beta_{an,w}}{1+\beta_{an,w}} W_{an,out} \tag{6.40}$$

Dynamic two-phase change

$$W_{an,v,phase} = \frac{dm_{an,v,phase}}{dt} = \frac{V_{an}}{R_v T_{an}} \frac{1}{\tau(T_{an}, P_{an,v}, \alpha_{an})} (P^{T_{an}}_{an,v,sat} - P_{an,v}) \tag{6.41}$$

6.2.2 Fuel Cell Body

While the energy conservation of the fuel cell body control volume is strongly coupled with another three control volumes (anode channel, cathode channel, micro channel of cooling water) and surrounding ambient, the species conservation is mainly involved in MEA, where the species diffusion and electrochemical reaction take place. As the fuel cell body is treated as a solid control volume, the mass conservation is trivial. However, the hydration condition of the membrane plays a critical role in gas diffusions and proton transportation and, consequently, has a significant effect on the electrochemical reaction and fuel cell performance. In the present study, membrane hydration will be incorporated into the system model and elaborated in what follows.

6.2.2.1 Membrane Hydration

It is well known that the Nafion membrane needs to be appropriately hydrated in order to efficiently conduct the hydrogen protons and prevent the localized hot spots. To achieve enough hydration, water is introduced into the cell by humidifying the inlet fuel/gas. The water transport across the membrane is generally recognized as the combined effects of two distinct mechanisms [4, 5, 10, 13, 19]. One is that the water molecules are dragged from the anode side to the cathode side by the hydrogen proton transport drag force which is referred to as the electro-osmotic drag force. Another mechanism is the back-diffusion from the cathode side to the anode side due to the water content difference between these two channels. It should be mentioned that the membrane hydration model has been well developed and simulated in such as [10] and [19]. In what follows we present the same equations and parameters given in [10] and [19], so the current modeling equations can be complete and self-contained. The mole flux of the water transported through the membrane can be written as [13, 19],

$$J_w = -D_w \frac{\partial C}{\partial x} + n_d \frac{I}{F} \tag{6.42}$$

where n_d is the electro-osmotic drag coefficient, D_w the membrane diffusion coefficient, and C the water concentration in the membrane. The transport direction from the anode side to the cathode side is designated as the positive direction in the above equation.

The coefficients n_d and D_w are nonlinearly related to the membrane water content [4, 5, 13, 19]. The water content in the membrane can be estimated using the water activity a_a at the anode side [2, 3, 8, 11],

$$\lambda = \begin{cases} 0.043 + 17.81a_a - 39.85a_a^2 + 36.0a_a^3 & 0 < a_a \le 1 \\ 14 + 1.4(a_a - 1) & 1 < a_a \le 3 \end{cases} \tag{6.43}$$

The electro-osmotic drag coefficient is calculated as [10, 13, 19],

$$n_d = 0.0029\lambda^2 + 0.05\lambda - 3.4 \times 10^{-19} \tag{6.44}$$

Using the membrane water content estimation given in Equation (6.43), we may obtain the water diffusion coefficient as [4, 10, 13, 19],

$$D_w = D_\lambda \exp(2416(\frac{1}{303} - \frac{1}{T_{body}})) \tag{6.45}$$

where

$$D_\lambda = \begin{cases} 10^{-10} & \lambda < 2 \\ 10^{-10}(1 + 2(\lambda - 2)) & 2 \le \lambda \le 3 \\ 10^{-10}(3 - 1.67(\lambda - 3)) & 3 < \lambda < 4.5 \\ 1.25 \times 10^{-10} & \lambda \ge 4.5 \end{cases} \tag{6.46}$$

The water activity in the anode/cathode channels is defined as the ratio of the actual vapor partial pressure and the saturation vapor pressure under the same conditions [4, 10, 13, 19],

$$a_i = \frac{P_{i,v}}{P_{i,v,sat}}, \quad i \in [an, ca] \tag{6.47}$$

In order to calculate the water flux across the membrane in Equation (6.42), we need to have the water concentration boundary conditions at the anode and cathode surfaces of the MEA, which can be estimated as [13, 19],

$$c_i = \begin{cases} \frac{\rho_{m,dry}}{M_{m,dry}}[0.043 + 17.81a_i - 39.85a_i^2 + 36.0a_i^3] & 0 < a_a \le 1 \\ \frac{\rho_{m,dry}}{M_{m,dry}}[14 + 1.4(a_i - 1)] & 1 < a_i \end{cases} \tag{6.48}$$

where $i \in [an, ca]$, and $\rho_{m,dry}$ and $M_{m,dry}$ are the dry membrane density and the dry membrane mole mass, respectively.

The water concentration in the membrane is generally assumed to change linearly over the membrane thickness. The water transport rate through the membrane is calculated as [10, 19]

$$W_{v,mem} = M_v A_{mem} N(n_d \frac{I}{F} - D_w \frac{(c_{ca} - c_{an})}{t_m}) \tag{6.49}$$

where M_v is the vapor mole mass, A_{mem} the membrane effective area, and t_m the thickness of the membrane.

6.2.2.2 *Fuel cell body temperature*

The fuel cell body is treated as a solid control volume. The mass of the fuel cell body is assumed to be a constant, and thus the mass conservation is trivial. The only concerned dynamic behavior of the fuel cell body is the temperature. Using the energy conservation, we have

$$\begin{aligned} m_{body} c_{body} \frac{dT_{body}}{dt} &= M_{H_2} \frac{NI}{2F} c_{p,H_2} T_{an} + M_{O_2} \frac{NI}{4F} c_{p,O_2} T_{ca} - M_{H_2O} \frac{NI}{2F} c_{p,H_2O} T_{body} \\ &+ W_{v,mem} c_{p,H_2O} [\{\text{sgn}\}_{mem} T_{an} - T_{body} + (1 - \{\text{sgn}\}_{mem}) T_{ca}] \\ &+ NI^2 R_{ohm} + \Delta H (M_{H_2} \frac{NI}{2F}) - NV_{cell} I \\ &+ (hA)_{an} (T_{an} - T_{body}) + (hA)_{ca} (T_{ca} - T_{body}) + (hA)_{\infty} (T_{\infty} - T_{body}) \\ &+ (hA)_{cw} (T_{cw} - T_{body}) \end{aligned} \tag{6.50}$$

where m_{body} and c_{body} are the fuel cell body mass and mixture specific heat, respectively. The right-hand side of Equation (6.50) shows all the contributing factors to the fuel cell temperature dynamics. The first three terms are related to the heat transfer due to species transport, which includes fuel diffusion from the anode channel to MEA, oxygen diffusion from the cathode channel to MEA, and vapor produced by the electrochemical reaction transported from MEA to the cathode channel. Depending on the direction of the water transport within the membrane, the channel temperature T_{an} and T_{ca} might affect the fourth term. If the water is transported from the anode side to the cathode side, *i.e.* $\{\text{sgn}\}_{an} = 1$, the anode channel temperature has a contribution to the fuel cell temperature change. Otherwise, $\{\text{sgn}\}_{an} = 0$, and it is the cathode channel temperature T_{ca} that will affect the fuel cell temperature. The fifth to seventh terms show the energy conversion effect on T_{body}. The fifth term is the heat produced by protons transported through the membrane. The sixth term is the total electrochemical energy in the fuel hydrogen, part of which is converted into electrical energy (the seventh term) while the rest is converted into heat energy. The last four terms represent the convective heat energy transferred into the fuel cell body from various components including the anode channel, cathode channel, ambient air, and cooling water, respectively.

6.2.3 Cooling Water

In order to control the fuel cell temperature within a certain range (normally less than 100^{0}C), two control strategies might be used: the air-cooling using fans and the water-cooling through the micro cooling channels [9]. To include this effect, here in this research we incorporate the water cooling mechanism into the PEMFC model. Since the liquid cooling water is incompressible, the mass conservation simply leads to that the inlet flowrate is equal to the outlet flowrate. The heat transfer between the fuel cell body and the cooling water is a convection process, and the energy balance equation for the cooling water is given as,

$$m_{cw}c_{p,cw}\frac{dT_{cw}}{dt}=W_{cw}c_{p,cw}T_{cw,in}-W_{cw}c_{p,cw}T_{cw}+(hA)_{cw}(T_{body}-T_{cw}) \tag{6.51}$$

where m_{cw} is the water mass in the microcooling channel, $c_{p,cw}$ is its specific heat, and W_{cw} is the flowrate. At the right-hand side of Equation (6.51), the first two terms indicate the heat transfer effects due to the mass (cooling water) transport, while the third term is the heat convection between the cooling water and the fuel cell body.

6.2.4 Electrochemical Reaction

In previous studies, empirical modeling has been developed for the MEA, which is built upon a phenomenological relationship between the fuel cell voltage and various contributing variables [6–8]. This approach has been employed in various system-level fuel cell modeling [9–12]. Typically, the cell voltage is assumed to depend upon state variables including the partial pressure of hydrogen and oxygen, the temperature of the cell body, and load current *etc.*, *i.e.*

$$V_{\text{cell}}=f(P_{\text{H}_2},P_{\text{O}_2},T_{\text{s}},i,\alpha) \tag{6.52}$$

where P_{H_2} is hydrogen partial pressure, P_{O_2} is oxygen partial pressure, T_{s} denotes the body temperature, i is load current, and α represents other possibly related parameters. In this study we use the same approach to perform the electrochemical modeling, which is based upon a curve-fitting method [12]. It is worth noting that, in classical empirical MEA modeling [6–8], the effective surface partial pressures are used in Equation (6.13). In this study, in order to simplify the empirical MEA modeling for large fuel cell systems, we use the hydrogen/oxygen partial pressures within the corresponding channels to obtain the empirical MEA model. While such direct utilization of hydrogen/oxygen partial pressures within the channels could lead to error in thermodynamic potential, such an error can readily be compensated for in the activation overvoltage calculation.

Generally, fuel cell voltage is the summation of three effects, the thermodynamic potential E_{Nernst}, the activation overvoltage η_{act} resultant from anode overvoltage and cathode overvoltage, and the ohmic overvoltage η_{ohmic}. For hydrogen/oxygen fuel cells the thermodynamic potential is [20],

$$E_{\text{Nernst}}=1.229-(8.5\times10^{-4})(T-298.15)+(4.308\times10^{-5})T[\ln(P_{\text{H}_2})+0.5\ln(P_{\text{O}_2})] \tag{6.53}$$

Here we will use the channel hydrogen/oxygen partial pressure instead of the effective partial pressure used in [12]. The activation overvoltage can be written as

$$\eta_{\text{act}} = \xi_1 + \xi_2 T + \xi_3 T \ln(i) + \xi_4 T \ln(c_{\text{O}_2}) \tag{6.54}$$

The parameters ξ_i in Equation (6.15) can be obtained from experimental data using a linear regression technique. While various empirical expressions or activation overvoltage exist, one can always use curve fitting to obtain a proper mathematical model [6, 8, 12, 20].

The ohmic overvoltage is caused by the electron flow resistance, proton flow resistance, and other contact resistances. One model being used is based upon the linear regression of experimental data under various cell temperature and load current [6, 8, 9]. Another model takes into account that the membrane conductivity is strongly dependent on the membrane water content and fuel cell temperature [4, 5, 13, 19],

$$\sigma_m = (0.00514\lambda - 0.00326)\exp(1268(\frac{1}{303} - \frac{1}{T_{body}}))\times 10^2 \tag{6.55}$$

The ohmic resistance of the membrane is proportional to the membrane thickness and inversely proportional to the membrane conductivity [4, 13, 10],

$$\eta_{ohmic} = -I\frac{t_m}{\sigma_m A_{mem}} \tag{6.56}$$

6.3 Model Validation and Analyses

The PEMFC dynamic model consists of Equations (6.6)-(6.20) and (6.25)-(6.56). The system is highly nonlinear. In what follows, we validate this model using experimental data from the open literature [9], and carry out analyses to explore the interactive dynamic behaviors. All simulations are performed under the MATLAB® environment. As some parameters such as channel volumes are small, the differential equation solver ODE15s is employed.

6.3.1 Validation with Respect to Experimental Data and Comparison

The PEMFC model presented in this chapter is first validated using the experimental data in [9]. Here the inlet and outlet flowrate coefficients of the anode and cathode channels are adjusted so that the inlet fuel/gas flowrates are consistent with the experimental setup. The temperatures, pressures and convective heat transfer coefficients are taken to be the same as given in [9]. The stack has 35 cells. For the electrochemical submodel, the empirical formulation of the total activation overvoltage is used [8, 9]. Since some of the physical parameters used in the experiment are not directly available, the values of these parameters are assumed from other literature. The parameters associated with the activation overvoltage are adjusted so that the fuel cell output voltage is the same as the experimental result at the same temperature. All the physical parameters used in this simulation are listed in Table 6.1. The comparison between the model-predicted results and the experimental results is given in Table 6.2. Clearly, the predicted outlet gas flowrates (at both anode and cathode outlets) are virtually the same as the experimental results. The predicted temperatures of the four control volumes and the fuel cell voltage and power output also match very well with their corresponding experimental values.

Table 6.1. Parameters used in the model validation and simulation

Parameters	Value	Reference	Parameters	Value	Reference
$K_{an,in}$	$1.23\cdot10^{-10}$ kg/(Pa·s)	a	$K_{an,out}$	2.10–5 kg/(Pa.s)	a
P_s^{an}	35psig	[9]	P_{atm}	$1.013\cdot10^5$ Pa	[9]
$T_{an,in}$	23.5°C	[9]	V_{an}	0.005 m^3	[10]
φ_{an}	0.9	b	$K_{ca,in}$	$9.148\cdot10^{-10}$ kg/(Pa·s)	a
$K_{ca,out}$	$2\cdot10^{-5}$ kg/(Pa·s)	a	P_s^{ca}	35psig	[9]
$T_{ca,in}$	23.5°C	[9]	V_{ca}	0.01 m^3	[10]
φ_{an}	0.0	b	MC	35000 J/K	[9]
W_{cw}	0.0331568 kg/s	[9]	$T_{cw,in}$	23.5°C	[9]
m_{cw}	2.5 kg	b	N	35	[9]
$(hA)_{an}$	2W/K	[9]	$(hA)_{ca}$	10W/K	[9]
$(hA)_{\infty}$	17W/K	[9]	$(hA)_{cw}$	50W/K	[9]
T_{∞}	23.5°C	[9]	I	20A	[9]
t_m	$1.275\cdot10^{-4}$ m	[10]	A_{cell}	0.028 m^2	[10]
$P_{m,dry}$	0.002 kg/cm^3	[13]	$M_{m,dry}$	1.1 kg/mol	[13]
T	30s	b	ζ_1	-0.944	[9]
ζ_2	$3.54\cdot10^{-3}$	[9]	ζ_3	$4.68\cdot10^{-4}$	c
ζ_4	$-1.96\cdot10^{-4}$	[9]			

[a] Coefficients are so chosen that the fuel/gas inlet flowrates are the same as those of the experimental setup.
[b] Parameters are assumed.
[c] Parameter is adjusted to fit the experimental result.

Table 6.2. Comparison between model predictions and experimental results [9]

Stream	Parameter	Predicted value	Experimental value
Inlet anode gas	$W_{H_2,in}$	0.0078958 mol/s	0.0078 mol/s
Outlet anode gas	$W_{H_2,out}$	0.004162 mol/s	0.0041 mol/s
	T_{an}	30.1°C	25.3°C
Inlet cathode gas	$W_{O_2,in}$	0.004 mol/s	0.004 mol/s
Outlet cathode gas	$W_{O_2,out}$	0.002189 mol/s	0.002 mol/s
	T_{ca}	40.7°C	38.8°C
Outlet water	T_{cw}	25.4°C	23.9°C
Other data	T_{body}	30.9°C	38.0°C
	V_{cell}	0.833 V	0.817 V
	V_{stack}	29.2 V	28.6 V
	P_{stack}	584 W	572 W

The model developed by Amphlett *et al.* [9] is in fact a special case of that developed in [12]. Xue *et al.* gave a model validation in [12] under the same experimental operation conditions. In both [9] and [12], however, the water vapor was simply treated as saturated in a steady-state manner. The current study improves the dynamic modeling presented in [12] by incorporating systematically the two-phase change process between liquid water and vapor involved in PEMFC operations. To highlight the significant effects of the dynamic two-phase change, here we compare the simulation results between the model given in [12] and the present model under the same parametric setup. We apply a step change of load current to the system, *i.e.* the external load current changes from 0 A to 20 A at the 200th second, then decreases from 20A to 10A at the 5000th second. It is worth noting that a constant time parameter is assumed for both condensation and evaporation procedures (shown in Table 6.1). More experimental studies are needed to determine this time parameter precisely. The simulation results are illustrated in Figure 6.2 (without phase-change model) and Figure 6.3 (with phase-change model), where the temperature responses of the stack body, cathode channel, anode channel, and cooling water are plotted. Clearly, after taking into account the phase-change effect, the temperature distribution within the fuel cell system becomes very different. When the phase change is not considered, the heat generated is only due to the electrochemical reaction in the cathode catalyst layer, which, based upon the control volume definition, will increase the fuel cell body temperature. After we incorporate the phase change effect, the condensation of water vapor in the cathode channel takes place under the current parametric setup, and the latent heat released from the phase change increases the gas temperature of the cathode channel. The combined effect of latent heat due to phase change in the cathode channel and the heat generation due to the electrochemical reaction in the cathode catalyst

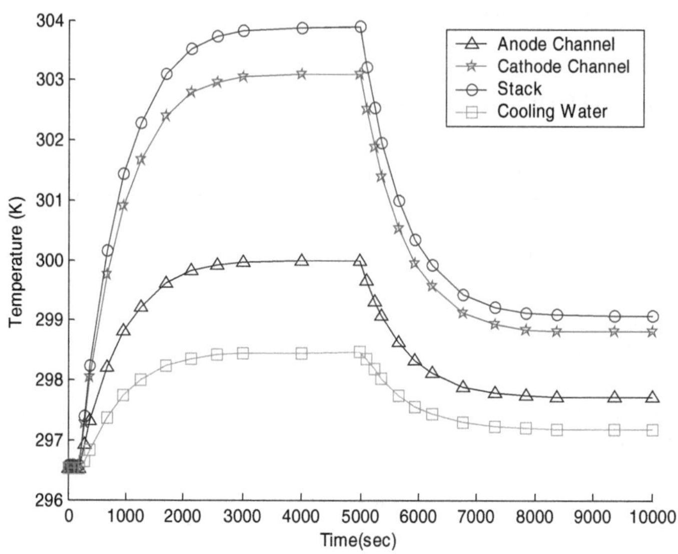

Figure 6.2. Temperature histories of fuel cell components without phase change

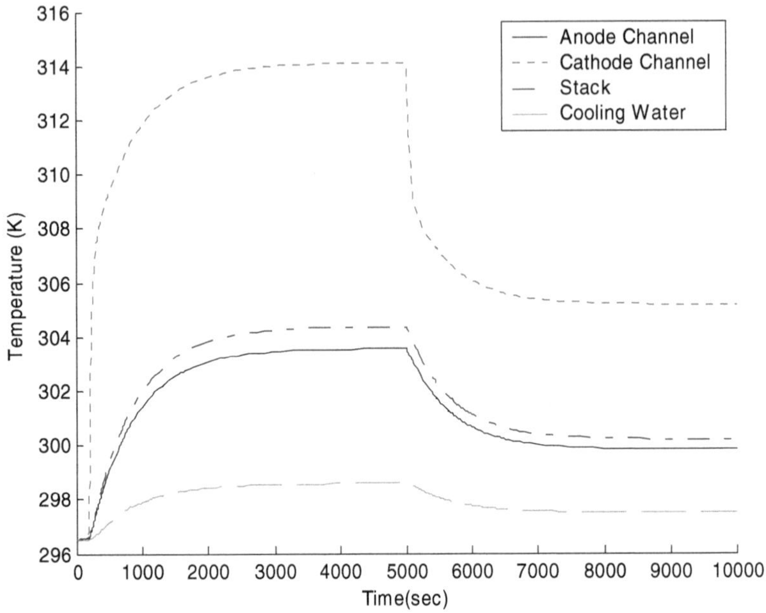

Figure 6.3. Temperature histories of fuel cell components with phase change

layer leads to the temperature distribution shown in Figure 6.3. Clearly, the cathode channel now has the highest temperature, which is different from that shown in Figure 6.2 where the fuel cell body has the highest temperature. The difference between Figure 6.2 and Figure 6.3 demonstrates the necessity of including phase-change in the dynamic modeling.

6.3.2 System-level Dynamic Analyses

In order to thoroughly study the complicated dynamic and especially transient behaviors within the fuel cell stack, a series of step changes in load current (Figure 6.4), which simulates the procedure of start up and shut down, are applied to the fuel cell system. Here we again use the initial conditions and system parameters listed in Table 6.1. The time length of the simulation is 20 000 seconds. The simulation results are illustrated in Figure 6.4-10.

Figure 6.4 shows the stack output voltage change due to the load current change. At the 1000th second, the load current rises from 0 to 20 A and, correspondingly, the stack voltage drops from 39 V to 29 V. The increase of the load current from 20 to 40 A at the 6000th second leads to further decrease of the stack voltage from 29 to 25 V. At the 12000th second, the external load shuts down, and the stack voltage goes up from 25 V to the original voltage 39 V. One interesting observation is that, while the respective load current remains constant at each stage, the voltage keeps changing. The varying stack voltage is due to the time-varying stack temperature, species partial pressure in the channels, and membrane humidity conditions, *etc.*

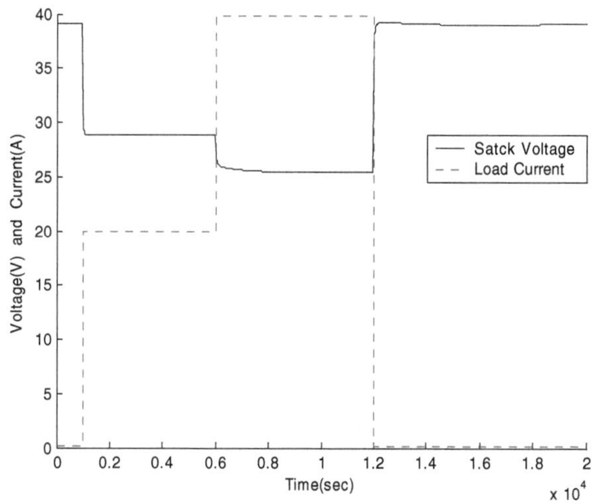

Figure 6.4. Stack output voltage and load current history

Figure 6.5 illustrates the temperature histories of the stack, anode/cathode channels and the cooling water. After the load current rises from 0A to 20 A at the 1000th second, the temperatures of the stack system need at least half an hour to reach the steady-state. While all the temperature histories of the four fuel cell components exhibit similar trends (Figure 6.5), the temperatures of the cathode channel and stack body are consistently higher than those of the anode channel and cooling water. Such a temperature difference comes from two effects: the electrochemical reaction at MEA that belongs to the stack body control volume, and the water

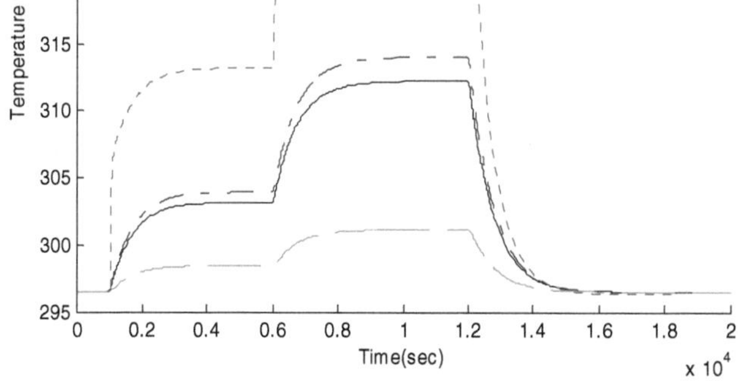

Figure 6.5. Temperature histories of fuel cell components with phase change

phase change in the cathode channel. One will see in what follows that the water phase change (mainly condensation) will take place in the cathode channel. The latent heat generated by the vapor condensation in the cathode channel and the heat generated by the electrochemical reaction both contribute to the temperature difference significantly. The temperature difference becomes even larger when the load current rises from 20 to 40 A. At the 12000th second, the stack load current goes down from 40 to 0 A, and the four temperatures decrease slowly and finally approach to the same value.

Figure 6.6 shows the species partial pressure histories in the anode and cathode channels. In the anode channel, we are concerned with the hydrogen partial pressure, vapor partial pressure, and the in-situ saturation vapor partial pressure that the anode channel can reach if liquid water exists. The hydrogen partial pressure shows a notable change at the 6000th second, which corresponds to the load current change from 20 to 40 A. After the load current rises to 40 A, the consumed hydrogen due to the electrochemical reaction increases, leading to the decrease of the hydrogen partial pressure in the anode channel. Once the load current shuts down at the 12 000th second, the hydrogen pressure goes up again. The vapor partial pressure in the anode channel is relatively low as compared to the hydrogen partial pressure, and it never surpasses the in-situ saturation vapor pressure. Therefore, no liquid water is generated in the anode channel even though the inlet hydrogen is humidified with relative humidity 90%. On the other hand, the species partial pressures in the cathode channel show very different trends from those in the anode channel. At each instant of load current change, *i.e.*, the 1000th second, the 6000th second, and the 12 000th second, the increase of the load current is followed by the decrease of the oxygen pressure, and vice versa. The oxygen partial pressure and the vapor partial pressure time histories have similar profiles. Clearly, this is because the consumed oxygen and the generated water vapor have the same trend of

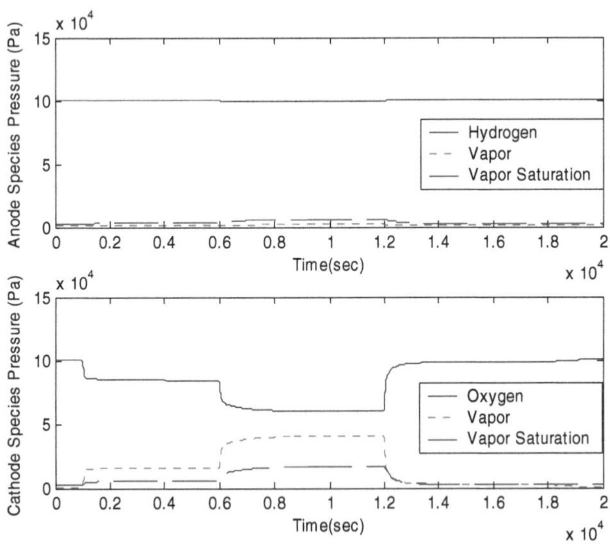

Figure 6.6. Species partial pressures

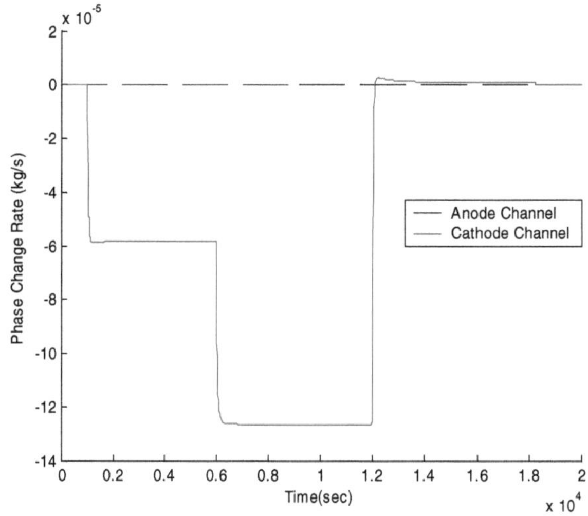

Figure 6.7. Phase change rates

change when load current changes. One can see that before the 1000th second, the actual vapor pressure is lower than the in-situ saturation vapor pressure, so no liquid water is generated. After that, the actual vapor in the cathode channel becomes over saturated, and the liquid water starts to be formed by condensation. The condensation procedure continues until the load current shuts down at the 12 000th second. After that, the actual vapor pressure becomes lower than the in-situ saturation vapor pressure, and the evaporation takes place. The evaporation procedure continues until the liquid water in the cathode channel is completely evaporated at approximately the 18 000th second.

Figure 6.7 shows the water phase-change rates in the anode and cathode channels. It is obvious that, under the current conditions, the phase change in the anode channel is zero, *i.e.*, no condensation/evaporation takes place and the liquid water is zero, which is shown in Figure 6.8. In the cathode channel, after the 1000th second, the generated over-saturated vapor leads to the condensation. The condensation rate goes up following the further increase of the load current at the 6000th second. After the load current shuts down and the actual vapor pressure goes down below the in-situ saturation vapor pressure at the 12 000th second, the phase change reverses its direction from condensation to evaporation. The evaporation rate becomes zero at approximately the 18 000th second when the liquid water mass becomes 0 (as shown in Figure 6.9).

Figure 6.8 and Figure 6.9 show the species mass histories in the anode and cathode channels. Generally, the species mass changes are similar to the species partial pressure changes. It is worth noting that a pseudo-humidity ratio is defined in Equation (6.14) to account for the outlet liquid water flow dragged by the gas phase species flow. In practical situations, the liquid water in the channel may flow out under gravitational force, depending on the geometry configuration of the channels, which is not considered in this study.

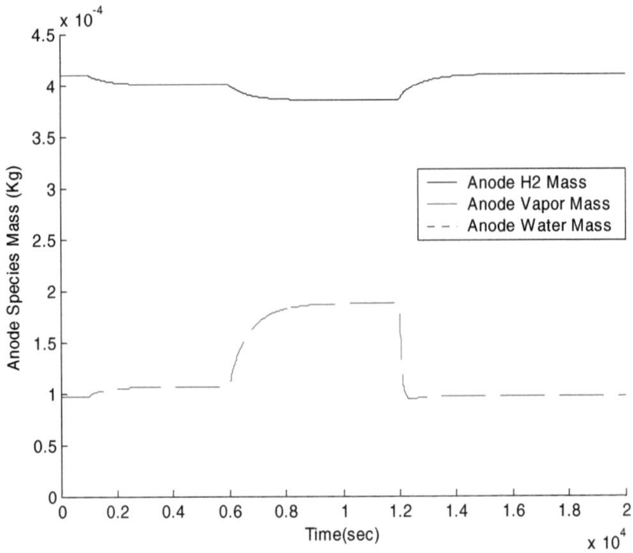

Figure 6.8. Species mass in anode channel

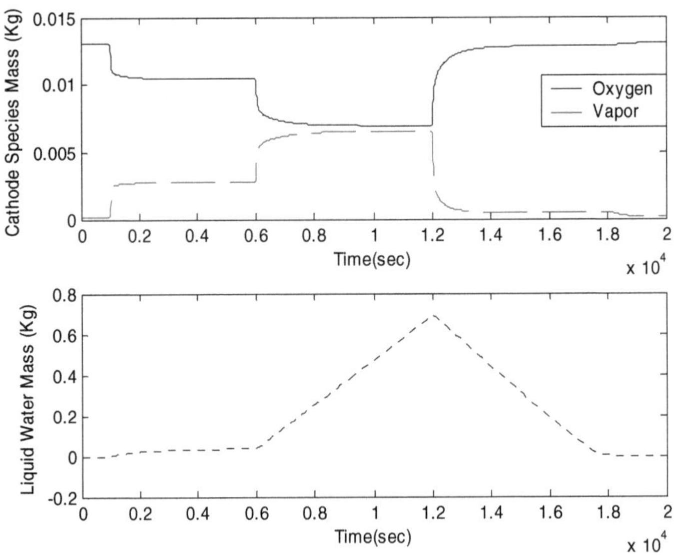

Figure 6.9. Species mass in cathode channel

Figure 6.10 illustrates the membrane relative humidity history during the fuel cell operation. As explained in Section 6.2.2.1, membrane hydration depends on the water concentrations in the anode and cathode channels, load current, and fuel cell temperature *etc.* After the load current rises from 0 A to 20 A at the 1000th second, the membrane relative humidity decreases. Due to the load current increase, the diffusion rate of water molecules from the anode side to the cathode side goes up. In the meantime, the back diffusion rises due to the increase of both fuel cell temperature and water concentration at the cathode side. Under these conditions, the vapor in the anode channel is not saturated while it is over saturated in the cathode. The overall result is that the diffusion rate due to the proton drag force overcomes the back diffusion rate, and the membrane relative humidity shows a decrease. When the load current further goes up at the 6000th second, the combined effect of the above two diffusion rates might reverse, and the membrane relative humidity increases. At the 12 000th second, a similar mechanism leads to the membrane relative humidity change as shown in Figure 6.10.

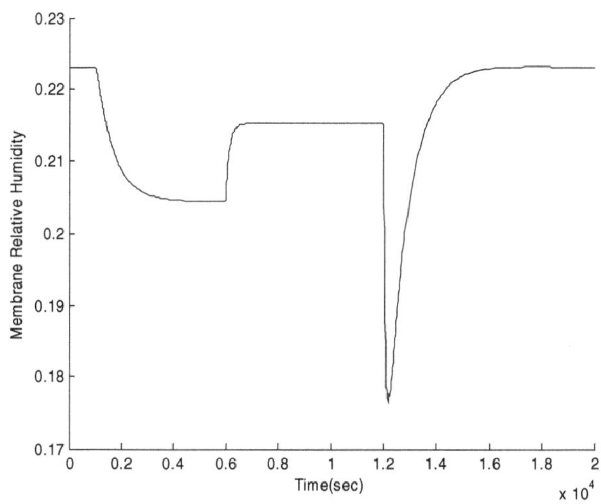

Figure 6.10. Membrane relative humidity history

6.4 References

1. Costamagna, P., and Srinivasan, S., 2001. Quantum jumps in the PEMFC science and technology from the 1960s to the year 2000, Part II, engineering, technology development and applications aspects. Journal of Power Sources, V102, pp253–269.
2. Bernardi, D.M., and Verbrugge, M.W., 1991. Mathematical model of a gas diffusion electrode bonded to a polymer Electrolyte. AIChE Journal, V37(8), pp1151–1163.
3. Bernardi, D.M., and Verbrugge, M.W., 1992. A Mathematical model of the solid polymer electrolyte fuel cell. Journal of Electrochemical Society, V139(9), pp2477–2491.
4. Springer, T.E., Zawodzinski, T.A., and Gottesfeld, S., 1991. Polymer electrolyte fuel cell model. Journal of Electrochemical Society, 138(8), pp.2334–2342.
5. Springer, T.E., Wilson, M.S., and Gottesfeld, S., 1993. Modeling and experimental diagnostics in polymer electrolyte fuel cells. Journal of Electrochemical Society, V140 (12), pp3513–3526.
6. Amphlett, J.C., Baumert, R.M., Mann, R.F., Peppley, B.A., Roberge, P.R., and Rodrigues, A., 1994. Parametric modeling of the performance of a 5-kW proton-exchange membrane fuel cell Stack. Journal of Power Sources, V49, pp349–356.

7. Amphlett, J.C., Baumert, R.M., Mann, R.F., Peppley, B.A., Roberge, P.R., and Harris, T.J., 1995. Performance modeling of the Ballard Mark IV solid polymer electrolyte fuel cell, I. Mechanistic model development. Journal of Electrochemical Society, V142(1), pp1–9.
8. Amphlett, J.C., Baumert, R.M., Mann, R.F., Peppley, B.A., Roberge, P.R., and Harris, T.J., 1995. Performance modeling of the Ballard Mark IV solid polymer electrolyte fuel cell, II. Empirical model development. Journal of Electrochemical Society, V142(1), pp10–15.
9. Amphlett, J.C., Mann, R.F., Peppley, B.A., Roberge, P.R., and Rodrigues, A., 1996. A model predicting transient responses of proton exchange membrane fuel cells. Journal of Power Sources, V61(1–2), 183–188.
10. Pukrushpan, J.T., Peng, H., and Stefanopoulou, A.G., 2004. Control-oriented modeling and analysis for automotive fuel cell systems. Transactions of the ASME, Journal of Dynamic Systems, Measurement, and Control, 126, pp.14–25.
11. Iqbal, M.T., 2003. Modeling and control of a wind fuel cell hybrid energy system. Renewable Energy, V28, pp223–237.
12. Xue, X., Tang, J., Smirnova, A., England, R., and Sammes, N., 2004. System level lumped-parameter dynamic modeling of PEM fuel cell. Journal of Power Sources, 133, pp.188–204.
13. Nguyen, T.V. and White, R.E., 1993. A water and heat management model for proton-exchange-membrane fuel cells. Journal of Electrochemical Society, 140(8), pp.2178–2186.
14. Wang, Z.H., Wang, C.Y., and Chen, K.S., 2001. Two-phase flow and transport in the air cathode of proton exchange membrane fuel cells. Journal of Power Sources, 90, pp. 40–50.
15. You, L., and Liu, H., 2002. A two-phase flow and transport model for the cathode of PEM fuel cells. International Journal of Heat and Mass Transfer, 45, pp.2277–2287.
16. Yerramalla, S., Davari, A., Feliachi, A., and Biswas, T., 2003. Modeling and simulation of the dynamic behavior of a polymer electrolyte membrane fuel cell. Journal of Power Sources, V124, pp104–113.
17. White, F.M., 1991, Viscous Fluid Flow, 2nd Edition, McGraw-Hill.
18. Sonntag, R.E., Borgnakke, C., and Van Wylen, G.J., 2003, Fundamentals of Thermodynamics (Sixth Edition), John Wiley and Sons.
19. Dutta, S., Shimpalee, S., and Van Zee, J.W., 2001. Numerical predication of mass-exchange between cathode and anode channels in a PEM fuel cell. International Journal of Heat and Mass Transfer, 44, pp.2029–2042.
20. Mann, R.F., Amphlett, J.C., Hooper, A.I., Jensen, H.M., Peppley, B.A., and Roberge, P.R., 2000. Development and application of a generalized steady-state electrochemical model for a PEM fuel cell. Journal of Power Sources, V86, pp173–180.

7

New Generation of Catalyst Layers for PEMFCs Based on Carbon Aerogel Supported Pt Catalyst (CASPC)

Alevtina Smirnova, Xing Dong, Hiro Hara, and Nigel M. Sammes

7.1 Introduction

The mass production of PEMFC power generators requires a price reduction and, thus, a decrease in the amount of noble metals present in the cathode and anode catalyst layers. Automotive, residential, military, and small scale applications require PEMFC stacks with a Pt-specific power density of at least 0.2 gPt/kW at cell voltages of about 0.65 V. However, existing PEMFC performance corresponds to approximately 0.85–1.1 g Pt/kW. Thus, at least a five-fold reduction of the amount of noble metal in the PEMFC catalyst layer is required for large scale manufacturing [1].

Developed over the past ten years, carbon aerogels are porous materials having a network structure of interconnected nanosized primary particles and in addition to a large surface area [2] include well-developed and controlled mesoporosity. Their tunable properties, such as surface area, pore volume, and pore size distribution, determined by the synthesis and processing conditions, allow for a wide spectrum of materials with unique properties and broad area of application, such as electrochemistry, fuel cells, catalysis, optics, and microelectronics [3, 4, 5, 6].

The carbon aerogel supported Pt catalyst (CASPC) structures differ significantly from those of presently available commercial catalysts produced by such companies as Tanaka, E-TEK, and Alfa Aesar. The carbon aerogel synthesized under supercritical conditions in the presence of CO_2, has a narrow pore size distribution and the size of the single pore that can be easily adjusted through synthesis modifications to the size of the polymer electrolyte molecule. Penetrating inside the aerogel pore, the molecule is trapped inside the pore and maintains excellent physical contact with the small platinum particles (approximately 1 nm in diameter) that are evenly distributed on the internal surface of the aerogel pore (Figure 7.1). In comparison to structurally disorganized commercial catalysts having noble metal particles randomly distributed on the surface of the spherical carbon particles, aerogel particles have less tendency for agglomeration, allow for the polymer molecule to be incorporated inside the single pore, maintain a good contact with Pt particles, and allow for efficient use of the whole surface of the noble metal. Correspondingly, the electrochemical surface area

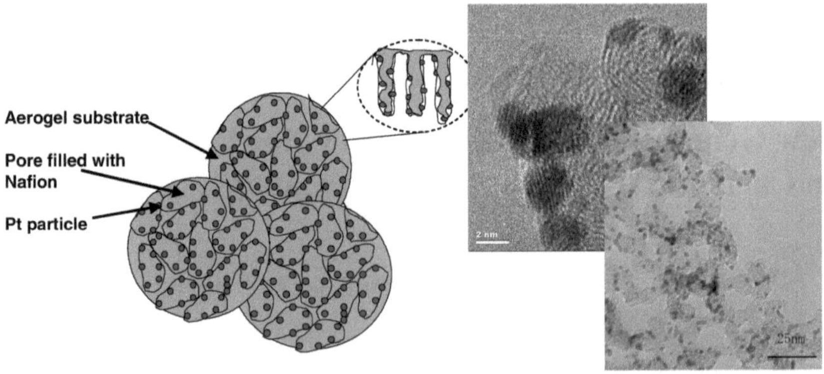

Figure 7.1. Schematic representation and HRTEM image of the CASPC catalyst. Pt particles are evenly distributed on the interior of the aerogel pores filled with polymer electrolyte. The shape and dimensions of the CASPC particles mainly depend on grinding/homogenization procedure. HRTEM image shows Pt particles on the surface of the carbon aerogel support

(ESA) of the catalyst, estimated by cyclic voltammetry demonstrates values at least twice those [7] in comparison to the ESA of existing catalysts. In addition, the homogeneous distribution of the platinum particles at the nanometer scale on the interior surface of mesopores [7,8] prevents their migration and growth, thus increasing the long term endurance of these materials. Thus, aerogel supported noble metal catalysts provide new opportunities for achieving the mass production of PEMFC power generators.

This chapter summarizes the CASPC catalytic properties and performance data of the PEMFCs with the low loaded ($\leq 0.1\,mg/cm^2$) CASPC layers. The influence of the electrolyte content (Nafion 1100) on the properties of the PEMFCs is discussed in terms of cell temperature, gas flowrates, and backpressure followed by the long-term endurance data.

7.2 Experimental

7.2.1 Characterization of the Aerogel Supported Pt Catalysts

Aerogel-supported Pt catalysts with 22 nm pore size distribution estimated by the Brunauer-Emmett-Teller (BET) method have been used for the cathode catalyst layers. Adsorption and desorption isotherms of nitrogen were measured using a Sorptomatic 1990, a fully automatic Gas Adsorption Analyzer from Horiba. The mesopore size distributions and mesopore volumes were estimated for the pore diameter range 2–50 nm by applying the Dollimore-Heal method to the desorption isotherm of nitrogen. The Pt mean diameter and the Pt surface area in the aerogel-supported Pt catalysts were measured by hydrogen adsorption. The catalysts differ by Pt content that corresponds to 20% for the samples 1–4 and 30% for sample #5, Table 7.1.

High Resolution TEM (HRTEM) was used for the estimation of the size of Pt particles in the aerogel supported Pt samples. The aerogel-supported Pt catalyst with and without Nafion was deposited on Quantifoil® (Quantifoil Micro Tools GmbH , Jena, Germany) gold grids covered by a holey carbon film with ~1.2 µm circular holes. TEM analyses were performed in a Philips EM-420 TEM equipped with a Kevex-Noran Be-window energy dispersive spectrometer (EDS), and operating at an accelerating voltage of 100 kV. Morphological characterization of the electrode surfaces has been made using a SEM JEOL JSM-5600LV.

7.2.2 Preparation of Catalyst Pastes and Membrane Electrode Assemblies

All the membrane electrode assemblies (MEAs) were made using carbon supported catalyst from Tanaka Inc. (Pt-Ru 53.5%) for the anodes and various carbon aerogels for cathodes. The anode catalyst was mixed under nitrogen with DI water, 5% Nafion 1100 solution from Solution Technology and appropriate solvents. The total amount of dry Nafion content was 25 wt.% in relation to total weight of the catalyst. Prepared mixtures were homogenized for few hours using an Ultra-Terrux homogenizer with further evaporation of the excess amount of the solvent in order to obtain the appropriate viscosity. The corresponding catalyst anode loadings did not exceed 0.3 mg/cm^2. Aerogel based cathode inks with ultra low Pt loadings (≤0.1 mg/cm^2) and different amounts of electrolyte corresponding to 25, 35, and 45wt. % (Table 7.1) of Nafion, were prepared from preliminary ground catalyst rods in the way similar to the preparation of the inks for the anode.

Teflon films used for decals were weighed before and after application of the catalyst ink. The ink was screen-printed using a Systematic Automation Model 810 Series Screen Printer. The polyester screen from SAATI Print™ was used for the deposition of ultra thin catalyst layers onto the surface of the Teflon film. To adjust the necessary CASPC loadings either a 110 or 280 mesh screen was used. The appropriate decals were placed on either side of the Nafion 112 membranes purchased from DuPont. After drying and hot pressing at 130°C and 70 kg/cm^2 for 3 min with Nafion 112, the Teflon supported films were peeled off from the cathode and anode sides of the MEAs.

Table 7.1. Compositions and properties of the PEMFC cathodes based on 22 nm pore size CASPC catalyst

Cell #	Wt.% Nafion	Wt. % Pt in CASPC	Catalyst loading, mg/cm^2	Estimated ESA of the CASPC catalyst, m^2/g
1	*25*	*20*	*0.10*	*80*
2 **3**	*35*	*20*	*0.05*	*120*
4	*45*	*20*	*0.10*	*175*
5	*35*	*30*	*0.08*	*175*

Due to very low cathode catalyst loadings that did not exceed 0.1 mg/cm^2 (Figure 7.2) the accuracy of the catalyst layer deposition has been statistically evaluated. The estimated error did not exceed 10% of the catalyst by weight. Further

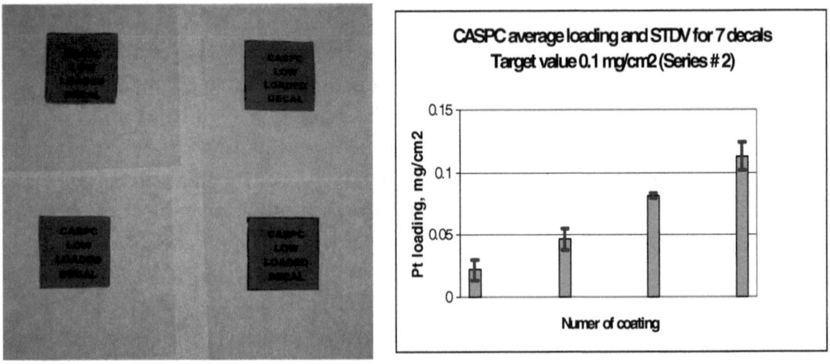

a) b)

Figure 7.2.a) The photograph of the Teflon substrates with ultra thin CASPC layers. The printed text “CASPC low loaded decal” can be read through the decal. b) Statistical evaluation of the catalyst loading gives a linear correlation between the number of depositions and the catalyst loading. Four depositions are needed to achieve the targeted value of 0.1 mg/cm^2 using the polyester screen with 280 mesh from SAATI print

fuel cell testing was performed in hardware with a 6.25 cm^2 active area from ElectroChem Inc. The cells included gas diffusion layers from SGL Technologies GmbH and Teflon gaskets with a thickness of 0.25 mm, which corresponded to a total pitch of 0.2 mm.

7.2.3 Cyclic Voltammetry Measurements

The measurements of Pt surface area in contact with Nafion and thus available for the electrochemical reaction were made using a cyclic voltammetry (CV) technique in a four-point probe configuration. CV plots were obtained in the range of 0.01–0.8 V at room temperature using a Princeton potentiostat Model 273. In order to avoid the presence of oxygen, pure nitrogen was supplied to the cathode as a working electrode and pure hydrogen to the anode as a counter electrode at a 200cc/min flowrate. The calculation of the electrochemical surface area of the catalyst was estimated from CV plots using the equation: A_{ESA}=A/(K × L× S). In this equation A is the averaged area of the reduction and oxidation peaks of the current-voltage plot; K = 0.21 mA/cm^2; S is scan rate (mV/sec), and L Pt loading in the cathode catalyst layer (mg/cm^2). It is necessary to note, that an averaging procedure applied to a number of CV plots revealed the known fact that the calculated area of oxidation peak is usually at least 10% greater than the corresponding area of the reduction peak, which is believed to be the result of hydrogen evolution at low voltages. However, for some of our CV plots we did not notice a significant difference between these two numbers, which can be attributed to relatively low hydrogen flowrates (0.2 L/min) and scan rate (20 mV/sec).

7.2.4 PEM Fuel Cell Testing Procedure

The cell performance was evaluated using a Teledyne Medusa™ RD test station. The test station was equipped with humidification chambers, mass flow and temperature controllers, and 50 A load box 890C from Scribner Associates, Inc that allowed simultaneous correction of the cell potential for the iR-drop. In order to avoid water condensation inside the cell, the feed lines from the cathode and anode side were heated 10°C higher than the cell temperature, which was equal to the temperature of both saturators.

7.3 Results and discussion

7.3.1 CASPC Morphology and BET Measurements

Morphological evaluation of the CASPC based cathode in comparison to an anode made from commercial Tanaka carbon-supported catalyst was made using MEAs with "partially-peeled-off" surfaces. The SEM image of the CASPC catalyst layer reveals its significant difference from commercially available Tanaka catalyst (Figure 7.3) and indicates that the size of the different in shape CASPC particles ranges from 1 to 10 micrometers, which is much bigger than the size of the particles in anode catalyst layer, which possesses fine microporous structure. Taking into account the macroporous structure of CASPC catalyst, we can assume that the observed size of the CASPC particles should be sufficient to achieve high electrochemical surface area (ESA) and could facilitate mass-transport through the catalyst layer [9,10]. However further homogenization and diminishing of the CASPC particle size in the cathode catalyst layer could be essential when using thinner (10–15μm) reinforced Nafion membranes. In this case the size of the CASPC particles could be extremely critical in avoiding possible MEA short circuit.

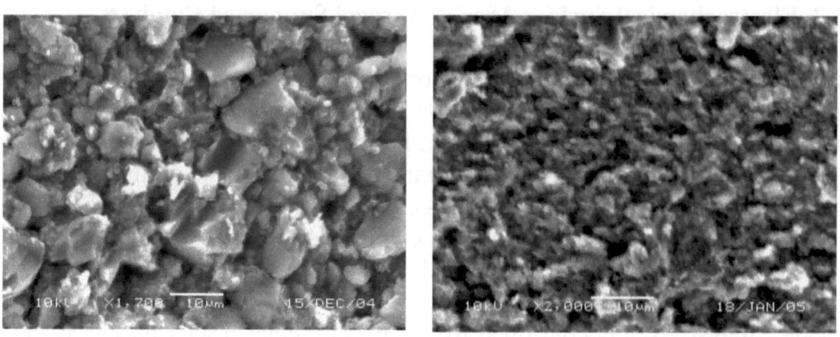

a) b)

Figure 7.3. SEM images of a) CASPC catalyst layer and b) Tanaka catalyst layer. Comparison shows much bigger size of the aerogel particles that cannot be attributed to the effect of agglomeration

It is well known that a tendency for agglomeration in commercial catalysts could decrease fuel cell performance because of the limited availability of Pt particles for polymer electrolyte molecules [9, 10, 11]. We believe that the observed shape of the CASPC particles is the result of inefficient grinding and cannot be considered as a result of agglomeration, since while preparing catalyst ink no agglomeration was detected. Furthermore, this structure could be beneficial for oxygen/air permeability in comparison to the conventional catalyst structure, where small homogeneous-resembling domains are surrounded by gas pores [12, 13].

The narrow pore size distribution (Figure 7.4a) in comparison to commercial catalyst (Figure 7.4b) is one of the major advantage of the CASPC that allows for efficient utilization of the Pt surface area and cannot be achieved for commercial carbon supported catalysts [7]. As a result of well defined mesoporous structure and thus, enhanced contact between polymer electrolyte and catalyst particles, the ESA of the carbon aerogel based catalyst could be significantly influenced by the Nafion loading.

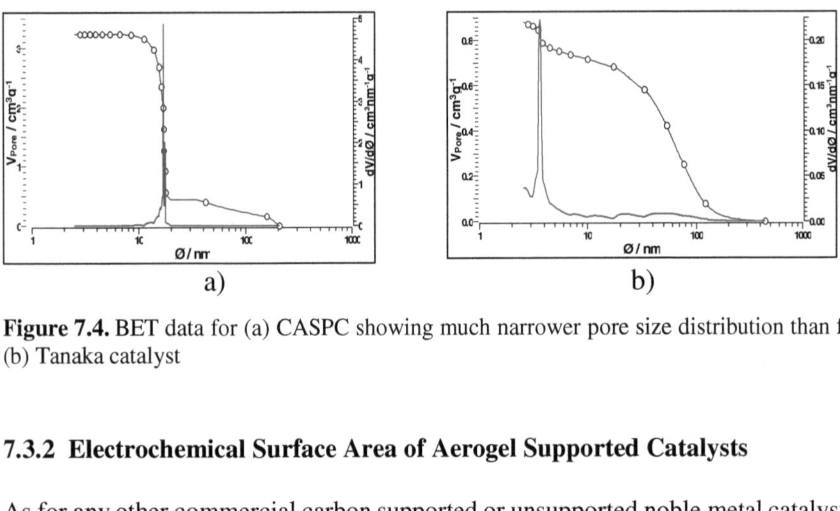

Figure 7.4. BET data for (a) CASPC showing much narrower pore size distribution than for (b) Tanaka catalyst

7.3.2 Electrochemical Surface Area of Aerogel Supported Catalysts

As for any other commercial carbon supported or unsupported noble metal catalysts, the CASPC should be mixed with Nafion as an ionic conductor to provide proton conductivity in the PEMFC catalyst layer. Formation of the thin film on the surface of noble metal particles improves ionic transport and thus ionic catalyst layer resistance. Based on various publications, it is evident that the ESA of the catalyst could be improved with an increase of Nafion content. However, mass transport limitations acting in the opposite direction could significantly impede this [12].

The assumption regarding the increase of ESA with Nafion content in the PEFC cathode catalyst layer is confirmed by the experimental results presented in Table 1. The ESA values obtained from CV data for different Nafion loadings, *e.g.* 25, 35, and 45%, correlate well with the general idea that consider an increase of the ESA with electrolyte Nafion loading.

It can be seen (Table 7.1) that an increase in Nafion content significantly enhances the ESA of the aerogel supported catalyst that can be explained by the consequent microstructural changes. Specifically, due to the better penetration of

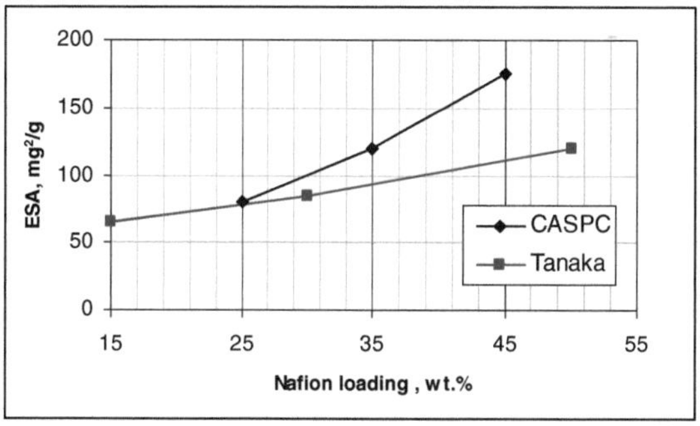

Figure 7.5. Electrochemical surface area (ESA) vs. Nafion loading in commercial (Tanaka) and CASPC estimated for 22 nm CASPC catalyst with 20% of Pt (cells # 1–4, Table 7.1)

Nafion inside the macroporous CASPC structure and better contact of Pt particles with the ionic conductor Nafion, the amount of Pt participating in the electrochemical reaction increases gradually, thus giving rise to the electrochemical surface area of the catalyst. Comparison of the electrochemical surface area of aerogel and commercial catalyst (Figure 7.5) revealed higher values of CASPC ESA especially at higher Nafion loadings (>35wt.%) that can be explained by the CASPC structure, specifically by higher Pt dispersion and better penetration of Nafion molecule inside the macropore fitting the size of the Nafion molecules.

7.3.3 Evaluation of PEMFC Performance at Elevated and Room Temperatures

The PEMFC performance measurements of voltage-current characteristics have been made at different operating conditions and were followed by the calculation of OCVs, Tafel slopes, membrane resistances, and power densities.

7.3.4 Catalytic Activity of CASPC: Open Circuit Voltage (OCV) and Tafel Slope

Our experiments revealed high OCVs for the MEAs with cathode layers based on CASPC catalysts. They ranged from 950 up to 1050 mV depending on the catalyst layer composition, operating conditions, and the gas supplied to the cathode, *e.g.* air, oxygen, or "synthetic air" with 21.4% O_2 balanced helium. Taking into account very low catalyst loadings (Table 7.1), we can assume that CASPC can be considered as a highly effective catalyst with wide prospective for PEMFC and DMFC applications.

The OCVs for the cell # 5 at 50°C can be seen in Figure 7.6a. Considering that each point is the average value from 20 measurement obtained during 5 minutes, the corresponding OCV values in Air, 20wt. %O_2 in He , and O_2(100%) at 3 bar absolute pressure correspond to: OCV_{air}=985±5 mV; $OCV_{20\%O2}$=998±2 mV, and OCV_{O2}=1040±5 mV.

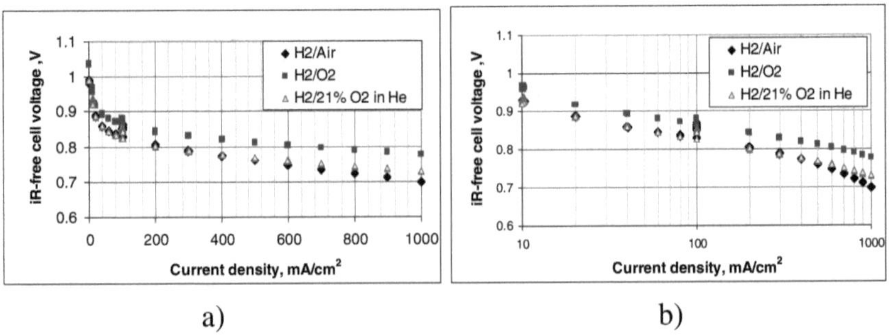

Figure 7.6. Compensated cell voltage vs. a) current density and b) logarithm of current density for the cell #5 (Table 7.1) at 50°C cell temperature and 3 bars absolute pressure in H_2/Air; H_2/O_2 , and H_2/20% O_2 in He. Anode flowrate (AFR)=288cc/min; cathode flowrate (CFR)=866 cc/min. Temperature of anode and cathode humidifiers is 60°C

The strong dependence of Nafion polymer electrolyte ohmic resistance and oxygen permeability on relative humidity (RH) that is established by the temperature of the humidifiers vs. cell temperature could play a significant role while estimating Tafel slopes in the low current density region. The decrease of the temperature of humidifiers vs. cell temperature could impede Tafel slopes due to the fact that both proton conductivity and oxygen permeability in Nafion [14] decrease with decreasing RH thus limiting proton migration in the active catalyst layer and oxygen diffusion inside agglomerates [9,10]. Therefore, Tafel slope measurements for CASPC based MEAs were performed together with corresponding membrane resistances evaluated by a current interrupt technique.

Evaluation of Tafel slopes revealing the electrode kinetics has been performed using current-voltage data obtained with oxygen, air, and 21% O_2 balanced helium and compensated for membrane resistance. It can be seen (Figures 7.6 b), that in

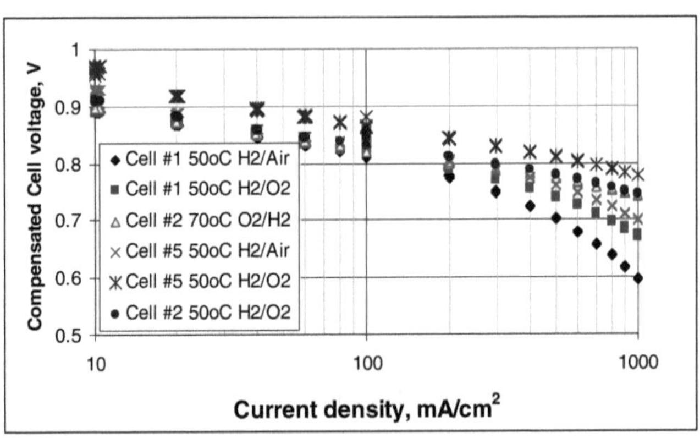

Figure 7.7. Current-voltage plots indicating Tafel slopes for MEAs with different amount of Nafion 1100 in the cathode CASPC layer (Table 7.1) at 50°C cell temperature and 3 bars absolute pressure in H_2/Air; H_2/O_2. Anode flowrate (AFR)=288cc/min, cathode flowrate (CFR)=866 cc/min. Temperature of humidifiers is 10°C higher than the cell temperature. No correction for hudrogen crossover has been made

the case of oxygen and an oxygen –helium mixture, containing the same amount of oxygen as air, the current-voltage plots are linear in the current density range up to 1 A/cm^2 where Tafel slopes remain constant and close to the theoretical values (70 mV/decade). On the contrary, in the case of air the linearity of the plots remains only in the lower current density region, that indicates existing mass transport limitations, which are much more pronounced for the cell # 4 (Table 7.1) containing higher concentration of Nafion (45 wt.%) even though the catalyst layer is very thin.

At a cell temperature 10°C lower than the temperature of the humidifiers, membrane resistances (Figure 7.8a) are almost constant and corresponding Tafel slopes are close to theoretical (Figure 7.7a). However, when the temperatures of the cell and both humidifiers are the same (Figure 7.8b), the membrane resistance in the low current density region is much higher, which is attributed to insufficient membrane humidification that causes an increase in Tafel slopes. For this reason all the experiments regarding Tafel slopes and cell performances have been carried out at a cell temperature 10°C lower than the temperature of saturators.

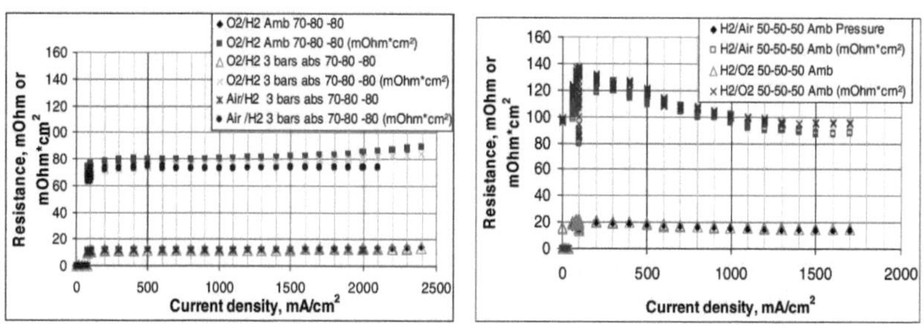

a) b)

Figure 7.8. Membrane resistances for the cell #2 (Table 7.1) at a) T_{cell}=70°C, $T_{saturators}$=80°C, and 3 bars absolute or ambient pressure. b) T_{cell}=50°C, $T_{saturators}$=50°C and ambient pressure

Influence of Nafion Content in CASPC Cathode Catalyst Layer

Electrocatalytic activity of PEMFC electrodes fabricated by mixing carbon supported catalyst with ionic conductive electrolyte is mainly determined by the electrolyte-catalyst ratio. In the case of Nafion, the catalyst-electrolyte ratio influences water management in the membrane and three-phase contact between the reactant gases, electrolyte, and catalyst [15]. Thus, in order to improve PEMFC performance many studies have been reported on the effect of Nafion content in the catalyst layer [16, 17, 18, 19]. However no studies have been done yet regarding the influence of Nafion content on the performance of CASPC based catalyst layers.

Our comparison of the cells with 25 and 35 wt. % (Figure 7.9a,b and Table 7.1) shows that the cell #1 with 25% Nafion demonstrates better performance both in high and low current density region. At low currents, specifically at 200mA/cm^2, corresponding voltages are higher for MEA containing 35wt.% Nafion ($E_{\#1}$=740 mV and $E_{\#2}$=670 mV). In a high current density regions, specifically at 1500mA/cm^2, the corresponding cell voltages for the MEA with 35wt. % Nafion

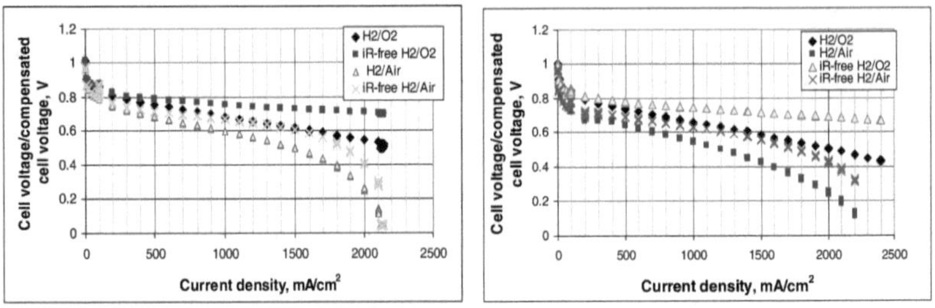

a) b)

Figure 7.9. Cell performance for a) Cell #1 with 25wt.% Nafion and b) Cell #2 with 35wt.% Nafion in H_2/ O_2 and H_2/air at 50°C and 3 bar absolute pressure

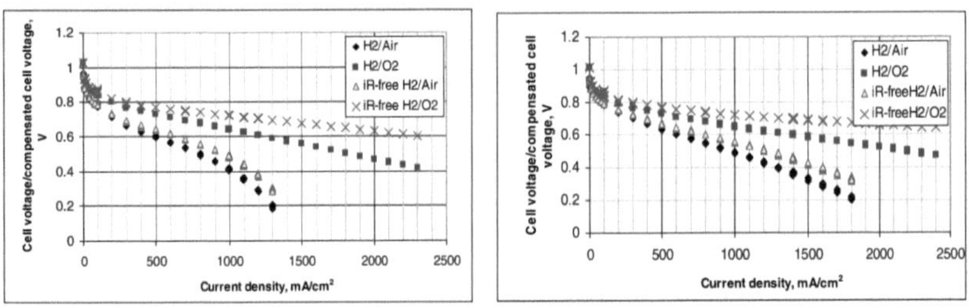

a) b)

Figure 7.10. Performance in H_2/air and H_2/O_2 for Cell # 4 (45wt.% Nafion) at the cell temperature a) 50°C and b) 80°C

also demonstrate higher performance: $E_{\#1}$=490mV and $E_{\#2}$=410mV. The result can be possibly explained by the fact that the cell #2 has half the catalyst loading in the cathode catalyst layer of cell # 1. However, this data demonstrate that even at 0.05 mg/cm^2 loading in the CASPC cathode layer the CASPC based cell still has reasonable performance with a power density of 0.6 W/cm^2 at 0.6 V.

An increase in Nafion content up to 45 wt.% further reduces the cell performance in comparison to the cells with 25 and 35wt.% Nafion (Figure 7.10a). However, the decrease in the cell performance has been detected only in the higher current density region that is due to mass transport limitations in the cathode catalyst layer especially in the presence of air. This data is consistent with earlier published data for carbon supported catalysts [20] and can be explained by the excessive thickness of Nafion film coating aerogel/Pt particles and thus, impeding mass transport inside the electrode structure. The cell performance at higher cell temperature (80°C) was improved (Figure 7.10b), especially in a high current density region. However, it was still much lower than the corresponding performance of the cells with lower Nafion content. The experiment showed that an increase in Nafion content for CASPC with 20wt.%Pt did not provide any improvement in cell performance, even though a significant enhancement in the ESA has been detected from cyclic voltammetry studies made for these cells (Figure 7.5).

Influence of Pt Loading in CASPC

Further improvement of the cell performance has been achieved using CASPC with noble metal concentration increased up to 30wt.% with addition of 35wt.% of Nafion (Figures 7.11 a, b and 7.12) and Table 7.2.

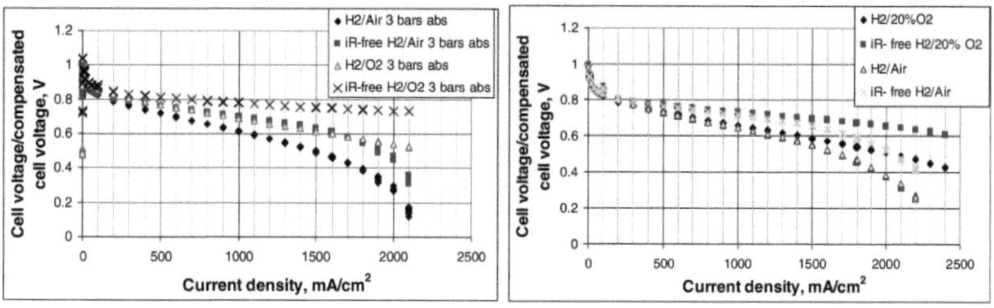

a) b)

Figure 7.11. Performance of the cell #5 cell at a) 50°C in H_2 /air and b) 80°C in H_2/Air/ 21%O_2in He

Comparison of the Cell Performance Under Pressurized and Non-pressurized Conditions

The influence of absolute pressure was assessed with cell #5 (Table 7.1) containing the highest Nafion concentration in the cathode catalyst layer. Because of the high Nafion content, we expected a relatively high influence of back pressure on the cell performance in the whole range of current densities. It can be seen (Figure 7.12), that the data correlates with the results published elsewhere [21] indicating that the cell performance can be significantly improved under pressurized conditions. Our data shows that pressurization of the anode does not influence the cell performance as much as anode pressurization, which is reasonable in terms of different reaction stoichiometry and mass transport limitations for anode and cathode gases. The difference in power densities in H_2/O_2 (Figure 7.1b) at ambient and pressurized conditions at 1.5 A/cm^2 is about 0.2 W/cm^2 and demonstrates the significant impact of the performance in the absence of pressurization.

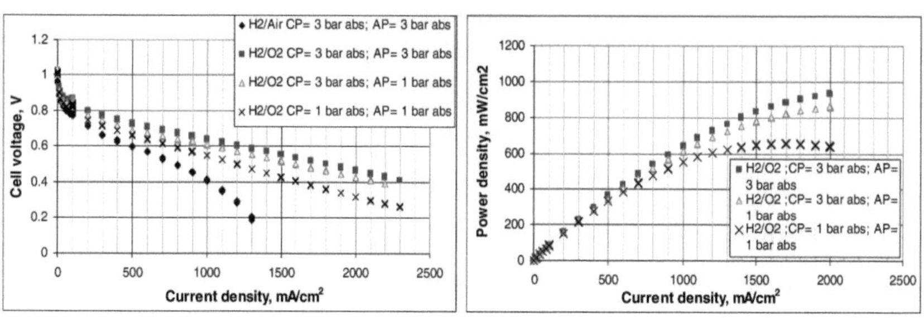

a) b)

Figure 7.12. The PEMFC performance (cell #4, Table 7.1) a) Cell voltage vs. current density b) cell voltage vs. power density under pressurized and non-pressurized conditions at T_{cell}=50°C and $T_{saturators}$=50°C

PEMFC Long-term Stability

The cells used in this work (Table 7.1) have been tested for at least 200 hours in scanning mode from OCV to 2.2 A/cm^2 and in cycling mode from 0.2 mA/cm^2 to 1.5 A/cm^2. The results based on performance and CV data did not indicate any decrease in cell performance or ESA of the catalyst during 200 hours of operation. The performance of cell #5 after 200 hours of operation is presented in Figures 7.13a, b). The cell voltages in air/H_2 at 50°C under pressurized conditions corresponded to approximately 0.82V at 200 mA/cm^2 and 0.6V at 1.5 A/cm^2, which in comparison to the data available in the literature, seems to be the highest among the earlier reported data [16]. The corresponding cell voltages compensated for membrane resistance were about 20 mV higher at 0.2 A/cm^2 (Table 7.2), however, at high current density the corresponding difference exceeded 100 mV, which was due to the resistive membrane losses. However, these characteristics can be further increased by optimization of the cathode catalyst layer morphology and composition.

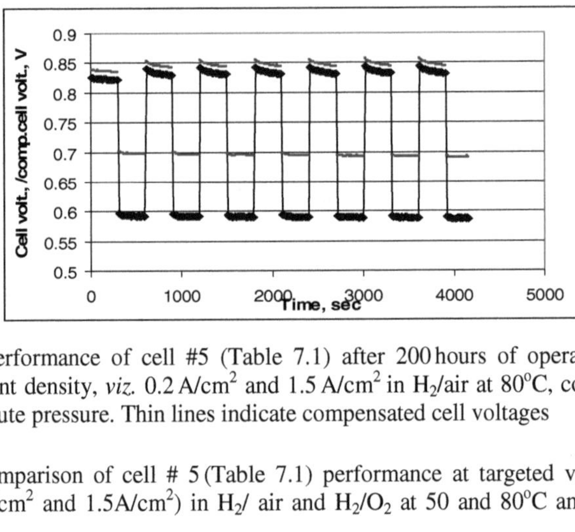

Figure 7.13. Performance of cell #5 (Table 7.1) after 200 hours of operation at targeted values of current density, *viz.* 0.2 A/cm^2 and 1.5 A/cm^2 in H_2/air at 80°C, constant flowrate, and 2 bar absolute pressure. Thin lines indicate compensated cell voltages

Table 7.2. Comparison of cell # 5 (Table 7.1) performance at targeted values of current density (0.2 A/cm^2 and 1.5A/cm^2) in H_2/ air and H_2/O_2 at 50 and 80°C and 2 bar absolute pressure

Current density, mA/cm^2	**Cell Voltage for the cell #5**			
	50°C		80°C	
	O_2	Air	20%O_2in He	Air
200	*828*	*790*	*788*	*796*
200 (iR-free)	*845*	*806*	*802*	*810*
1500	*613*	*498*	*588*	*576*
1500 (iR-free)	*751*	*631*	*697*	*656*

7.4 Conclusion

Aerogel supported Pt catalysts have been evaluated for possible application in polymer electrolyte fuel cell cathode layers. Morphological investigation revealed a relatively large size of the catalyst particles in comparison to commercially produced carbon supported catalyst samples. On the other hand, CV data demonstrated much higher electrochemical surface area for CASPC in comparison to commercial catalyst. The highest ESA has been obtained for the CASPC mixed with 45wt.% Nafion. However, this composition did not show a high power density in comparison to the other cells because of significant mass transport limitations, and thus, reduced performance in a high current density region.

The best performance in both low and high current density regions has been obtained for the 30wt.% Pt in aerogel based catalyst with 35wt.% Nafion (Figure 7.14). After 200 hours of stable operation, the compensated cell voltages in the low current density region at 80°C and 3 bars absolute pressure were close to the value of 0.85 V. In the high current density region (1.5 A/cm^2) the compensated for ohmic resistance cell voltage was in-between 0.65 and 0.7 V depending on the air flowrate. Uncompensated cell voltages were lower than corresponding iR-free data, especially at the current density due to the membrane resistance.

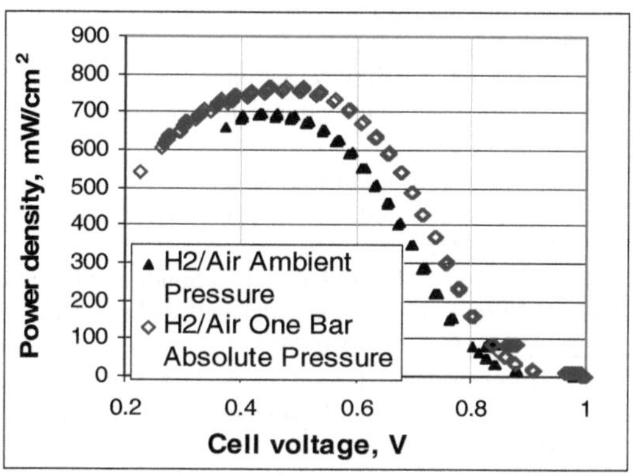

Figure 7.14. Power density vs. cell voltage for the cell with 0.1 mg/cm^2 CASPC in cathode catalyst layer

The H_2/Air PEMFC performance of the membrane electrode assemblies (MEAs) with low Pt loadings (0.1 mg/cm^2) in the CASPC cathode catalyst layer (Figure 7.14) shows 600mW/cm^2 at 0.6V and ambient pressure and thus, demonstrates a significant reduction in the amount of platinum compared to the existing PEMFC technologies [22, 23, 24, 25]. In terms of specific power density, the estimated loading for this type of MEA (cathode only) is about 0.16gPt/kW that is very close to the DOE target [1]. The cell performance evaluated at different temperatures (25–80°C) and pressures (1–3 bars absolute) demonstrates stable performance during 200 hours of operation. However, further investigation is necessary for the optimization of electrolyte content in the PEMFC catalyst layers based on CASPC, as well as long-term endurance tests that will reveal the actual stability of the Pt and carbon phases.

7.5 References

1. Gasteiger H A, Kocha S S, Sompalli B and Wagner F T. Activity benchmarks and requirements for Pt, Pt-alloy, and non-Pt oxygen reduction catalysts for PEMFCs. Applied Catalysis B: Environmental, 2005;56:9–35.
2. Moreno-Castilla C and Maldonado-Hódar F J. Carbon aerogels for catalysis applications: An overview. Carbon, 2005;4(3):455–465.
3. Wei Y-Z, Fang B, Iwasa S and Kumagai M. A novel electrode material for electric double-layer capacitors, Journal of Power Sources, 2005;141(2):386–391.
4. Maldonado-Hódar F J, Moreno-Castilla C and Pérez-Cadenas A F. Catalytic combustion of toluene on platinum-containing monolithic carbon aerogels. Applied Catalysis B: Environmental, 2004;54(4):217–224.
5. Baker W S, Long J W, Stroud R M and Rolison D R. Sulfur-functionalized carbon aerogels: a new approach for loading high-surface-area electrode nanoarchitectures with precious metal catalysts. Journal of Non-Crystalline Solids, 2004;350(15):80–87.
6. Marie J, Berthon-Fabry S, Achard P, Chatenet M, Pradourat A and Chainet E. Highly dispersed platinum on carbon aerogels as supported catalysts for PEM fuel cell-electrodes: comparison of two different synthesis paths. Journal of Non-Crystalline Solids, 2004;350(15):88–96.
7. Smirnova A, Dong X, Hara H, Vasiliev A, Sammes N, Novel carbon aerogel-supported catalysts for PEMFC application. International Journal of Hydrogen Energy, 2005;30: 149–158.
8. Saquing C D, Kang D, Aindow M and Erkey C. Investigation of the supercritical deposition of platinum nanoparticles into carbon aerogels. Microporous and Mesoporous Materials, 2005;80(1–3):11–23.
9. Ihonen J, Jaouen F, Lindbergh G, Lundblad A, and Sundholm G. Investigation of mass-transport limitations in the solid polymer fuel cell cathode. Mathematical model. J. Electrochem. Soc.2002;149(4):A437–A447.
10. Ihonen J, Jaouen F, Lindbergh G, Lundblad A, and Sundholm G. Investigation of Mass-Transport Limitations in the Solid Polymer Fuel Cell Cathode. Experimental. J. Electrochem. Soc.2002;149(4):A448–A454.
11. Uchida M, Fukuoka Yu, Sugawara Ya, Eda N, and Ohta A. Effects of microstructure of carbon support in the catalyst layer on the performance of polymer electrolyte fuel cells. J. Electrochem. Soc. 1996; 143: 2245–2252.
12. Perry M.L., Newman J, and Cairns J. Mass transport in gas-diffusion electrodes: A diagnostic tool for fuel cell cathodes. J. Electrochem. Soc. 1998;145:5–15.
13. Broka K, Ekdunge P. Modelling the PEM fuel cell cathode. J. Appl. Electrochem, 1997;27:281–289.
14. Broka K, Ekdunge P. Oxygen and hydrogen permeation properties and water uptake of Nafion® 117 membrane and recast film for PEM fuel cell.J. Appl. Electrochem, 1997;27(2):117–123.
15. Sasikumar G, Ihm J W and Ryu H. Dependence of optimum Nafion content in catalyst layer on platinum loading. Journal of Power Sources, 2004;132:11–17.
16. Gamburzev S and Appleby A J. Recent progress in performance improvement of the proton exchange membrane fuel cell (PEMFC). J. Power Sources, 2002;107:5–12.

17. Siroma Z, Sasakura T, Yasuda K, Azuma M, Miyazaki Y, Effects of ionomer content on mass transport in gas diffusion electrodes for proton exchange membrane fuel cells. J. Electroanal.Chem., 2003;546:73–78.
18. Qi Z and Kaufman A. Low Pt loading high performance cathodes for PEM fuel cells. J. Power Sources, 2003;107:37–43.
19. Lee S J, Mukerjee S, McBreen J, Rho Y W, Kho Y T and Lee T H. Effects of Nafion impregnation on performances of PEMFC electrodes. Electrochimica Acta, 1998;43(24):3693–3701.
20. J. M. Song, S. Y. Cha and W. M. Lee, Optimal composition of polymer electrolyte fuel cell electrodes determined by the AC impedance method. J. Power Sources, 2001; 94(1):78–84.
21. Uribe FA and Zawodzinski TA. A study of polymer electrolyte fuel cell performance at high voltages. Dependence on cathode catalyst layer composition and on voltage conditioning. Electrochimica Acta, 2002;47(22–23): 3799–3806.
22. Litster S and McLean G. PEM fuel cell electrodes. J. Power Sources, 2004;130(1–2): 61–76.
23. Haile S M. Fuel cell materials and components. Acta Materialia, 2003; 51(19):5981–6000.
24. Ch. Song, Fuel processing for low-temperature and high-temperature fuel cells: Challenges, and opportunities for sustainable development in the 21st century, Catalysis Today, 2002;77(1–2):17–49.
25. Mehta V, Cooper JS. Review and analysis of PEM fuel design and manufacturing. J. Power Sources, 2003;114:32–53.

8

Power Conditioning and Control of Fuel Cell Systems

Elias W. Faraclas, Syed S. Islam, A. F. M. Anwar

8.1 Introduction

The development of Fuel Cell technology, an alternative energy source, involves direct conversion of chemical energy to electrical energy through a controlled chemical reaction. This has attracted attention due to the following attributes:

- This is inherently clean (water and CO_2 byproducts).
- Absence of moving parts makes the system inherently reliable.
- Also very quiet as there is no combustion
- All fuel cells are based on hydrogen, which is an abundant natural resource.

The application of fuel cell-based technology may be in a wide variety of systems, namely:

- Electrical power generation for homes and businesses is attractive mainly for the reduction of pollutants as well as an end to dependency on the limited supply of fossil fuels.
- Fuel cells to power transportation: trains, busses, automobiles - similar motivation.
- Also a huge interest in using small-scale integrated fuel cells to power portable electronics. Its advantages, over traditional [1] batteries are the much higher available power densities and the near instantaneous refueling / recharging time.

As evident from the list of possible applications the development of Fuel Cells covers the entire spectrum of power generation ranging from mega-watts (MW) to watts. This in turn requires the development of different methods of power generation as evident form the different technologies, namely, Proton Exchange Membrane Fuel Cell, Phosphoric, Acid Fuel Cell, and has been discussed in detail in previous chapters.

The reliability of the generated power, however, has yet to be addressed. It is to be noted that the output power of the fuel cell is highly unregulated resulting in a drastic drop in the output voltage with increasing load. Moreover, reliable operation also requires consistent startup, ability to deliver instantaneous power among other things. In this chapter, we present the control electronics required to obtain reliable power out of fuel cells. The sections are organized as follows. In Section 8.2, a brief overview of the generation of electricity is presented. This is followed by a section on power conditioning that provides the details of different control circuit architecture for reliable operation of fuel cell systems. In Section 8.4, the special application towards low power, portable systems is presented. The chapter ends with a conclusion and relevant references.

8.2 Fuel Cell Basics

8.2.1 Physics

While there are several specific fuel cell variations, all are based on the same basic principles of electrochemistry to create electrical power through controlled chemical reactions. The schematic of a typical *Proton Exchange Membrane* (PEM) fuel cell is shown in Figure 8.1. It is to be noted that the discussion to follow is strictly valid for PEM fuel cells; however, similar chemical processes are present in other types of fuel cells as well. Fuel, hydrogen in some form, is applied to the anode under controlled pressure where it is dissociated into ions by an electrolytic membrane according to the following reaction:

$$H_2 \leftrightarrow 2H^+ + 2e- . \tag{8.1}$$

The electrons generated in the above reaction are then free to travel to the cathode through an electrical load connecting the two terminals. The H^+ ions, which are nothing more than protons, travel to the cathode through the *Proton Exchange*

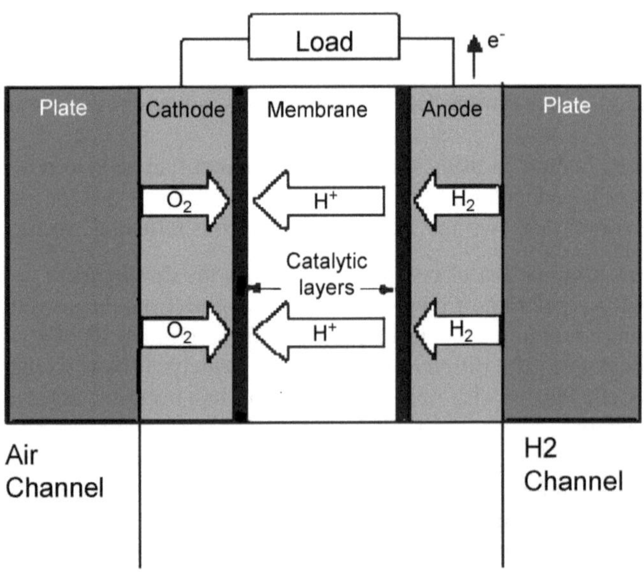

Figure 8.1. Schematic of basic PEM Fuel Cell operation

Membrane (PEM). At the cathode, oxygen is supplied in addition to the incoming protons and electrons that allow them to recombine as water:

$$2H^{+} + \frac{1}{2}O_2 + 2e^{-} \leftrightarrow H_2O\,. \tag{8.2}$$

Thus, in general, a fuel cell operates on chemical reactions involving Hydrogen and Oxygen whose only byproduct is water according to the complete reaction:

$$H_2 + \frac{1}{2}O_2 \leftrightarrow H_2O\,. \tag{8.3}$$

8.2.2 Power Generation

While the maximum voltage of the cell can be determined from underlying thermodynamics, there are also three types of polarizations inherent in fuel cells that degrade the output voltage as a function of the operating current density: activation, Ohmic, and concentration polarizations [2-4]. A typical polarization curve is seen in Figure 8.2 [2, 5]. Activation polarization is due to the limited rates of the chemical reaction at the cathode, which is attributed primarily to catalyst inefficiency. Ohmic losses are the IR losses experienced during transport through the membrane. At very high current densities, concentration polarization occurs where the necessary concentrations of fuel cannot be maintained to support the reactions [2, 3].

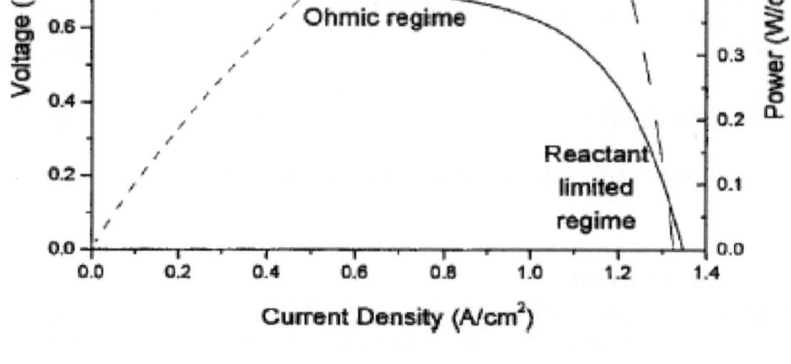

Figure 8.2. Typical polarization curve for H_2/O_2 fuel cell. Helen L. Maynard and Jeremy P. Meyers, "Miniature fuel cells for portable power: Design considerations and challenges", *J. Vac Sci. Technol. B,* 20(4), 2002, pp. 1287–1297

8.2.3 Loss Mechanism

The total voltage across the fuel cell can be written as a combination of the open circuit voltage V_{OC} and the polarization losses V_{act}, V_{ohm}, and V_{conc} as [4]:

$$V_{fc} = V_{OC} - V_{act} - V_{ohm} - V_{conc} . \tag{8.4}$$

The open circuit voltage, V_{OC}, which is the voltage across the FC under no-load, is due to the free energy of the reactants, namely hydrogen fuel and oxygen, and is empirically expressed as [4]:

$$V_{OC} = 1.229 - 8.5\times10^{-4}\left(T_{fc} - 298.15\right) + 4.3085\times10^{-5} T_{fc}\left[\ln\left(p_{H_2}\right) + \tfrac{1}{2}\ln\left(p_{O_2}\right)\right] \tag{8.5}$$

where T_{fc} is the temperature of the fuel cell (K) and p_{H2} and p_{O2} are the partial pressures (atm) of hydrogen and oxygen, respectively. The activation voltage losses are mainly due to the energy required to break and form chemical bonds [4] and can be expressed as a function of the current density by an approximation of the Tafel equation as:

$$V_{act} = v_0 + v_a\left(1 - e^{-ci}\right), \tag{8.6}$$

where, v_o, v_a, and c_1 are constants extracted from experimental data. The ohmic component accounts for the voltage drop across the distributed internal resistance of the FC and is expressed as:

$$V_{ohm} = i \cdot r_{cell} , \tag{8.7}$$

where r_{cell} is the lumped internal resistance.

The concentration losses occur at extremely high current densities where fuel flow cannot keep up with the rate of the required reaction. As this portion of the operating curve is less efficient, concentration losses are often neglected as large scale systems generally operate at the point of maximum efficiency. However, small power supplies, for portable electronics, will operate in this portion of maximum power density [6]. The concentration losses are approximated as [4]:

$$V_{conc} = i\left(c_2 \frac{i}{i_{\max}}\right)^{c_3}, \tag{8.8}$$

where i_{max}, c_2, and c_3 are empirically determined constants.

8.2.4 Equivalent Circuit

It is important to note that, from an electronics viewpoint, a fuel cell is a type of current source whose output voltage is a function of the free energy of the fuel used and the thermodynamics of the reaction. Figure 8.3 shows the equivalent circuit model of the fuel cell as a voltage-dependent current source. Here r_{cell} is the previously defined internal resistance of the membrane, V_{OC} is the open circuit voltage, and C is the internal capacitance.

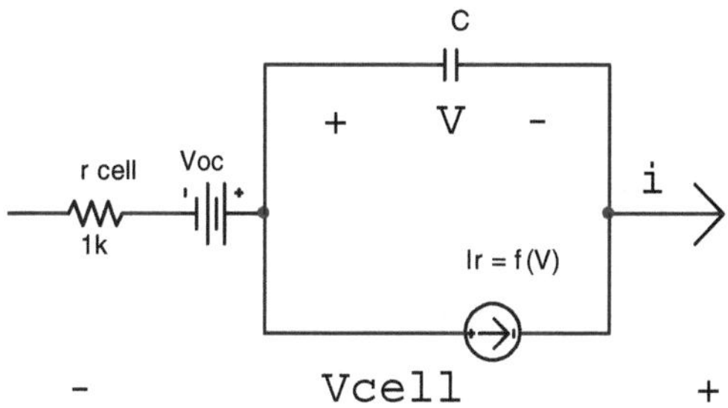

Figure 8.3. Equivalent circuit of a fuel cell

8.3 Power Conditioning

The output power of a fuel cell is highly unregulated and stability is a serious concern [7, 8]. The output voltage of a fuel cell has a very low magnitude and varies greatly with increasing load. Bernay *et al.* [7] reported a decrease in cell voltage from 0.82 V to 0.62 V as the current density increased from $10mA/cm^2$ to $350mA/cm^2$ for a phosphoric acid fuel cell (PAFC). Similar variations in cell voltage with increasing operating current densities are reported for proton exchange membrane (PEMFC), molten carbonate (MCFC), solid oxide (SOFC), alkaline, direct methanol (DMFC) and protonic ceramic (PCFC) fuel cells. Argyropoulos *et al.* [8] analyzed the dynamic response of the direct methanol fuel cell under variable load conditions. Under pulsating load conditions the reported cell voltage variation was from 600 mV to 180 mV with an applied load pulse of $100mA/cm^2$. A typical V-I characteristics is shown in Figure 8.2.

Limited gas exchange efficiency has been identified as the cause of poor voltage regulation in fuel cell systems, as shown in Figure 8.3. Significant cell voltage variation during no load period was also observed. Thus power conditioning circuitry to accommodate for the cell to cell inconsistencies, load-driven output voltage variation, and redundant power supply is a critical component in any fuel cell system.

8.3.1 Fuel Cell Systems

The theoretical limit of output voltage from a single cell is of the order of 1.2 V [5] and a number of fuel cells can be connected in parallel/series to reach the required power/voltage levels. An electronic power conditioning system is essential to supply power from the fuel cells to customers as well as to the power system grids. A conceptual fuel cell system for portable low power applications such as cell phone or laptop computers is shown below:

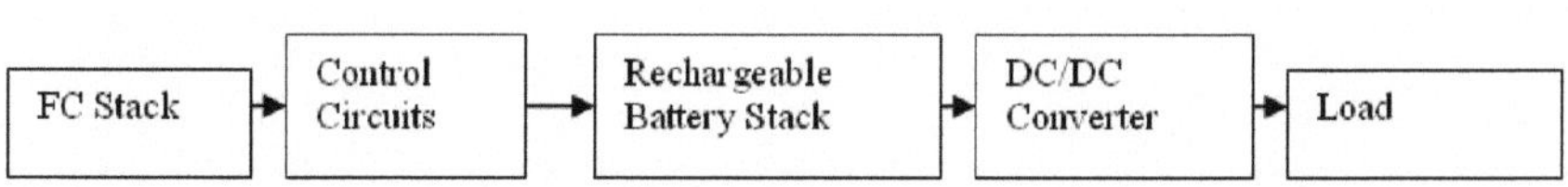

The output of fuel cells is DC. In this application, the purpose of the fuel cell is to charge the rechargeable battery stacks used to drive the portable devices. Control circuits are essential to prevent overcharging. DC/DC converters may be used to extend/adjust the required driving DC voltage across the load. If isolation between load and fuel cell is required, then the DC/DC converter can be replaced by DC/AC transformer-rectifier combination. The output of the fuel cell is converted to an intermediate high frequency AC voltage. A small transformer is used to convert the AC voltage to the required level that is converted to DC by using a rectifier.

While the electronics in the feedback control system operate extremely rapidly, there is a significant time delay in the fuel cell itself between the increase of fuel and the corresponding increase in output voltage. Additionally, a portable power supply also needs to have the capability of supplying the necessary power for self-starting. Thus, for reliable operation, storage facilities should be integrated with a fuel cell system to accommodate load initiated transients as well as the initial electrical energy for self startup. Solmecke *et al.* [9] have reported a solar-hydrogen storage system where energy produced by a photovoltaic cell was partly used for the generation and storage of H_2 before supplying to a fuel cell. Such a storage system is expensive and needs more space. A better solution is the integration of a battery and charging circuitry into the electrical subsystem [7]. As shown in Figure 8.4, the charging circuit connects a battery to the fuel cell system to store energy during off-load period for later use during peak load. Such an energy storage facility is compact and relatively inexpensive. Additionally, a comparator circuit monitors the output voltage of the fuel cell so that additional fuel may be supplied if necessary.

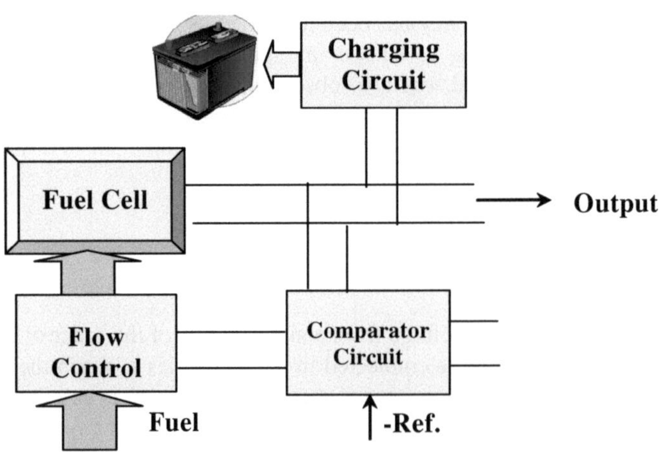

Figure 8.4. Schematic of the fuel cell integrated with the battery subsystem, charging circuitry, and comparator circuit for voltage regulation

8.3.2 Storage System

The redundant storage system consists of a battery and a charging system. The battery charging circuit is shown in Figure 8.5. During off-load periods when the fuel cell voltage goes above the battery voltage then charging of the battery takes place through diode D1 and inductor L1. D1 directs the current, while L1 eliminates any current discontinuity to limit harmonic generation. During the peak load period the fuel cell voltage decreases and when the fuel cell voltage falls below the battery voltage, the charging circuit is disconnected and energy supply from the battery takes place through diode D. The L-C filter is required to suppress harmonics by eliminating any voltage or current discontinuities.

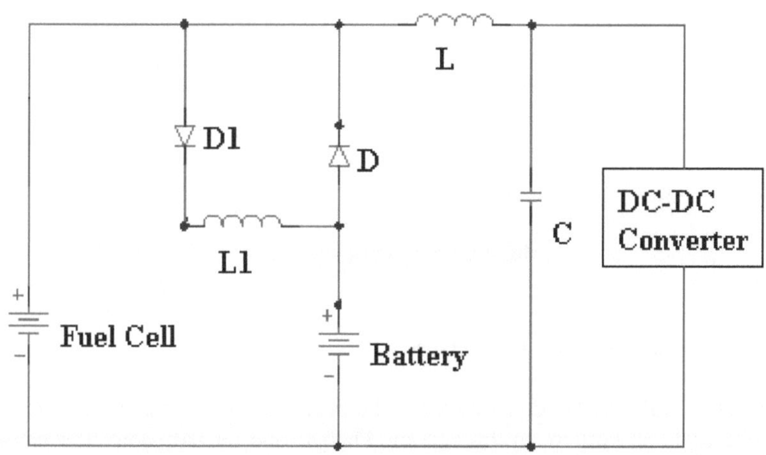

Figure 8.5. Battery Charging Circuit

8.3.3 Voltage Regulation

The load voltage control circuit is triggered automatically to adjust the fuel intake of the fuel cell when the battery storage system becomes inadequate to supply power to the load. At high system load, the fuel cell voltage drops due to poor regulation and polarization effects. The charging system comes out of the charging mode and enters into the discharging mode to improve voltage regulation. If the load demand is even higher than what can be supplied by battery and fuel cell jointly, the load voltage drops that triggers the load-voltage control circuits.

At the front end of the load-voltage control system, a comparator circuit is placed to detect the magnitude of load voltage deviation from a prespecified reference. The comparator circuit is shown in Figure 8.6. V_{out} is the voltage across the load. The output voltage of the operational amplifier (opamp) is given by,

$$V_{comparator} = \left(-V_{out} + V_{REF}\right)\frac{R_f}{R}, \tag{8.9}$$

with $R_f = R$ and $V_{comparator} = V_{REF} - V_{out}$, with a single bias voltage connection as shown in Figure 8.6, the comparator circuit gives 0 V at the output when $V_{REF} < V_{out}$. With $V_{REF} > V_{out}$, the difference appears across the output. The output of the comparator controls the openings of the valve to the fuel intake system to adjust the fuel intake proportional to the deviation of V_{out} from V_{REF}.

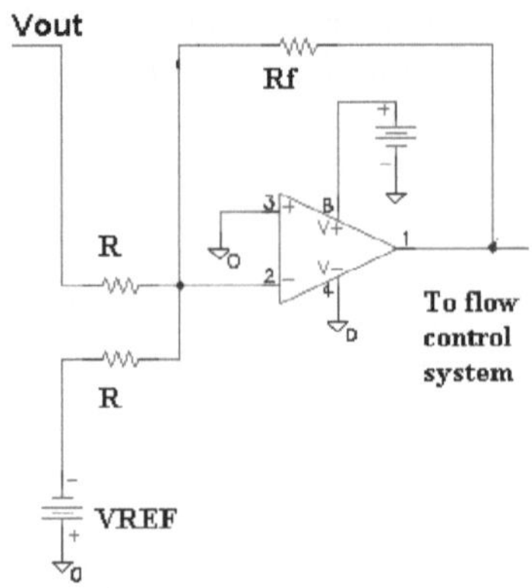

Figure 8.6. Comparator circuit

8.3.4 DC/DC and DC/AC Converters

The DC/DC and DC/AC converters are the principal power electronic circuitries used in a fuel cell system. In this section, DC/DC and DC/AC converter architectures are described following a discussion on possible power transistors.

8.3.5 Power Transistors

The most commonly used power transistors in power conversion circuits are:

1) Bipolar junction transistors (BJTs)
2) Metal-oxide-semiconductor field-effect-transistors (MOSFETs)
3) Static induction transistors (SITs)
4) Insulated-bipolar transistors (IGBTs)

Conventional converters are implemented using MOSFETs, power transistors or IGBTs as switches [1]. Most of the commercially available devices are mostly Si-based and are subjected to lower power handling capabilities. Recently, GaN and SiC based devices have emerged as promising devices for high power circuits. Because of higher breakdown voltage, higher carrier saturation velocity and the possibility to grow on higher thermal conductivity materials, these devices can deliver high power at high temperature and high frequency. GaN based devices have been demonstrated to operate at 9.8 W/mm output power at 8 GHz [10] and operating temperature is reported up to 750°C [11]. SiC-based devices have been demonstrated to operate at a power density of 700 W/cm^2 with a reverse blocking voltage as high as 1800 V [12]. Thus, while Si-based devices are still the most prevalent, GaN/SiC based devices have been identified as suitable candidates to be used as power electronic switches and should be expected soon to be the *de facto* material families for power transistors.

8.3.6 DC/DC

Often, a fuel cell is used as a DC power generator with applications ranging from electric motors in automobiles to laptop computers and other portable electronics. As mentioned earlier, the primary obstacles for integrating fuel cells with modern electronics are the low output voltage of the cell combined with its instability over the range of electrical loading. Furthermore, the low output voltages (~1 V) are not high enough to power typical DC applications (12–48 V). A step up DC/DC converter is used to amplify the voltage in conjunction with a feedback fuel-control system to maintain the appropriate DC level, as seen in Figure 8.7.

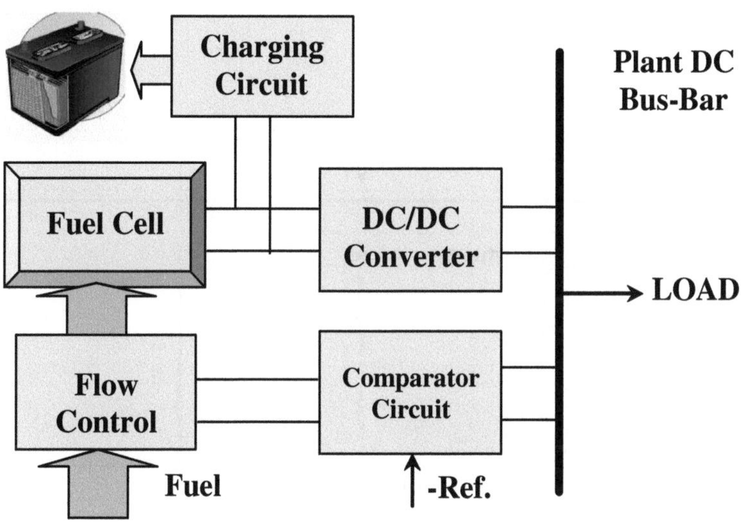

Figure 8.7. Complete system including the DC/DC converter

8.3.6.1 Step Down DC/DC converter

Step down or buck type DC/DC converters are used for direct down conversion of DC voltage levels. A step-down DC/DC converter with resistive load and the timing diagrams are shown in Figure 8.8.

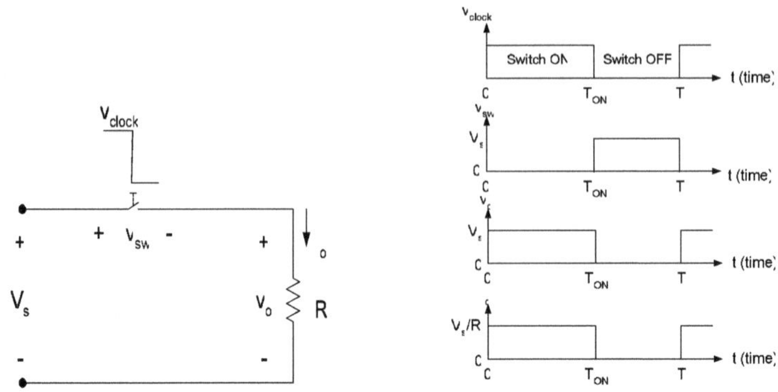

(a) (b)

Figure 8.8. Step-down DC/DC converter (a) circuit schematics and (b) timing diagrams

When the switch is ON ($0 \leq t \leq T_{ON}$), the voltage across the switch is ideally zero and the total DC supply voltage is across the load resistance R. The output voltage reduces to zero as the total supply voltage appears across the ideal switch

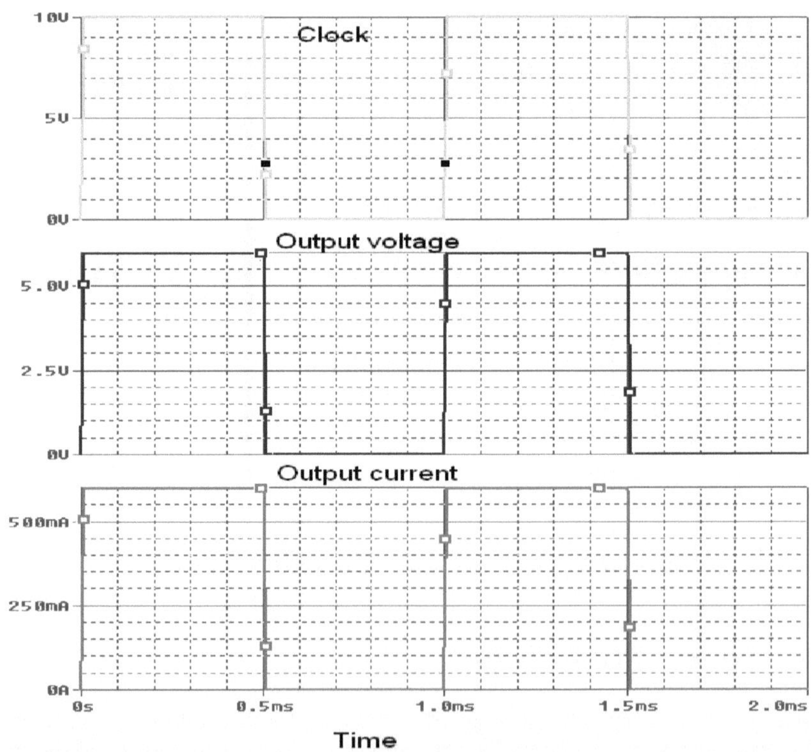

Figure 8.9. SPICE simulation results of a step-down converter ($V_s = 6\,V$, $k = 0.5$, $T = 1ms$ and $R = 10\,\Omega$)

when the switch is OFF ($T_{ON} \le t \le T$). The average value of the output voltage is given by:

$$V_{o,avg} = \frac{1}{T}\int_0^T v_o(t)\cdot dt = \frac{1}{T}\int_0^{T_{ON}} V_s \cdot dt = V_s \cdot \frac{T_{ON}}{T} = V_s \cdot k\,, \tag{8.10}$$

where, k is the duty cycle (T_{ON}/T). The average output power can be expressed as:

$$P_{o,avg} = \frac{1}{T}\int_0^T v_o(t)\cdot i_o(t)\cdot dt = \frac{1}{T}\int_0^{T_{ON}} V_s \cdot \frac{V_s}{R}\cdot dt = \frac{V_s^2}{R}\cdot\frac{T_{ON}}{T} = \frac{V_s^2}{R}\cdot k\,\cdot \tag{8.11}$$

Thus, both the average DC output voltage and power can be controlled by the duty cycle of the clock. As the duty cycle is varying from 0 to 1, $V_{o,avg}$ and $P_{o,avg}$ vary from 0 to V_s and from 0 to V_s^2/R, respectively. Figure 8.9 shows the SPICE simulation results of a step-down converter with 50% duty cycle. Here a MOSFET is used as the switch.

8.3.6.2 Step Up DC/DC converter

A transistor operated boost type DC/DC converter [13] is outlined below. The use of a single transistor makes the converter highly efficient. Compared to thyristor driven DC/DC converters, the transistor driven converters are simple as the latter do not require complex control circuitry to turn the device OFF. Additionally, this topography avoids the size and weight penalties inherent in using a transformer based converter. A single-level of the converter is shown in Figure 8.10. The circuit operation can be divided into two modes. In mode 1, with $\mathbf{Q_1}$ ON, the input current increases and flows through the inductor *L* and transistor $\mathbf{Q_1}$. In mode 2, with $\mathbf{Q_1}$ OFF, the input current flows through the inductor *L*, diode D_m, capacitance *C*, and load. The load current starts falling and the energy stored in the inductor *L* is transferred to the load boosting the output voltage. The process continues until the transistor is turned ON again at the beginning of the next cycle.

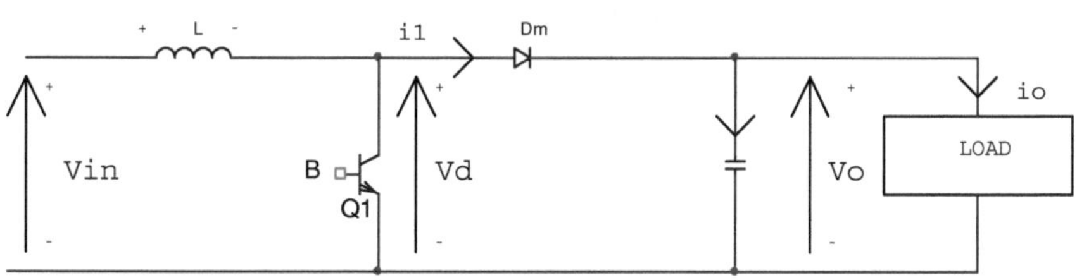

Figure 8.10. Transistor operated DC/DC boost converter

The smooth control of the output voltage of the single-level converter is achieved by changing the pulse width of the input signal applied at the input of the driving transistor. The output voltage (V_o) as a function of the input voltage (V_{in}) can be expressed as:

$$V_o = \frac{V_{in}}{1-k}, \tag{8.12}$$

where k is the duty cycle of the input signal to the transistor. For practical implementation, a smooth variation of k can be achieved by using micro-controller ICs. Moreover, the input current is continuous, which ensures minimum harmonic generation.

Figure 8.11 and Figure 8.12 show the simulation waveforms with k = 0.3 and k=0.6, respectively. In this simulation, L = 1 mH, C = 220 μF and load resistance R_L = 100 Ω. The switching frequency is 20 kHz. The output voltage increases from 16 V to 27 V as the duty cycle increases from 30% to 60%. In addtion, a slight improvement in settling time is observed with increasing duty cycle. With k = 0.3, an overshoot in the output voltage is observed before settling down to 16 V. The overshoot is eliminated for k = 0.6.

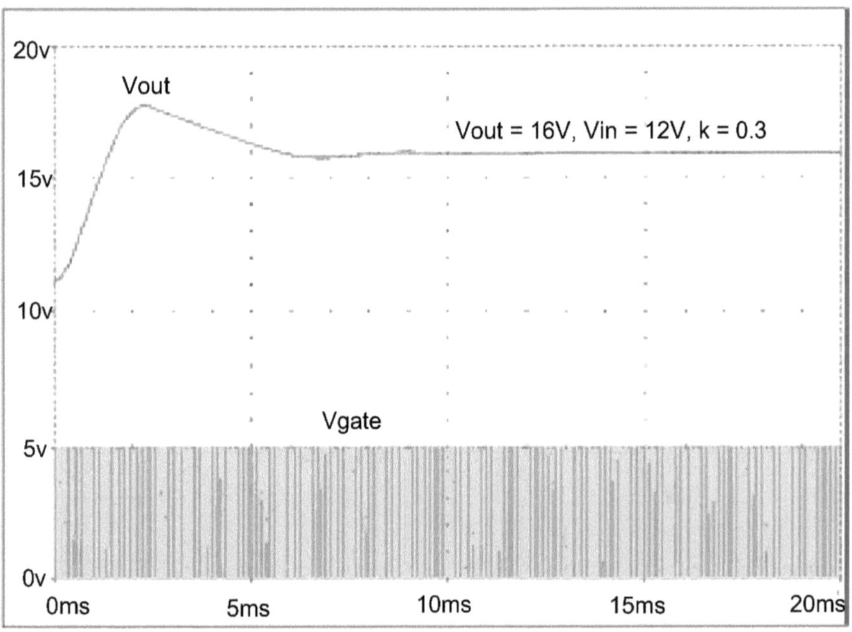

Figure 8.11. Output voltage and gate pulse waveforms with k = 0.3 (L = 1mH, C = 220 μF, R =100 Ω)

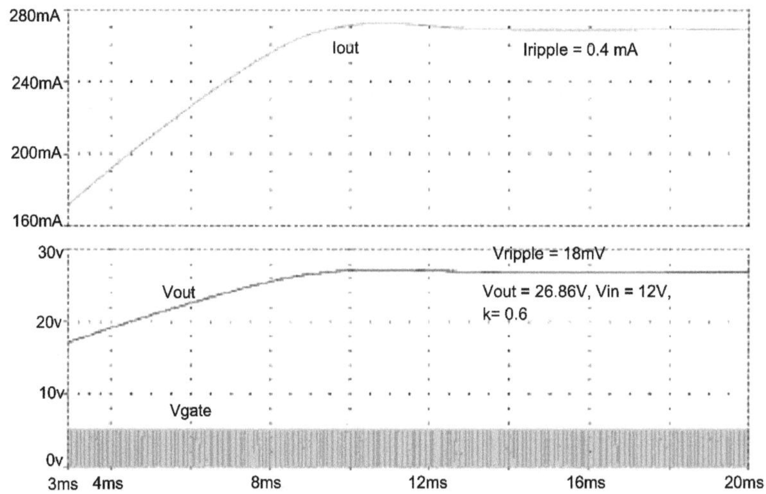

Figure 8.12. Output voltage and gate pulse waveforms with $k = 0.6$ (L = 1mH, C = 220 μF, R =100 Ω)

8.3.7 DC/AC

Fuel cells are also capable of generating electrical power for homes and businesses. In this case it is necessary to connect the output of the fuel cell directly to the main power grid. A DC/AC converter or inverter is used to convert DC fuel cell voltage into AC voltage to drive AC loads or to supply power to the system grid. The circuit diagram of the DC/AC converter is shown in Figure 8.13. While the following discussion is for generation of a single phase, it can be easily extended to 3-phase power generation, shown schematically in Section 8.3.10 in Figure 8.21 and Figure 8.20.

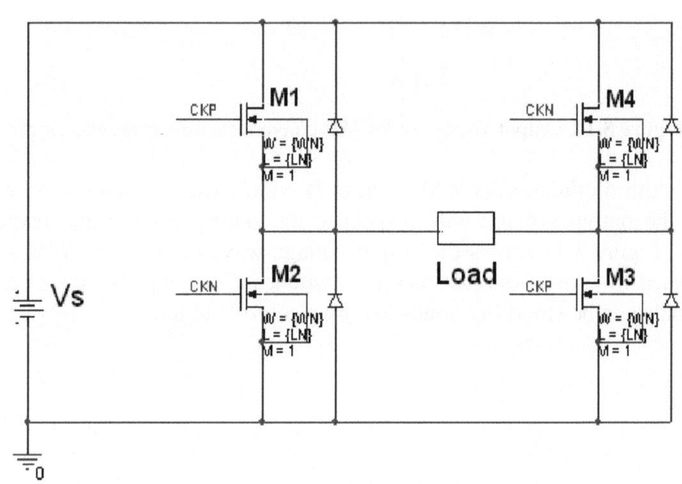

Figure 8.13. Circuit diagram of a single-phase DC/AC converter

For single-phase implementation, two nonoverlapping clocks are required. The frequency of the clock determines the frequency of the output AC signal. On one half cycle, M1-M3 switches are ON. On the following half cycle, M2-M4 switches are ON which applies voltage across the load in the opposite direction compared to the previous half cycle. SPICE simulation results of a single-phase inverter are shown in Figure 8.14.

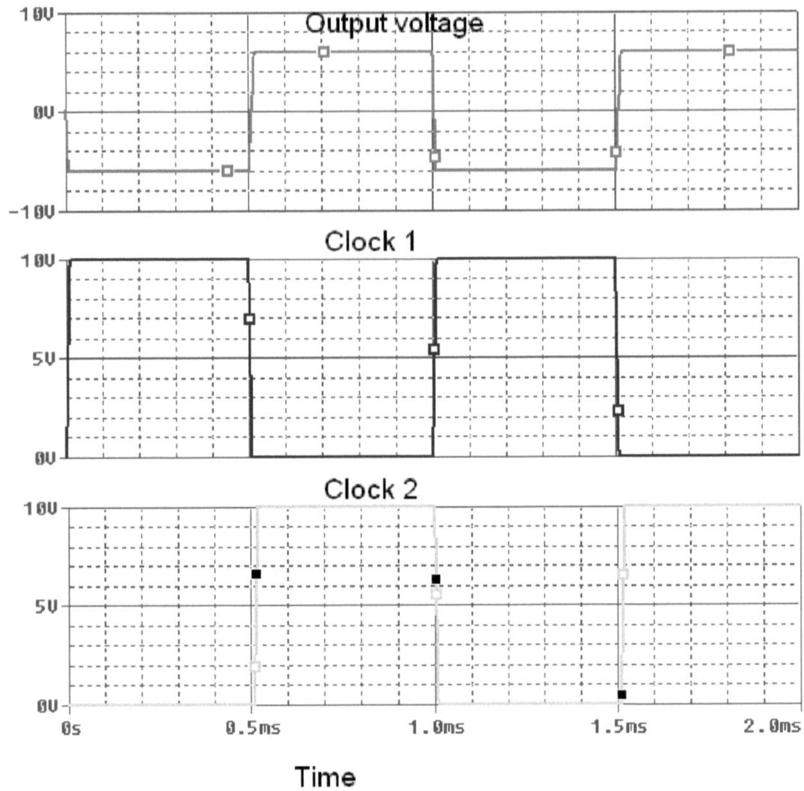

Figure 8.14. Output voltage of DC/AC converter with rectangular clocks

Pulse-width modulation (PWM) control is widely used to achieve a better control over the output voltage with respect to the supply voltage and inverter load variations. Figure 8.15 shows the output voltage waveform when PWM clock signals are used to control operation of the switches. The pulse widths of the PWM clock signals can be smoothly controlled by a control signal.

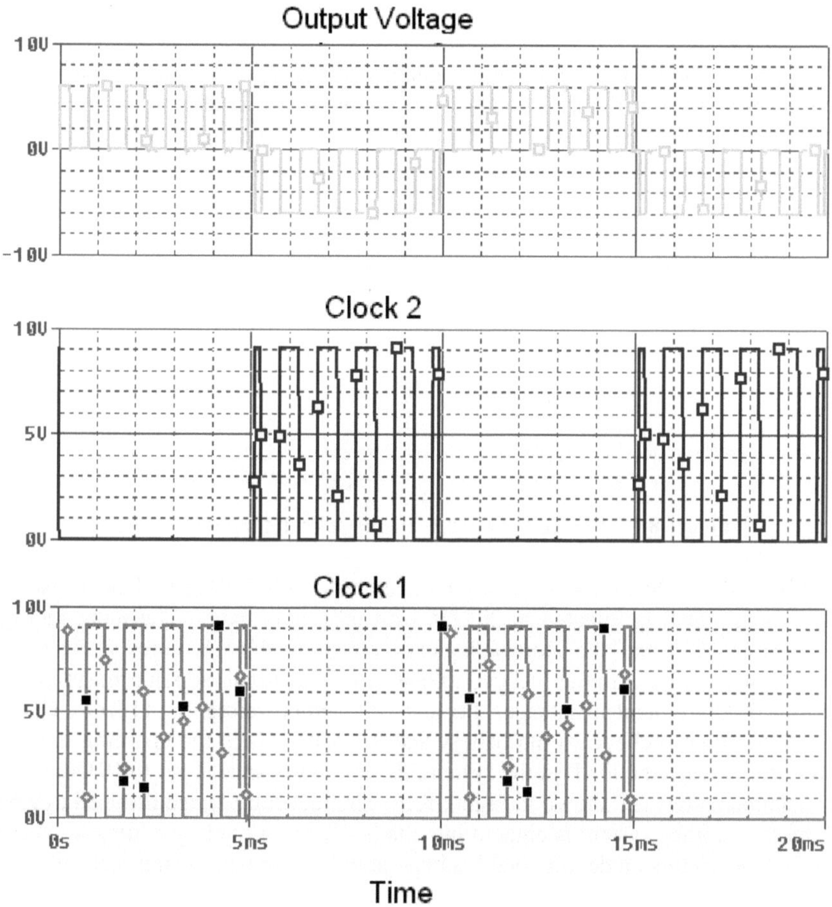

Figure 8.15. Output voltage of a DC/AC converter with PWM clocks

Following the DC/AC converter, a low pass filter is used to ensure the elimination of any harmful harmonics. The complete system diagram is seen in Figure 8.16.

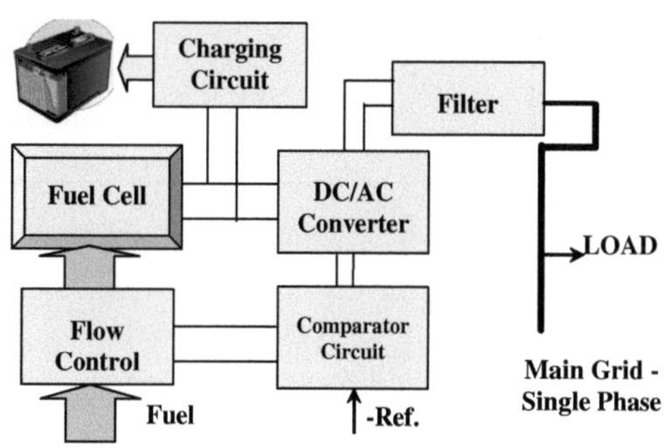

Figure 8.16. Schematic of fuel cell system for AC power generation

8.3.8 Simulation of Fuel Cell Power Conditioning Systems

Figure 8.17, shows a possible implementation of the fuel cell power conditioning system described in Figure 8.7. The system consists of a battery charging circuit, see Figure 8.5, and a step-up DC/DC converter circuit, see Figure 8.10.

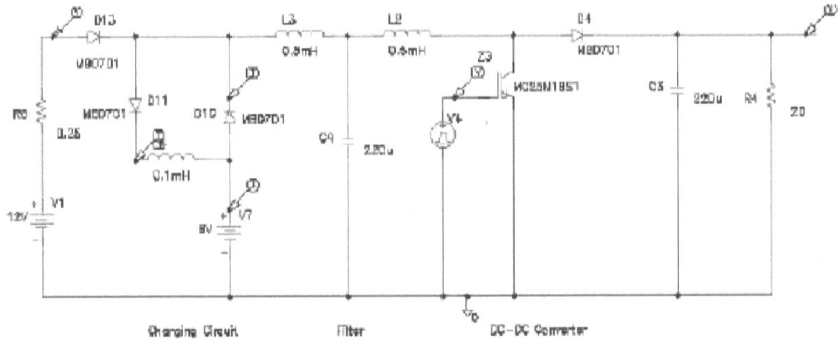

Figure 8.17. Fuel cell power conditioning circuit for portable devices

Figure 8.18, shows the simulation waveforms with 100 Ohms load resistance. Switching signal is selected at 20 kHz with 80% duty cycle. The input fuel cell voltage is 12 V and the storage battery voltage is 9 V. A series resistance of 0.25 Ohms is used with the fuel cell to realize voltage regulation during simulation. With 100 Ohms load resistance, the fuel cell output voltage is 11.21 V. Current flows from the fuel cell to the battery and the battery is in the charging mode. In order to increase the load on the system, the load resistance is set to 10 Ohms. The simulation results are shown in Figure 8.19. The fuel cell voltage dropped to 9.6 V. A battery storage system is entered into the discharging mode and supplies current to the load. In this mode, the total load is shared by both battery and fuel cell.

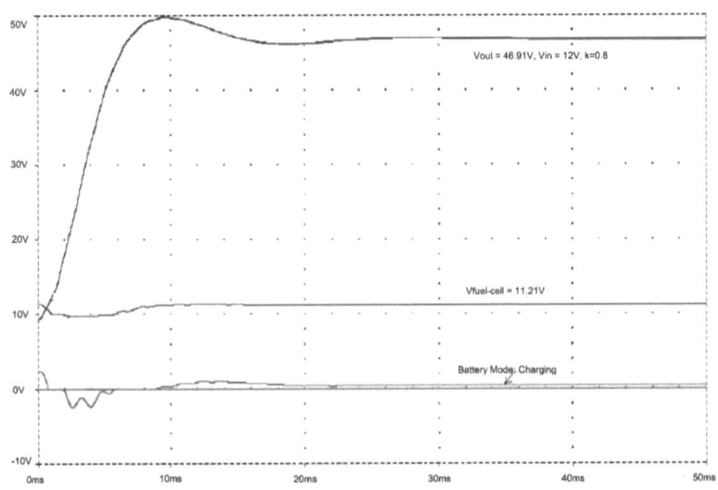

Figure 8.18. Output waveforms with 100 Ohms load resistance (Battery is in charging mode)

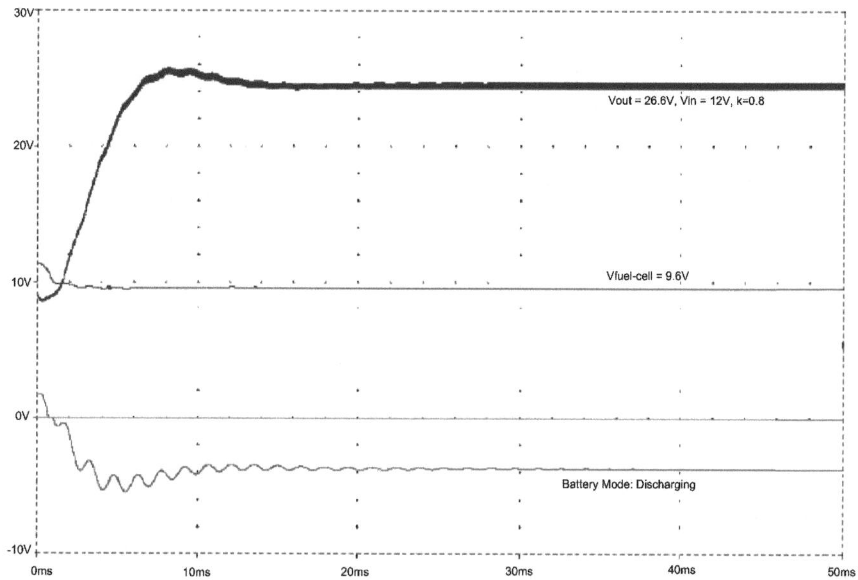

Figure 8.19. Output waveforms with 10 Ohms load resistance (Battery system is in discharging mode)

8.3.9 Low Power Applications

The DC/DC converter shown in Figure 8.7 is appropriate when a large stack of fuel cells can be created to provide an initial output of several volts. However, for extremely low supply voltage, it can be difficult to obtain the necessary voltage amplification with a practically sized inductor. In this case, the previously discussed DC/DC converter is fed into a Full-Bridge Inverter. The small AC signal is then stepped up by a transformer that is then rectified by a Bridge Rectifier to obtain the desired DC output. This is seen schematically below.

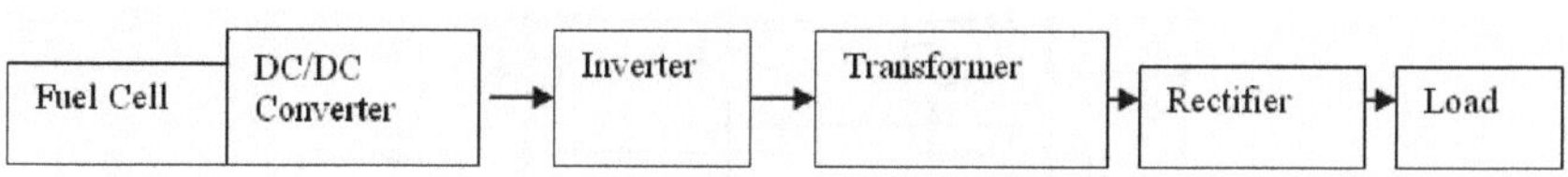

8.3.10 Multi-level DC/DC and DC/AC Converter

Conventional DC/AC converters (inverters) are required to connect AC loads or power systems to the DC bus-bars necessitating the use of transformers that, at the power supply frequency, are rather bulky and account for 50% of the total loss of the system. Additional transformers are required to upgrade the AC voltage to the required level and to control the reactive power generation for system stabilization, while maintaining the required supply voltage. Multi-level inverters, using a number of switches, can successfully achieve DC/AC conversion with efficiency greater

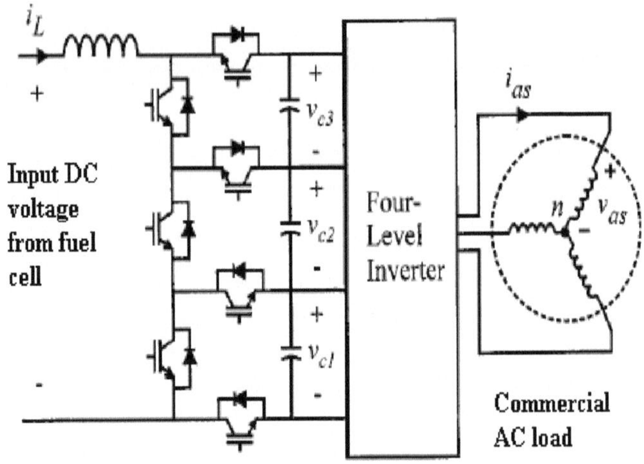

Figure 8.21. Four level DC/DC converter from a single DC level into a four level inverter and a 3-phase AC load

than 90% and total harmonic distortion less that 5% at low power. Its application is restricted at high power due to the complexity of the circuit as well as the requirement of a rather large number of capacitors and switches.

The multi-level converters are an extension of the concept reported by Tolbert *et al.* [14] and Corzine *et al.* [15] where a voltage from a battery was used as input with a limited number of levels. In the four-level DC/DC converter shown in Figure 8.21, the input voltage is divided before applying across each switch. The switches are operated at low voltage and are less expensive. Varying the switching sequence and duty cycle of the driving signal can vary the output voltage across the capacitors. The four-level DC/AC converter, shown in Figure 8.20, gets balanced

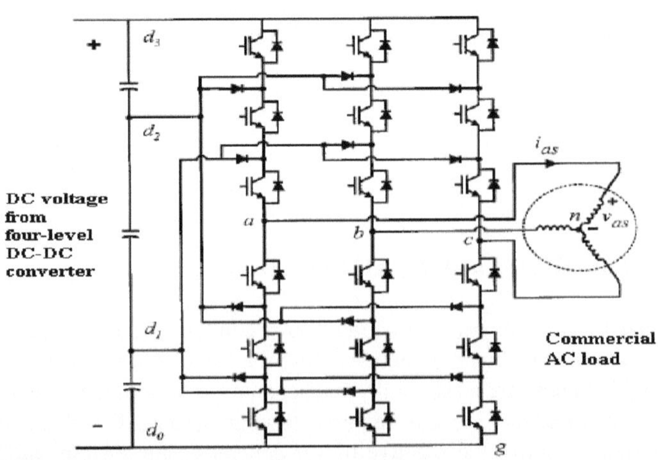

Figure 8.20. Four level DC/DC converter from multiple DC levels into a four level inverter and a 3-phase AC load

DC input voltages across the capacitors from the DC/DC converter. Increasing the number of stages controls the magnitude of the output AC voltage. The output voltage waveform increases in small steps, as shown in Figure 8.22, and harmonic distortion is very low even without the use of a filter.

Figure 8.23 shows the feedback control loop for static reactive power generation and output voltage stabilization scheme. The magnitude and phase angle of the output voltage of the DC/AC converter are sensed by the magnitude and phase detector. The output signals of the detector determine the timing angles of the switching signals of the DC/AC converter to control the static reactive power generation and output voltage magnitude. The switching signal generators are often programmable and may be micro-controller based.

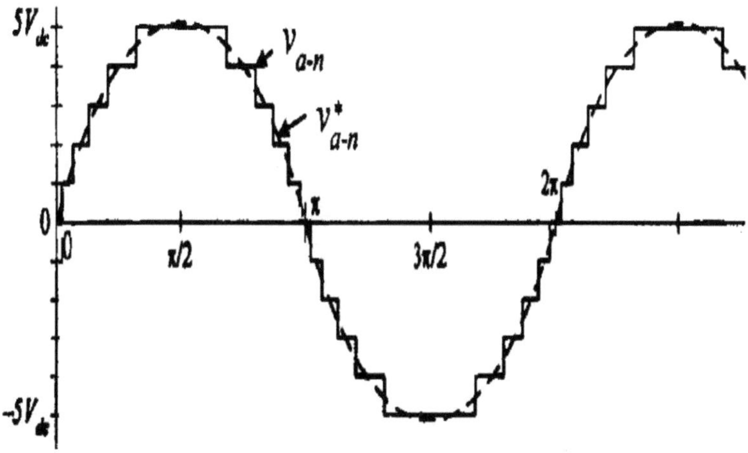

Figure 8.22. Output waveform of multi-level DC/AC converter

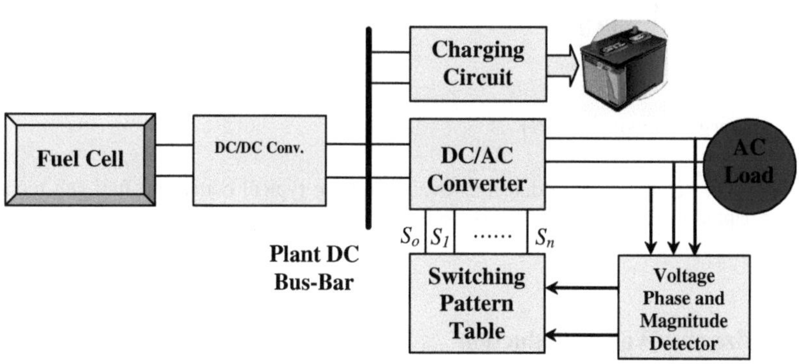

Figure 8.23. Static reactive power generation and output voltage stabilization loop

8.4 Small Scale Systems

8.4.1 Increased Available Power

As portable electronics continue to be ever more power hungry, the power supplied by traditional batteries becomes more and more inadequate. It is clear, from Figure 8.24, that fuel cell technology promises much higher power density (Wh/A) than current battery technologies.

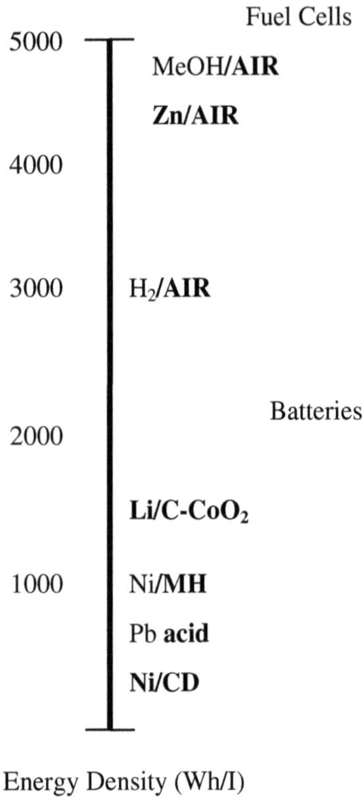

Figure 8.24. Relative energy densities available from typical batter and fuel cell technologies. Adapted from C.K. Dyer, "Fuel Cells for Portable Applications", *J. of Power Sources*, vol. 106, 2002, pp. 31–34

8.4.2 Size and Weight Reduction

In addition to the requirement for higher power densities, portable electronics continuously suffer from the increased size and weight of the battery systems used to power them. Not only is the battery large and heavy, but the battery charger, which can be eliminated even in a hybrid application using the fuel cell as the charger, contributes significantly to this. As seen in the Figure 8.25. Fuel Cell based systems

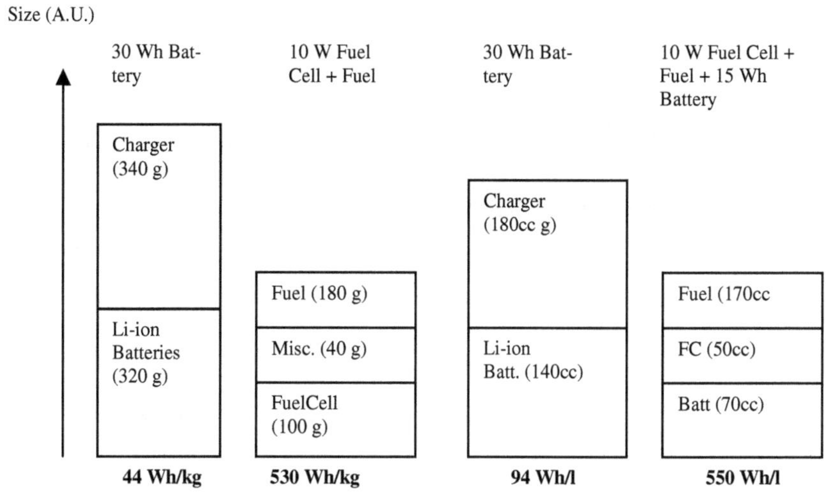

Figure 8.25. Figure comparing the power per weight and power per size of battery and fuel cell systems. Adapted from C.K. Dyer, "Fuel Cells for Portable Applications", *J. of Power Sources*, vol. 106, 2002, pp. 31–34

simultaneously increase the available power density significantly, while dramatically reducing the size and weight of the power supply.

8.4.3 Difference in Philosophy

For large scale systems, the emphasis is on the creation of a high powered, efficient stack of cells. The layout of the stack and the many interfaces of the fuel cell with the power grid or the environment are secondary considerations. However, fuel cells for powering portable devices must be designed in conjunction with the entire system, not just optimizing the performance of the stack.

Since the output voltage of a fuel cell is typically around 1 V, large fuel cell systems employ a "stack" where multiple fuel cells are placed on top on one fanother, with the contacts connected in series, to increase the output voltage. However, this method is unrealistic for portable power supplies due to size limitations in portable systems. Instead, architectures for planar stacks need to be developed, see Figure 8.26.

O'Hayre has promoted PCB (printed circuit board) technology for the integration of small-scale fuel cells with electronic systems [16]. Advantages over other methods include "increased design flexibility, potentially higher power densities, ease of device integration, and improved packaging form factors" that are inherent strengths of PCB technology. Maynard and Meyers have proposed two small-scale fuel cell systems grown directly on a silicon substrate, for seamless integration with existing electronics. One design uses two separate Si wafers for the anode and cathode contacts while the other is a monolithic design incorporating both contacts on a single Si wafer [2].

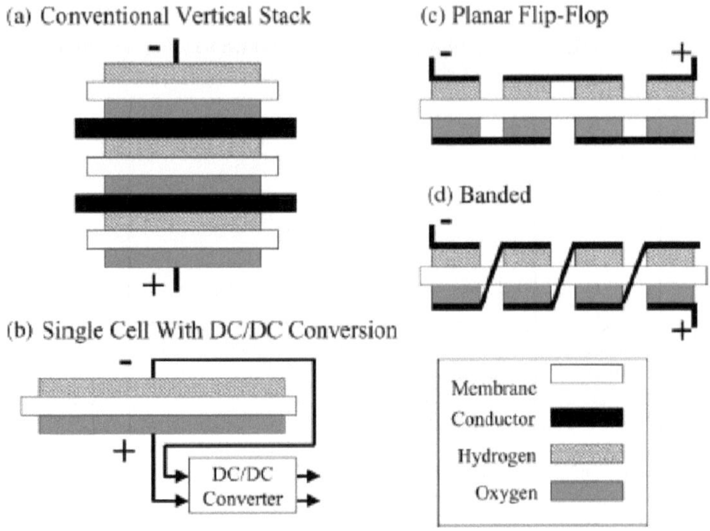

Figure 8.26. Stack designs for membrane fuel cells: (a) conventional vertical (bipolar) stack; (b) single cell with DC/DC conversion; (c) planar flip-flop configuration; (d) planar banded configuration. From R. O'Hayre *et al.* "Development of portable fuel cell arrays with printed-circuit technology", *J. of Power Sources,* 124 (2003), 459–472

Also, a rechargeable battery system can be considered completely isolated from the user and the environment. This is not true in fuel cell systems that depend on the environment for operation. Moreover, especially for portable electronics, the user must be adequately shielded from both the fuel supply as well as the thermal and gaseous byproducts [2, 6].

8.5 Conclusion

This chapter has outlined the power conditioning and control electronics required to obtain reliable power out of Fuel Cells. It is important to realize that only with the proper development and adaptation of these circuits can fuel cells be used effectively in the broad spectrum of applications that are being pursued. When designing a power conditioning circuit, it is important to combine state of the art semiconductor technologies with optimal circuit topologies.

8.6 References

1. G.J. Ball, D.C.B. Jr., W. Gish, W.A. Guro, E.W. Kalkstein, A. Pivec, P.W. Powerll, F. Soudi, J.W. Stevens, M. Yalla, and R. Zavadil, "Summary of static power converters of 500 kW or less serving as the relay interface for non-conventional generators". *IEEE Trans. Power Delivery*, 1994. **9**(3): p. 1325–1331.
2. H.L. Maynard and J.P. Meyers, "Miniature fuel cells for portable power: Design considerations and challanges". *J. Vac. Sci. Technol. B*, 2002. **20**(4): p. 1287–1297.

3. S.K. Mazumder, K. Acharya, C.L. Haynes, R. Wiilliams, M.R.v. Spakovsky, D.J. Nelson, D.F. Rancruel, J. Hartvigsen, and R.S. Gemmen, "Solid-Oxide-Fuel-Cell Performance and Durability: Resolution of the Effects of Power-Conditioning Systems and Application Loads". *IEEE Trans. on Power Electronics*, 2004. **19**(5): p. 1263–1278.
4. J.T. Pukrushpan, H. Peng, and A.G. Stefanopoulou, "Modeling and Analysis of Fuel Cell Reactant Flow for Automotive Applications". *J. of Dynamic Systems, Measurment and Control*, 2004. **126**(1): p. 14–25.
5. F. Blaabjerg, Z. Chen, and S.B. Kjaer, "Power Electronics as Efficient Interface in Dispersed Power Generation Systems". *IEEE Trans. on Power Electronics*, 2004. **19**(5): p. 1184–1194.
6. C.K. Dyer, "Fuel Cells for Portable Applications". *J. of Power Sources*, 2002. **106**: p. 31–34.
7. C. Bernay, M. Marchand, and M. Cassir, "Prospects of different fuel cell technologies for vehicle applications". *J. of Power Sources*, 2002. **108**: p. 139–152.
8. P. Argyropuolos, K. Scott, and W.M. Taama, "Dynamic response of the direct methanol fuel cell under variable load conditions". *J. of Power Sources*, 2000. **87**: p. 153–161.
9. H. Solmecke, O. Just, and D. Hackstein, "Comparison of solar hydrogen storage systems with and without power electronic DC-DC converters". *Renewable Energy*, 2000. **19**: p. 333–338.
10. Y.-F. Wu, D. Kapolnek, J. Ibbetson, P. Parikh, B. Keller, and U.K. Mishra, "Very high power density AlGaN/GaN HEMTs". *IEEE Trans. on Electron Dev.*, 2001. **48**(3): p. 586–590.
11. I. Daumiller, C. Kirchner, M. Kamp, K. Ebeling, L. Pond, C.E. Weitzel, and E. Kohn, "Evaluation of AlGaN/GaN HFET's up to 750°C". *Device Res. Conf. Dig.*, 1998: p. 114–115.
12. D. Peters, R. Schorner, P. Friedrichs, J. Volkl, H. Mitlehner, and D. Stephani, "An 1800V triple implanted vertical 6H-SiC MOSFET". *IEEE Trans. on Electron Dev.*, 2001. **46**(3): p. 542–545.
13. M.H. Rashid, *Power Electronics: circuits, devices and applications, Second Edition*. 1993, Eaglewood Cliffs, NJ: Prentice Hall.
14. L.M. Tolbert, F.Z. Peng, and T.G. Habetler, "Multi-level converters for large electric drives". *IEEE Trans. Industry Appl.*, 1999. **35**(1): p. 36–44.
15. K.A. Corzine and S.K. Majeethia, "Analysis of a novel four-level DC-DC boost converter". *IEEE Trans. Industry Appl.*, 2000. **36**(5): p. 1342–1350.
16. R. O'Hayre, D. Braithwaite, W. Hermann, S.-J. Lee, T. Fabian, S.-W. Cha, Y. Saito, and F.B. Prinz, "Development of portable fuel cell arrays with printed-circuit technology". *J. of Power Sources*, 2003. **124**: p. 459–472.

9

Microbial Fuel Cells

Ken Noll

9.1 Microbial Fuel Cells

9.1.1 Introduction

Cellular life exists at the interface between electrochemical extremes. The energy of most living cells depends on the transfer of electrons from intracellular, electrically reduced biochemicals to oxidized extracellular acceptors. For almost one hundred years investigators have tried to tap into these processes in microbes for electrical power generation. Efforts have been made to use microbes as complex catalysts to oxidize relatively inexpensive organic and inorganic substrates as fuels in compact spaces in microbial fuel cells (MFCs). However, natural selection does not favor microbial metabolism under such conditions. Evolution has shaped microbes to use their growth substrate efficiently to reproduce under changing environmental conditions. Consequently, they are not optimized for use in MFCs, where electrons derived from substrate oxidation go only to the anode and not to cell growth.

In contrast to conventional fuel cells, MFCs operate under relatively mild conditions, use a wide range of organic fuels, and do not use expensive precious metals as catalysts. However, MFC technology is still rudimentary and there are several areas for improvement [28]. For example, typical MFCs exhibit low coulombic efficiencies due to inefficient electron transfer between the microbial cells and the anode. This inefficiency results in incomplete oxidation of the fuel [7, 14, 50] and undesired assimilation of some of the fuel carbon into biomass [42, 50]. Power output from MFCs may also be limited by the rates of intracellular substrate oxidations [48]. Even though there has been much work to optimize MFC configurations [43, 44, 56], their physical and chemical operating conditions [19], and their choice of microorganisms [14, 19, 29, 32, 50], the optimization of microbial metabolism to increase electron production has received insufficient attention. In this chapter, we will summarize the current state of MFC research and describe our efforts to model the energy metabolism of the bacterium *Escherichia coli* (*E. coli*) to design methods to optimize electron output from this organism. We note that there is a growing interest in the use of redox-active enzymes as both anodic and cathodic catalysts in micro- or even nano-scale fuel cells [24, 51, 65]. Although there is great promise in the use of enzymes for specialized fuel cell applications, a review of that technology is outside the scope of this chapter.

9.1.2 Historical Perspective

For over 90 years investigators have known that microbes can catalyze the oxidation of substrates and donate the resulting electrons to the anode of MFCs [13, 46]. With the dawn of manned space travel, interest in MFCs was renewed by their potential use in converting biowaste to energy in spacecraft [55]. The electrical current generated by that generation of MFCs was low for their size and so limited their applicability. With recent advances in microelectronics, the power demand for devices has drastically decreased. Consequently, microbe-catalyzed small fuel cells have once again emerged as a tantalizing alternative to fuel cells employing inorganic catalysts.

Microbes can be used for fuel cell applications either directly as integral catalysts in the fuel cell or indirectly as catalysts for fuel synthesis. In direct applications, microbial cells are placed in contact with the anode of the MFC. In indirect applications, microbes generate fuel, typically hydrogen or methane, by decomposition of organic matter. Many kinds of microbes produce hydrogen or methane as a product of their metabolism. The indirect approach is especially attractive as a means to generate a useful product from decomposition of sewage, garbage, and agricultural wastes [38, 63]. Phototrophic organisms have also been reportedly used for electricity generation from light either directly [61] or indirectly [36]. Different kinds of bacteria and, less frequently, yeast have been used as catalysts in MFCs [23]. This summary of MFC research will only consider those MFCs that use organisms in direct contact with the anode since that is the system that we seek to optimize.

9.1.3 MFC Performance

(Table 9.1) summarizes information gleaned from recent publications describing MFC research. In many cases it is difficult to compare results reported in the literature because key experimental parameters are sometimes not provided or critical comparative measurements of electrical output are not reported. Of greater concern for those interested in improving the microbiological aspects of MFC performance is the fact that much of the reported work in MFCs has focused on the electrochemical aspects of MFC performance so that critical biological parameters are not reported.

The low coulombic efficiencies and low power outputs of MFCs are major limitations to their utility [31]. Significant improvements have been reported in both areas in recent years. As described in a later section, electrochemical mediator dyes are typically used to increase the efficiency of electron transfer in MFCs that employ pure cultures of microorganisms [12, 14, 28, 42–44, 50]. However, soluble mediators limit the use of MFCs to closed systems. To solve the problem of low coulombic efficiencies while avoiding the use of mediators, some investigators have used single species of microorganisms that use the anode directly as terminal electron acceptor [7, 10, 27]. The bacteria *Shewanella putrefaciens, Geobacter sulfurreducens,* and *Rhodoferax ferrireducens* have been shown to naturally donate electrons to external electrodes. High coulombic efficiencies have been reported in these studies reflecting the highly efficient conversion of fuel to measurable electron output. However, these studies do not report concomitant high power outputs, which are necessary for practical applications. These low power outputs may be due to the fact that these bacteria transfer electrons from their metabolism to the anode at midpoint potentials more positive than do bacteria that utilize mediators, so a significant fraction of their electrical capacity is not captured.

Other groups have achieved improved coulombic efficiencies by allowing natural, mixed populations of bacterial species in the fuel cell to select for those that can donate electrons to the anode. These open systems do not require the use of mediators [8, 10, 17, 22, 25, 27, 45]. Natural electrical mediators produced by the microbes provide electrical contact between intracellular redox biochemistry and the anode [47]. Coulombic efficiencies of over 80% have been reported in such systems [7, 10, 47, 48, 53]. The reported power densities are quite variable, though, with values ranging from 14 - 16 mW m^{-2} to about 4 W m^{-2} [6, 47, 48, 60]. These variations may be due to differences in the sources of inocula or the selective pressures to which the mixed culture were exposed during the experiments. The species compositions of these natural microbial populations are unknown. It is believed that the composition of these microbial communities depends upon the nature of the electrode, the carbon sources available, and any competing electron acceptors present in the natural environments surrounding the electrode.

To address the problem of low power outputs in mediator-less MFCs, investigators are exploring the immobilization of mediators on electrode surfaces [43, 44]. Electrical output in such systems is comparable to or even exceeds that of systems that use soluble mediators. A novel MFC electrode was constructed with a layer of the conductive polymer polyaniline covering a Pt electrode. Using *E. coli* as catalyst and glucose as fuel, the power output was 900 mW m^{-2} which is over ten-fold higher than any previously reported using MFCs containing pure cultures of any microbe [39, 54]. It is believed that the conductive polymer forms an electrochemically active, biocompatible layer on top of the electrode that serves as mediator for electron transfer from the bacterial cells to the anode while protecting the underlying Pt electrode from poisoning [54]. The same group reported even higher power output (18 W m^{-2}) using a similar electrode with a layer of the perfluorinated form of polyaniline and the bacterium *Clostridium beijerinckii* as catalyst with starch as the fuel [40]. It is not clear how these bound mediators connect with the intracellular redox biochemistry of cells.

9.1.4 MFC Applications

MFCs have been examined for use in different applications. Among the most common is the use of MFCs as biosensors for biological oxygen demand (BOD) estimation of available oxygen in wastewater [8, 9, 22, 25, 68]. MFCs used for

BOD estimations experience errors introduced by the presence of competing electron acceptors such as nitrate or oxygen in the wastewater [9]. This diminishes the sensitivity of the biosensor signal and decreases its reproducibility. Kim, *et al.* [26] reported that inhibition of the terminal oxidase of the microbes in the mixed culture to circumvent donation to dissolved oxygen did not affect current production from MFCs. To improve the performance of MFCs as BOD biosensors, inhibitors of terminal oxidases were added to the catalyst community resulting in coulombic efficiency increases of up to 90% [9].

Another MFC application that has seen increasing interest is in recovery of electricity from natural decomposition processes. Electricity production from MFCs using natural communities of microbes was demonstrated over 40 years ago [57, 66]. Recently, this process has been re-examined and may hold promise for uses ranging from electricity generation from wastes to power generation from sediments [6, 26, 33, 60]. The latter may provide an *in situ* power source for remotely placed environmental microelectronic sensors. Electricity recovery from wastewater treatment is very attractive because it provides the possibility of decreasing overall treatment costs while reducing the production of biomass. Other potential applications of MFCs include other types of biosensors [5, 20, 59] and as power sources for self-feeding robots [64]. Such environmental MFCs capitalize on microbial colonization of anodes placed in natural settings. MFC applications such as these are best powered by natural communities and not by engineered organism-driven MFCs.

It has been suggested that the rate of fuel oxidation by microbial metabolism is the limiting step in MFC performance [48]. The next step to improve MFC output lies in the improvement of electron production from the microbe. Advanced microbial genetic engineering methods are available to carry out this work with selected microbes, such as *E. coli*.

9.2 Microbiology Overview

9.2.1 Bacterial Structure

A bacterial cell is surrounded by a cell envelope that consists of two parts: an outer cell wall and an inner cytoplasmic membrane. Together they are responsible for maintaining the integrity of the cell and for controlling the movement of molecules into and out of the cell. The cell wall of *E. coli* is a bilayer membrane composed of hydrophobic molecules (lipids) attached to an inner, rigid network of sugars and polypeptides called peptidoglycan. Inside this layer is the cytoplasmic membrane that is also a hydrophobic bilayer of lipids. Within this is the cytoplasm of the cell that contains most of the biochemical catalysts (enzymes) and all the genetic information (in the form of deoxyribonucleic acid, DNA). An *E. coli* cell is rod-shaped and is typically 0.5 μm in diameter and 2 μm long when growing.

9.2.2 Nutrient Transport

The bacterial cytoplasmic membrane is a highly selective barrier between the cytoplasm and the external environment. This membrane is generally permeable to small molecules (like water), gases (carbon dioxide, oxygen) and hydrophobic molecules (hydrocarbons, protonated fatty acids). To obtain most organic nutrients necessary for growth, bacteria use different uptake mechanisms. Except in exceptional cases where the external concentration of a nutrient is much higher than the steady state internal concentration, nutrient uptake requires the cell to expend energy for concentrative transport.

9.2.3 Cellular Energy and Electron Carriers

In *E. coli*, the energy for transport and other cellular functions is derived from the oxidation of nutrients. *E. coli* can use a variety of organic compounds as growth substrates including sugars and fatty acids. Under fully aerobic conditions, these substrates can be oxidized to carbon dioxide *via* several enzyme-catalyzed reaction steps (Figure 9.1). When oxygen is limiting, less oxidized compounds are secreted by cells as products of their catabolism. Ethanol, organic acids (acetic, formic, succinic, and lactic acids) and gases (hydrogen and carbon dioxide) are products of *E. coli* catabolism under such conditions. Fatty acids (including acetic acid) can only be used as growth substrates when oxygen is available.

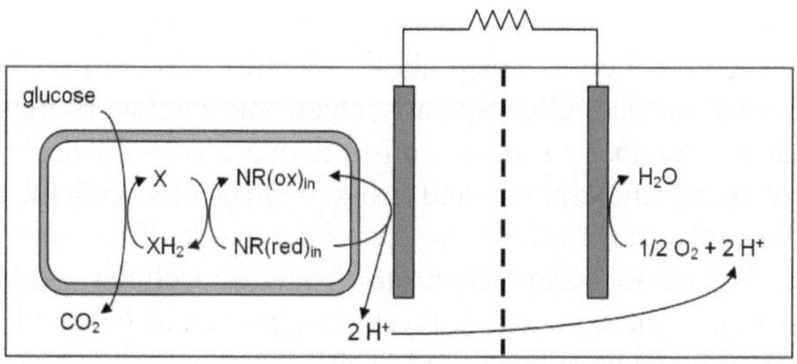

Figure 9.1. Schematic diagram depicting the complete oxidation of glucose coupled to electrical output in an MFC. The bacterium in the anode chamber catalyzes the oxidation of the glucose fuel to carbon dioxide *via* several enzyme-catalyzed reactions. Electrons from these reactions are passed to an intermediate intracellular carrier molecule (X) and then to a mediator dye molecule (NR, neutral red) that is freely diffusible across the bacterial cell's membrane. The reduced dye is reoxidized at the anode and electrons pass through an external circuit to the cathode. Protons released from the bacterium exchange with an ion exchange membrane that separates the two chambers. Protons in the cathode chamber recombine with electrons from the cathode and oxygen to form water

Energy released from intracellular oxidative reactions is captured in one of two forms: as covalent bond energy in phosphoesters and thioesters or as transmembrane ion gradients. Covalent bond energy is stored most commonly in phosphodiester bonds of the compound adenosine triphosphate (ATP). When an ATP molecule is hydrolyzed, usually to adenosine diphosphate (ADP) and phosphate, the released energy drives other endergonic reactions. Cellular energy is also stored in ion gradients established across the cytoplasmic membrane. Protons and in some cases sodium ions provide potential energy stores in the form of both chemical and electrical gradients. The energy in these gradients can be used directly by coupling ion uptake with uptake or export of molecules. Ion gradients are also used to synthesize ATP *via* a membrane-bound protein complex.

A transmembrane proton gradient is created in *E. coli* by passing the electrons harvested from nutrient oxidation to a series of membrane-bound carrier proteins. This "bucket brigade" proceeds from proteins with low midpoint potentials to those of higher potential. These protein carriers, called dehydrogenases and cytochromes, comprise an electron transport chain. The consequent loss of energy in electron transfer to each carrier is captured in part by the coupled export of protons from the cell. Ultimately the electrons are passed to a final acceptor molecule. *E. coli* uses an acceptor with the highest midpoint potential available to maximize energy conservation. The oxygen/water couple is the most electropositive couple *E. coli* can use. This process of electron transport is called respiration. Inside the cell, a number of compounds serve as electron carriers that can donate their electrons to the electron transport chain or be used as sources of reductant for different cell biosynthetic processes. Chief among these carriers are the nicotinamide adenosine dinucleotides (NAD and NADP, NAD containing an additional phosphate) and flavin adenine dinucleotide (FAD).

9.2.4 Coupling Cellular Electrochemistry to the Anode

Except in the exceptional cases noted above, electrons do not pass from inside the cell to the external surface. Consequently electronic mediators must be added to cells in the anode chamber to allow cell-to-anode electron transfer (Figure 9.1). Mediators have to be amphiphilic and cross hydrophobic cell membranes while remaining sufficiently water soluble to access the anode. They also need midpoint potentials low enough to accept electrons from intracellular oxidations without significant loss of electrical potential. Dyes have proven to be best for this purpose and many different kinds have been used (Table 9.1)..

As described previously, bacteria such as *Geobacter* and *Shewanella* species can donate electrons directly to insoluble electron acceptors and so do not require the use of mediators in MFCs [6, 27]. Although mediator-independence is desirable in environmental applications of MFCs, there is no advantage gained by using mediatorless MFCs in closed electrical devices. Indeed, these organisms donate their electrons from relatively high potential biochemical donors (an electron carrier called a c-type cytochrome, $E_0 = -170\,mV$) and so useful energy is lost during intracellular electron transfer to this cytochrome from lower potential redox reactions [34]. In addition, although some of these bacteria are amenable to simple genetic manipulations, gene replacement techniques are not as advanced as are those with *E. coli*. Much less is known about their central catabolism, so that designing metabolic engineering efforts is more difficult as well [37]. Thus, *E. coli* remains the best choice for developing an efficient, high output fuel cell catalyst for small, closed system electronic device MFCs.

9.3 A Theoretical Treatment of MFC Anode Reactions

9.3.1 What Limits MFC Electrical Output?

Many important questions remain unanswered regarding activities in the anode compartment of MFCs. Does the rate of microbial catabolism limit MFC output? If so, what process (or processes) within the bacteria can be changed to improve MFC performance? Is the flow of electrons from the microbes to the anode limiting? If so, is the limitation due to mass transport of the mediator or the oxidation rate of the mediator at the anode surface? To answer these questions we will examine the processes involved in intracellular electron generation and electron transfer from the microbial cells to the anode.

For this analysis, we will consider the case of an MFC with glucose as the fuel, *E. coli* as the catalyst, and the dye neutral red (NR) as the mediator (Figure 9.2, step A). Many dyes have been studied as mediators in MFCs (see (Table 9.1) and NR has been shown to have very good electrical properties and is compatible with the physiology of this organism [42]. We will compare the rate of oxidation of NR at the electrode surface with the rate of NR reduction by the bacterial cell. To simplify this analysis, we will consider the entire bacterial cell as a catalyst and will define this apparent rate of NR reduction as the composite of the rates of glucose uptake, the several rates of enzymatic hydrolysis and substrate oxidation reactions, and rates of intracellular electron transfer to NR.

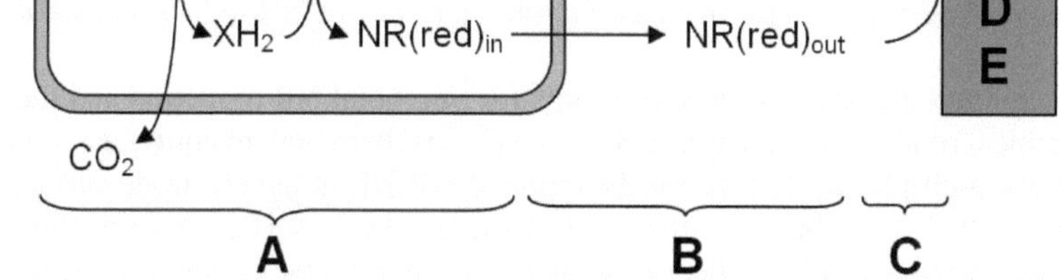

Figure 9.2. Schematic diagram depicting electron transfer from a bacterial cell to the anode. The reaction is the as same described in the legend of Figure 9.1. Transfer of the NR species from intracellular pools to extracellular pools is depicted. The electron transfer process is broken down into three components. A. Intracellular substrate oxidation and reduction of NR *via* intracellular carriers (X). B. Mass transfer of reduced NR [NR(red)] and oxidized NR [NR(ox)] through the cell membrane and solution by diffusion. C. Oxidation of reduced NR by the anode. The rates of each of these component reactions are compared in the text

We start with a consideration of the rate of NR oxidation at the anode (Figure 9.2, step C). The electrode surface reaction can be described using the Butler-Volmer equation for electrode kinetics:

$$i = nFk\left[C^{o}_{Ox} e^{\frac{-nF\alpha\eta}{R_i T}} - C^{o}_{Rd} e^{\frac{nF(1-\alpha)\eta}{R_i T}} \right] \quad (9.1)$$

where k is the standard rate constant (cm s^{-1}). The reported values of k for different compounds vary from less than 10^{-9} cm s^{-1} to more than 10 cm s^{-1} [3]. Since the molecular structure of NR is not significantly altered in the redox reaction, one can expect the value of k for NR to be among the larger values [3]. Since the structure of NR is similar to anthracene, its value of k (27 cm s^{-1}) will be used for NR (Figure 9.3) [30].

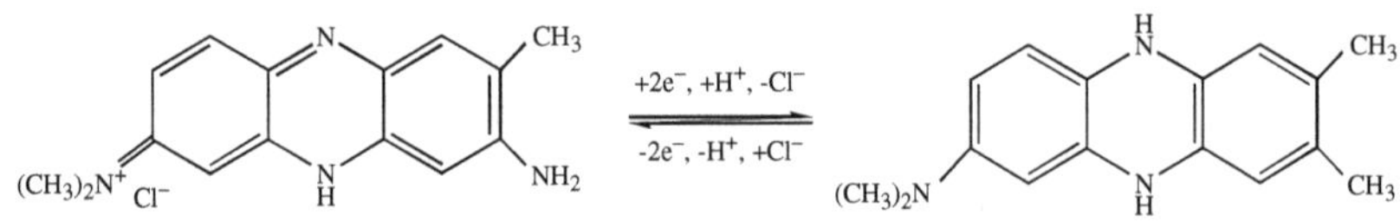

Figure 9.3. Redox chemistry of neutral red (NR), an electrochemical mediator dye commonly used in MFCs

Roller *et al.* [50] reported the rates of microbial reduction of MFC mediator dyes. They did not report the rate of NR reduction, but since dyes with similar chemical structures are reduced at similar rates, we will use the rate of reduction they reported for thionine, a dye similar in structure to NR. This dye had a specific reduction rate of 4.13 μmol (g dry wt)$^{-1}$ s^{-1}, the highest rate reported in their study [50]. If we assume in an MFC the bacterial cell concentration is 10^8 cells mL^{-1}, the volumetric rate of reaction would be 0.116 nmol mL^{-1} s^{-1}. Using the volumetric rate of thionine reduction and a typical thionine concentration (50 μM) [50], we obtain a value of 2.32×10^{-3} s^{-1} for the specific rate constant for NR reduction by bacterial cells.

To compare the standard rate constants for microbial NR reduction and the electrochemical oxidation of the dye, one must convert them both to equivalent units. To do so, we multiply the k value for the oxidation of NR at the electrode surface by a parameter A. We define A as the ratio of the electrode surface area to the anode chamber solution volume. There is no standard ratio, so we choose a typical value, one we use in our own MFCs, of 1.09 cm^{-1}. After this adjustment for units, the standard rate constant for the electrochemical oxidation reaction is 29.4 s^{-1}. This value is more than four orders of magnitude larger than the standard rate for the microbial dye reduction (29.4 s^{-1} *vs.* 0.00232 s^{-1}). In many MFCs, the anode is composed of a rough or woven graphite surface(Table 9.1), so the actual surface area is much larger than the value we used in these calculations. Consequently, our calculation underestimates a typical standard rate constant of anode oxidation of NR.

Mass transfer of the reduced dye to the anode must also be considered as a possible rate-limiting step in the MFC. We will compare electron output from the bacterial cell before the mass transport process (Figure 9.2, step A) with reported values measured in MFCs (Figure 9.2, steps A+B+C). Using the volumetric dye reduction rate from the previous calculations (0.116 nmol mL^{-1} s^{-1}) and assuming a 20 mL anode chamber and 2 moles of electrons carried per mole NR, we calculate the expected current output from the bacterial cells to be 448 μA. Reported values for current output using *E. coli* cells range from 120 μA to 4800 μA (Table 9.1). Since the measured current in MFCs is within the range of the theoretical output from the bacterial cells, we can conclude that mass transport is not limiting MFC performance.

From the calculations above, we have established that bacterial catabolism is the rate-limiting step in MFCs. To improve MFC performance, one can engineer the microbial catalyst to increase electrical output. As noted previously, this involves directing the catabolism of the microbe in a direction different from that resulting from billions of years of evolution. Many changes need to be made in the cellular biochemistry of the bacterial catalyst, so *E. coli* is an obvious choice for such efforts. *E. coli* is metabolically versatile and catabolizes a variety of compounds including sugars and fatty acids. There are many procedures available to genetically engineering its catabolism. It is also a proven catalyst for MFC applications.

9.4 Metabolic Engineering

Analysis of metabolic pathways in *E. coli* to increase electrical output can be aided through the use of metabolic engineering. Using metabolic engineering, it is possible to determine whether power enhancement is possible. If enhancement is possible, then metabolic engineering may be used to identify the pathways to be manipulated that are likely to provide the best results. One of the fundamental techniques for carrying out such an investigation is metabolic flux analysis [62].

The benefit of metabolic flux analysis or MFA is that it provides a snapshot of the physiological state of the cell. The purpose of MFA is to determine the intracellular concentration of metabolites and the flux of the metabolites through their metabolic pathways. Fluxes are calculated through the development of stoichiometric models of the metabolic reaction network. Generally the theory of reaction kinetics requires variation of intracellular metabolites over time to be described *via* a system of differential equations. However, if the experimental system can be manipulated to approximate or operate at steady state, such as by using a chemostat or working with resting cells, the variation with respect to time goes to zero. The model is then reduced to a system of algebraic equations. Knowing the metabolic network along with measured extracellular fluxes, it is possible to determine metabolic fluxes by performing mass balances on each metabolite.

MFA allows one to determine a number of other cellular features. As pointed out by Stephanopoulos *et al.* [58], features such as nodal rigidity, alternative pathways, values of non-measurable fluxes, and maximum theoretical yields may be determined. Nodal rigidity refers to how much or how little the flux distribution through a given branch point will change when operating conditions are changed. Alternative pathway analysis may be required when several different pathways appear feasible, but the actual pathway is unknown. Through MFA, it may be possible to show that some of the pathways are actually not used (zero flux) or impossible (negative flux).

Such an analysis was used to identify the correct pathway for citric acid fermentation by *C. lipolytica* [1]. Actual experimental measurement of some fluxes may simply be impossible. However, enough information from other fluxes may be available to uniquely identify what the unknown flux must be. Due to the knowledge of the stoichiometry of the reaction system, MFA also allows one to determine the maximum specific product yield for a given substrate, including power output for a microbial fuel cell.

The nature of MFA makes it amenable to analysis using linear algebra. Given substrate, metabolites, and products, the mass balance on the constituents may be written in differential form with respect to time as

$$\frac{d\overline{X}}{dt} = \overline{r} - \mu\overline{X} \tag{9.2}$$

where $\overline{X}$ is a vector of the metabolites of interest, $\overline{r}$ is the vector of rate expressions, and μ is the bacterial growth rate. The term $\mu\overline{X}$ represents the dilution of the metabolites as the cells grow. However, when dealing with MFCs, the cells are in a resting state. As a result μ is approximately 0 and the term may be neglected. Since the system is at a steady state, $\frac{d\overline{X}}{dt}$ may also be considered to be 0. All that remains then is

$$0 = \overline{r} \tag{9.3}$$

It is further possible to write the rate expression in terms of the stoichiometric coefficients and their associated fluxes, such that

$$\overline{r} = \overline{\overline{S^T}}\,\overline{v} \tag{9.4}$$

where $\overline{\overline{S}}$ is the matrix of stoichiometric coefficients and $\overline{v}$ is the vector of fluxes. By writing the reaction rate in this way, it is possible to easily couple reactions. Combining equations (9.3) and (9.4) results in

$$\overline{0} = \overline{\overline{S^T}}\,\overline{v} \tag{9.5}$$

To illustrate how MFA can be used to analyze *E. coli* metabolism for use in an MFC in a simple hypothetical example, consider the case of the resting bacterial cell metabolizing glucose. Potential pathways that may be used for glucose catabolism are the Embden-Meyerhof-Parnas pathway, the mixed acid fermentation pathways, and the TCA cycle, as shown in Figure 9.3 and Figure 9.4. Assuming that the system is at a steady-state, it is possible to set up mass balances based on each of the intracellular species based on simple mass balances. Specifically,

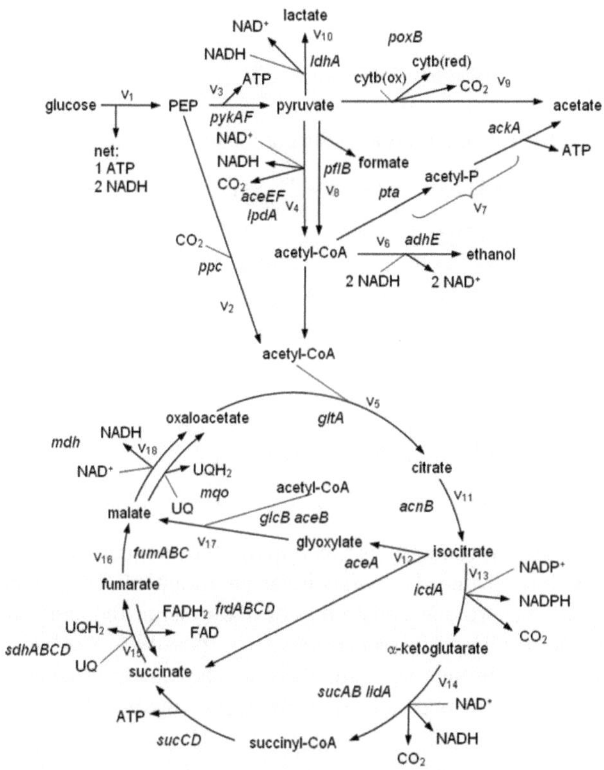

Figure 9.4. Glucose catabolism in *E. coli via* the Embden-Meyerhof-Parnas pathway, mixed acid fermentation pathways and the TCA cycle. The genes encoding the relevant enzymes are shown in italics. Fluxes for reactions 9.5–9.22 are also indicated. Flux v_7 includes reactions catalyzed by both Pta and AckA

$$PEP:\quad v_1 - v_2 - v_3 \quad = 0 \tag{9.5}$$

$$pyruvate:\quad v_3 - v_4 - v_8 - v_9 - v_{10} \quad = 0 \tag{9.6}$$

$$AcCoA:\quad v_4 - v_5 - v_6 - v_7 - v_{27} \quad = 0 \tag{9.7}$$

$$citrate:\quad v_5 - v_{11} \quad = 0 \tag{9.8}$$

$$isocitrate:\quad v_{11} - v_{12} - v_{13} \quad = 0 \tag{9.9}$$

$$\alpha - ketoglutarate:\quad v_{13} - v_{14} \quad = 0 \tag{9.10}$$

$$succinate:\quad v_{12} + v_{14} - v_{15} \quad = 0 \tag{9.11}$$

$$fumarate:\quad v_{15} - v_{16} \quad = 0 \tag{9.12}$$

$$malate:\quad v_{16} + v_{17} - v_{18} \quad = 0 \tag{9.13}$$

$$oxaloacetate:\quad v_2 - v_5 + v_{18} \quad = 0 \tag{9.14}$$

The flux for glucose substrate consumption is represented by

$$glucose: \quad -\frac{1}{2}v_1 \tag{9.15}$$

Cofactors and byproducts produced during metabolism are

$$NADH: \quad v_1 + v_4 - 2v_6 - v_{10} + v_{13} + v_{14} + v_{18} \tag{9.16}$$

$$FADH_2: \quad v_{15} \tag{9.17}$$

$$acetate: \quad v_7 + v_8 + v_9 \tag{9.18}$$

$$formate: \quad v_8 \tag{9.19}$$

$$ethanol: \quad v_6 \tag{9.20}$$

$$CO_2: \quad -v_2 + v_4 + v_{13} + v_{14} \tag{9.21}$$

$$ATP: \quad v_3 + v_7 \tag{9.22}$$

The cofactors and byproducts described in the above equations may be produced and secreted by the cell into the extracellular environment. These components are often measurable and provide constraints on the system, reducing the number of degrees of freedom. Given the constraints, it is possible to solve the equations above simultaneously through standard algebra. The system of equations may also be recast in matrix form and solved using linear algebra. For the intracellular metabolites

$$\begin{pmatrix} r_{PEP} \\ r_{pyr} \\ r_{AcCoA} \\ r_{cit} \\ r_{iso} \\ r_{\alpha-keto} \\ r_{suc} \\ r_{fum} \\ r_{mal} \\ r_{ox} \end{pmatrix} = \begin{pmatrix} 1 & -1 & -1 & 0 & 0 & 0 & 0 & 0 & 0 & 0 & 0 & 0 & 0 & 0 & 0 & 0 & 0 & 0 \\ 0 & 0 & 1 & -1 & 0 & 0 & 0 & -1 & -1 & -1 & 0 & 0 & 0 & 0 & 0 & 0 & 0 & 0 \\ 0 & 0 & 0 & 1 & -1 & -1 & -1 & 0 & 0 & 0 & 0 & 0 & 0 & 0 & 0 & 0 & -1 & 0 \\ 0 & 0 & 0 & 0 & 1 & 0 & 0 & 0 & 0 & 0 & -1 & 0 & 0 & 0 & 0 & 0 & 0 & 0 \\ 0 & 0 & 0 & 0 & 0 & 0 & 0 & 0 & 0 & 0 & 1 & -1 & -1 & 0 & 0 & 0 & 0 & 0 \\ 0 & 0 & 0 & 0 & 0 & 0 & 0 & 0 & 0 & 0 & 0 & 0 & 1 & -1 & 0 & 0 & 0 & 0 \\ 0 & 0 & 0 & 0 & 0 & 0 & 0 & 0 & 0 & 0 & 0 & 1 & 0 & 1 & -1 & 0 & 0 & 0 \\ 0 & 0 & 0 & 0 & 0 & 0 & 0 & 0 & 0 & 0 & 0 & 0 & 0 & 0 & 1 & -1 & 0 & 0 \\ 0 & 0 & 0 & 0 & 0 & 0 & 0 & 0 & 0 & 0 & 0 & 0 & 0 & 0 & 0 & 1 & 1 & -1 \\ 0 & 1 & 0 & 0 & -1 & 0 & 0 & 0 & 0 & 0 & 0 & 0 & 0 & 0 & 0 & 0 & 0 & 1 \end{pmatrix} \times \overline{v}$$

The number of intracellular metabolic reactions specified is 10. If it is assumed that the glyoxylate shunt is not functioning, the fluxes v_{12} and v_{17} will be 0. Measuring any six of the cofactors or byproducts will fix the system allowing one to calculate the remaining the fluxes.

Table 9.1. Literatuyre reports of MFC performance

Organism	Anode	Cathode	Fuel	Mediator[a]	OC Volt (V)	Current (mA)	Load (Ω)	Peak Current Density (mA/m^2)	Peak Power Density (mW/m^2)	Total Charge (C)	Coulombic Efficiency (%)	Comments[c]	Reference
E. coli	*disc of glassy carbon or packed bed*	*oxygen gas diffusion*	*glucose*	*HQN*			*0–10,000*	*1,300**				**Current density was measured under short circuit conditions*	*[56]*
	woven graphite, NR-linked graphite, Mn^{4+} graphite	*woven graphite or Fe^{3+} graphite electrode*	*glucose*	*(immobilized in electrode)*			*300*	*325**	*91**			**Density values are approximations calculated by the author*	*[44]*
	Graphite felt	*Graphite felt*	*glucose*	*NR*	*0.85*	*4.8*	*120*	*0.851**	*0.528**	*2,039*		** Calculated using authors' data*	*[42]*
	graphite plate	*graphite plate*	*acetate*	*NR*		*0.28*	*1,000*	*31*					*[41]*
	Pt plate or glassy carbon	*Pt*	*glucose*	*TH, RS*	*0.7*	*0.12*	*1,000*	*40**	*16**			** Calculated using authors' data*	*[4]*
	polyaniline-modified Pt electrode	*woven graphite*	*glucose*	*(immobilized in electrode)*	*0.895*	*150*	*10–1,000*	*15,000*	*900**			**The peak power was at 1.95 A/m^2, and 0.469 V (surface area of 1 cm^2)*	*[54]*
	Reticulated vitreous carbon	*Pt foil*	*glucose*	*various[b]*	*0.5–0.8*	*0.8*	*560*	*10**					

Alcaligenes eutrophus	*Reticulated vitreous carbon*	*Pt foil*	*glucose*	*various*[b]	*0.5–0.8*		*560*					**Current density calculated with approximate electrode surface area of 800 cm² given by the author.*	*[14]*
Bacillus subtilis	*Reticulated vitreous carbon*	*Pt foil*	*glucose*	*various*[b]	*0.5–0.8*	*0.8*	*560*	*10**					
	Reticulated vitreous carbon	*Pt foil*	*glucose*	*various*[b]	*0.5–0.8*	*0.8*	*560*	*10**			*32 - 62*		*[11]*
	Graphite felt	*Graphite felt*	*glucose*	*HQN*	*0.7*	*0.5*	*1000*	*1.06**				**Calculated using authors' data*	*[2]*
	Reticulated vitreous carbon	*Pt plate*	*glucose*	*TH*		*≈ 0.7*	*560*						*[11]*
Proteus vulgaris	*Reticulated vitreous carbon*	*Pt plate*	*sugars*	*TH*		*≈ 0.4*	*1,000*				*26 - 56*		*[28]*
	Reticulated vitreous carbon	*Pt plate*	*sugars*	*TH*		*≈ 0.4*	*1,000*				*12 - 63**	**The energy source was assumed to be completely consumed in the calculations.*	*[29]*
Mixed culture from anaerobic sludge	*Graphite electrode*	*Graphite electrode*	*glucose*	*none*			*100*		*3,600**	*2,991*	*89**	**Efficiency decreased at higher glucose concentrations; four MFCs were stacked.*	*[48]*
Mixed culture from marine sediments	*graphite discs*	*not specified*	*sediments*	*none*			*not specified*	*≈ 100**	*28 (average)*			**Potential was constant at 0.27 V*	*[60]*
Desulfuromonas acetoxidans	*graphite*	*not specified*	*acetate*	*AQDS*			*500*	*26**				**Potential was constant at 0.17 V*	

	graphite	not specified	acetate	none			500	20*			*Potential was constant at 0.13 V	
	graphite	graphite	acetate	none			500		14 (ave-rage)*		*AQDS increased power by 24%	[6]
Mixed culture from marine sediments	graphite	graphite	sediments	none			not speci-fied		16 (ave-rage)			
Rhodoferax ferrireducens	graphite rod, graphite foam or graphite felt	graphite rod, graphite foam or graphite felt	glucose	none		0.2	1,000	31	8.215*	83		[10]
Geobacter sulfurreducens	unpolished graphite	unpolished graphite	acetate	none		0.4	500	65	15.5*	96.8		[7]
Mixed culture from anaerobic sludge	woven graphite, NR-linked graphite, Mn^{4+} graphite	woven graphite or Fe^{3+} graphite electrode	glucose	(immobili-zed in electrode)			300	1,750*	787.5*		*The electro-des surface area used in the calcula-tions are an approximation from the author.	[44]
	woven graphite, NR-linked graphite, Mn^{4+} graphite	Fe^{3+} graphite electrode	lactate, pyruvate, acetate, glucose	(immobili-zed in electrode)		2.5*	1,000		10		*Current was measured in short circuit configuration.	[43]
Shewanella putrefaciens	graphite felt	graphite felt	lactate	none	0.5	0.04	1,000	8*			*An apparent electrode surface area of 50 cm^2 was used for calculations	[27]
Anabaena variabilis	carbon cloth	carbon cloth	sugars	HQN	0.8	0.5	1,000			32 - 38	*Photosyntheti c MFC; energy efficiency about 0.2%.	[67]

Mixed culture	*graphite felt*	*graphite felt*	*wastewater*	*none*		*1.6*	*10–1,000*				**Current was measured using 10 ohm resistance.*	*[17]*
Activated sludge	*graphite felt*	*graphite felt*	*wastewater*	*none*		*0.2*	*1,000*	*956**				*[26]*
Mixed culture from wastewater	*graphite rods*	*carbon/platinum catalyst*	*wastewater*	*none*			*16–5,000*	*125*	*26*	*< 12*	**Maximum power measured using 69 ohm resistor.*	*[33]*
Anaerobic sludge	*graphite*	*graphite*	*glucose*	*none*			*100*	*6,490*	*4,310*	*81*		*[47]*
Clostridium beijerinckii	*tetrafluoropolyaniline-modified woven graphite electrode*	*woven graphite*	*starch*	*(immobilized in electrode)*	*759*	*30*	*0.1 - 1,001*	*40,000**	*18,667**		**Density values calculated from cross sectional area; two fuel cells in series were used*	*[40]*
Activated sludge	*graphite felt*	*graphite felt*	*glucose and glutamate*	*none*			*100*	*9*	*1.3*			*[21]*

[a] Abbreviations: ABB, alizarin brilliant blue; AQDS, anthraquinone 2,6-disulfonate; BCB, brilliant cresyl blue; BV, benzyl viologen; DCPIP, 2,6-dichorophenol-indophenol; DMST2, N,N-dimethyl-disulphonated thionine; GC, gallocyanine; HQN, 2-hydroxy-1,4-naphtoquinone; MB, methylene blue; NMB, new methylene blue; NR, neutral red; PES, phenazine ethosulphate; PTZ, phenothiazinone; RS, resorufin; SF, safranine-O; TB, toluidine blue-O; TH, thionine;

[b] ABB, BCB, BV, DCPIP, DMST2, GC, NMB, PES, PTZ, RS, SF, TB, and TH.

[c] Values with asterisks were calculated for this table using assumptions described in the corresponding rows.

For certain metabolic systems, it may not be possible to reduce the number of degrees of freedom to 0. Under such circumstances, the MFA problem may be re-cast as an optimization problem at which point one may solve for the fluxes. However, the appropriate choice of an objective function is not trivial. Although the MFC operator would like to maximize power output, the cells will want to maxi mize their chances for survival and propagation. To determine accurate fluxes, it is critical to identify exactly what the cell is attempting to maximize and utilize that objective function in solving for the fluxes. Generally, maximization of growth rate has been used as the objective function. However, for cells forced into a resting state, growth rate may not be appropriate. Others have used maximization of metabolic energy, as represented by ATP, with some success [35, 49].

Having set up the model, it is possible to perform *in silico* experiments to determine how manipulation of different pathways will change levels of NADH and $FADH_2$ and the concomitant change in power output. Although the example shown above is for a small portion of *E. coli* metabolic pathways, it should be noted that modeling of the whole metabolic network of *E. coli* and the complete metabolism of other bacteria has been accomplished [15, 16, 18, 52].

9.5 References

1. Aiba, S., and M. Matsuoka. 1979. Identificationof metabolic model: Citrate production from clucose by *Candida lipolytica*. Biotechnol Bioeng 21:1373–1386.
2. Allen, R. M., and H. P. Bennetto. 1993. Microbial fuel cells: Electricity production from carbohydrates. App. Biochem. Biotechnol. 39/40:27–40.
3. Bard, A. J., and L. R. Faulkner. 2001. Electrochemical methods: fundamentals and applications, 2nd ed. Wiley, New York.
4. Bennetto, H. P., J. L. Stirling, K. Tanaka, and C. A. Vega. 1983. Anodic reactions in microbial fuel cells. Biotechnol. Bioeng. 25:559–568.
5. Bentley, A., A. Atkinson, J. Jezek, and D. M. Rawson. 2001. Whole cell biosensors--electrochemical and optical approaches to ecotoxicity testing. Toxicol. In Vitro 15:469–475.
6. Bond, D. R., D. E. Holmes, L. M. Tender, and D. R. Lovley. 2002. Electrode-reducing microorganisms that harvest energy from marine sediments. Science 295:483–485.
7. Bond, D. R., and D. R. Lovley. 2003. Electricity production by *Geobacter sulfurreducens* attached to electrodes. Appl. Environ. Microbiol. 69:1548–1555.
8. Chang, I. S., J. K. Jang, G. C. Gil, M. Kim, H. J. Kim, B. W. Cho, and B. H. Kim. 2004. Continuous determination of biochemical oxygen demand using microbial fuel cell type biosensor. Biosensors and Bioelectronics 19:607–13.
9. Chang, I. S., H. Moon, J. K. Jang, and B. H. Kim. 2004. Improvement of a microbial fuel cell performance as a BOD sensor using respiratory inhibitors. Biosensors and Bioelectronics:In Press, Corrected Proof.
10. Chaudhuri, S. K., and D. R. Lovley. 2003. Electricity generation by direct oxidation of glucose in mediatorless microbial fuel cells. Nature Biotechnol. 21:1229–1232.
11. Choi, Y., E. Jung, S. Kim, and S. Jung. 2003. Membrane fluidity sensoring microbial fuel cell. Bioelectrochem. 59:121– 127.
12. Choi, Y., J. Song, S. Jung, and S. Kim. 2001. Optimization of the performance of microbial fuel cells containing alkalophilic *Bacillus* sp. J. Microbiol. Biotechnol. 11: 863–869.
13. Cohen, B. 1931. The bacterial culture as an electrical half-cell. J. Bacteriol. 21:18.

14. Delaney, G. M., H. P. Bennetto, J. R. Mason, S. D. Roller, J. L. Stirling, and C. F. Thurston. 1984. Electron-transfer coupling in microbial fuel cells: 2. Performance of fuel cells containing selected microorganism-mediator-substrate combinations. J. Chem. Technol. Biotechnol. 34B:13–27.
15. Edwards, J. S., R. U. Ibarra, and B. O. Palsson. 2001. In silico predictions of Escherichia coli metabolic capabilities are consistent with experimental data. Nat Biotechnol 19:125–30.
16. Edwards, J. S., and B. O. Palsson. 2000. The Escherichia coli MG1655 in silico metabolic genotype: its definition, characteristics, and capabilities. Proc Natl Acad Sci U S A 97:5528–33.
17. Gil, G.-C., I.-S. Chang, B. H. Kim, M. Kim, J.-K. Jang, H. S. Park, and H. J. Kim. 2003. Operational parameters affecting the performance of a mediator-less microbial fuel cell. Biosensors and Bioelectronics 18:327–334.
18. Ibarra, R. U., J. S. Edwards, and B. O. Palsson. 2002. Escherichia coli K-12 undergoes adaptive evolution to achieve in silico predicted optimal growth. Nature 420:186–9.
19. Ikeda, T., and K. Kano. 2001. An electrochemical approach to the studies of biological redox reactions and their applications to biosensors, bioreactors, and biofuel cells. J. Biosci. Bioeng. 92:9–18.
20. Ikeda, T., T. Kurosaki, K. Takayama, K. Kano, and K. Miki. 1996. Measurements of oxidoreductase-like activity of intact bacterial cells by an amperometric method using a membrane- coated electrode. Anal. Chem. 68:192–198.
21. Jang, J. K., T. H. Phama, I. S. Changa, K. H. Kanga, H. Moona, K. S. Chob, and B. H. Kim. 2004. Construction and operation of a novel mediator- and membrane-less microbial fuel cell. Process Biochem. 39:1007–1012.
22. Kang, K. H., J. K. Jang, T. H. Pham, H. Moon, I. S. Chang, and B. H. Kim. 2003. A microbial fuel cell with improved cathode reaction as a low biochemical oxygen demand sensor. Biotechnol. Lett. 25:1357–61.
23. Katz, E., A. N. Shipway, and I. Willner. 2003. Biochemical fuel cells. *In* A. Lamm (ed.), Handbook of Fuel Cells-Fundamentals, Technology and Applications, vol. Volume 1: Fundamentals and Survey of Systems. John Wiley & Sons, Ltd.
24. Katz, E., and I. Willner. 2003. A biofuel cell with electrochemically switchable and tunable power output. J. Amer. Chem. Soc. 125:6803–6813.
25. Kim, B. H., I. S. Chang, G. C. Gil, H. S. Park, and H. J. Kim. 2003. Novel BOD (biological oxygen demand) sensor using mediator-less microbial fuel cell. Biotechnol. Lett. 25:541–545.
26. Kim, B. H., H. S. Park, H. J. Kim, G. T. Kim, I. S. Chang, J. Lee, and N. T. Phung. 2004. Enrichment of microbial community generating electricity using a fuel-cell-type electrochemical cell. Appl. Microbiol. Biotechnol. 63:672–681.
27. Kim, H. J., H. S. Park, M. S. Hyun, I. S. Chang, M. Kim, and B. H. Kim. 2002. A mediator-less microbial fuel cell using a metal reducing bacterium, *Shewanella putrefaciense*. Enz. Microbial Technol. 30:145–152.
28. Kim, N., Y. Choi, S. Jung, and S. Kim. 2000. Development of microbial fuel cells using *Proteus vulgaris*. Bull. Korean Chem. Soc. 21:44–48.
29. Kim, N., Y. Choi, S. Jung, and S. Kim. 2000. Effect of initial carbon sources on the performance of microbial fuel cells containing *Proteus vulgaris*. Biotechnol. Bioeng. 70:109–114.
30. Kojima, H., and A. J. Bard. 1975. Determination of rate constants for the electroreduction of aromatic compounds and their correlation with homogeneous electron transfer rates. J. Amer. Chem. Soc. 97:6317–6324.
31. Kosaric, N., and J. Velikonja. 1995. Liquid and gaseous fuels from biotechnology - challenge and opportunities. FEMS Microbiol. Rev. 16:111–142.

32. Lee, S. A., Y. Choi, S. H. Jung, and S. Kim. 2002. Effect of initial carbon sources on the electrochemical detection of glucose by *Gluconobacter oxydans*. Bioelectrochem. 57:173–178.
33. Liu, H., R. Ramnarayanan, and B. E. Logan. 2004. Production of electricity during wastewater treatment using a single chamber microbial fuel cell. Environ. Sci. Technol. 38:2281–2285.
34. Lloyd, J. R., C. Leang, A. L. Hodges Myerson, M. V. Coppi, S. Cuifo, B. Methe, S. J. Sandler, and D. R. Lovley. 2003. Biochemical and genetic characterization of PpcA, a periplasmic c-type cytochrome in *Geobacter sulfurreducens*. Biochem. J. 369:153–161.
35. Majewksi, R. A., and M. M. Domach. 1990. Simple constrained optimization view of acetate overflow in *Escherichia coli*. Biotechnol Bioeng 35:732–738.
36. Melis, A., and T. Happe. 2001. Hydrogen production. Green algae as a source of energy. Plant Physiol. 127:740–748.
37. Methe, B. A., K. E. Nelson, J. A. Eisen, I. T. Paulsen, W. Nelson, J. F. Heidelberg, D. Wu, M. Wu, N. Ward, M. J. Beanan, R. J. Dodson, R. Madupu, L. M. Brinkac, S. C. Daugherty, R. T. DeBoy, A. S. Durkin, M. Gwinn, J. F. Kolonay, S. A. Sullivan, D. H. Haft, S. J., T. M. Davidsen, N. Zafar, O. White, B. Tran, C. Romero, r. H. A. Forberge, J. Weidman , H. Khouri, T. V. Feldblyum, T. R. Utterback, S. E. Van Aken, D. R. Lovley, and C. M. Fraser. 2003. Genome of *Geobacter sulfurreducens*: metal reduction in subsurface environments. Science 302:1967–1969.
38. Nath, K., and D. Das. 2004. Improvement of fermentative hydrogen production: various approaches. Appl. Microbiol. Biotechnol. 65:520–529.
39. Niessen, J., U. Schröder, M. Rosenbaum, and F. Scholz. 2004. Fluorinated polyanilines as superior materials for electrocatalytic anodes in bacterial fuel cells. Electrochem. Commun. 6:571–575.
40. Niessen, J., U. Schröder, and F. Scholz. 2004. Exploiting complex carbohydrates for microbial electricity generation - a bacterial fuel cell operating on starch. Electrochem. Commun. 6:955–958.
41. Park, D. H., S. K. Kim, I. H. Shin, and Y. J. Jeong. 2000. Electricity production in biofuel cell using modified graphite electrode with Neutral Red. Biotechnol. Lett. 22:1301–1304.
42. Park, D. H., and J. G. Zeikus. 2000. Electricity generation in microbial fuel cells using neutral red as an electronophore. Appl. Environ. Microbiol. 66:1292–1297.
43. Park, D. H., and J. G. Zeikus. 2002. Impact of electrode composition on electricity generation in a single-compartment fuel cell using *Shewanella putrefaciens*. Appl. Microbiol. Biotechnol. 59:58–61.
44. Park, D. H., and J. G. Zeikus. 2003. Improved fuel cell and electrode designs for producing electricity from microbial degradation. Biotechnol. Bioeng. 81:348–355.
45. Pham, C. A., S. J. Jung, N. T. Phung, J. Lee, I. S. Chang, B. H. Kim, H. Yi, and J. Chun. 2003. A novel electrochemically active and Fe(III)-reducing bacterium phylogenetically related to *Aeromonas hydrophila*, isolated from a microbial fuel cell. FEMS Microbiol. Lett. 223:129–34.
46. Potter, M. C. 1911. Electrical effects accompanying the decomposition of organic compounds. Proc. Royal Soc. Lond. Series B 84:260–276.
47. Rabaey, K., N. Boon, S. D. Siciliano, M. Verhaege, and W. Verstraete. 2004. Biofuel cells select for microbial consortia that self-mediate electron transfer. Appl. Environ. Microbiol. 70:5373–5382.
48. Rabaey, K., G. Lissens, S. D. Siciliano, and W. Verstraete. 2003. A microbial fuel cell capable of converting glucose to electricity at high rate and efficiency. Biotechnol. Lett. 25:1531–1535.

49. Ramakrishna, R., J. S. Edwards, A. McCulloch, and B. O. Palsson. 2001. Flux-balance analysis of mitochondrial energy metabolism: consequences of systemic stoichiometric constraints. Am J Physiol Regul Integr Comp Physiol 280:R695–704.
50. Roller, S. D., H. P. Bennetto, G. M. Delaney, J. R. Mason, J. L. Stirling, and C. F. Thurston. 1984. Electron-transfer coupling in microbial fuel cells: 1. Comparison of redox-mediator reduction rates and respiratory rates of bacteria. J. Chem. Tech. Biotechnol. 34B:3–12.
51. Sasaki, S., and I. Karube. 1999. The development of microfabricated biocatalytic fuel cells. TIBTech 17:50–52.
52. Schilling, C. H., M. W. Covert, I. Famili, G. M. Church, J. S. Edwards, and B. O. Palsson. 2002. Genome-scale metabolic model of Helicobacter pylori 26695. J Bacteriol 184:4582–93.
53. Scholz, F., and U. Schroder. 2003. Bacterial batteries. Nature Biotechnol. 21:1151–1152.
54. Schröder, U., J. Nießen, and F. Scholz. 2003. A generation of microbial fuel cells with current outputs boosted by more than one order of magnitude. Angew. Chem. Int. Ed. 42:2880–2883.
55. Sell, D. 2001. Bioelectrochemical Fuel Cells, Biotechnology, vol. 10 Special Processes. Wiley-VCH.
56. Sell, D., P. Krämer, and G. Kreysa. 1989. Use of oxygen gas diffusion cathode and a three-dimensional packed bed anod in a bioelectrochemical fuel cell. Appl. Microbiol. Biotechnol. 31:211 - 213.
57. Sisler, F. D. 1962. Electrical energy from microbiological processes. J. Wash. Acad. Sci. 52:181–187.
58. Stephanopoulos, G. N., A. A. Aristidou, and J. Nielsen. 1998. Metabolic Engineering: Principles and Methodologies. Academic Press, San Diego.
59. Takayama, K., T. Kurosaki, T. Ikeda, and T. Nagasawa. 1995. Bioelectrocatalytic hydroxylation of nicotinic acid at an electrode modified with immobilized bacterial cells of *Pseudomonas fluorescens* in the presence of electron transfer mediators. J. Electroanalyt. Chem. 381:47–53.
60. Tender, L. M., C. E. Reimers, H. A. r. Stecher, D. E. Holmes, D. R. Bond, D. A. Lowy, K. Pilobello, S. J. Fertig, and D. R. Lovley. 2002. Harnessing microbially generated power on the seafloor. Nat. Biotechnol. 20:821–825.
61. Tsujimura, S., A. Wadano, K. Kano, and T. Ikeda. 2001. Photosynthetic bioelectrochemical cell utilizing cyanobacteria and water-generating oxidase. Enz. Microb. Technol. 29:225–231.
62. Varma, A., and B. O. Palsson. 1994. Stoichiometric flux balance models quantitatively predict growth and metabolic by-product secretion in wild-type *Escherichia coli* W3110. Appl. Environ. Microbiol. 60:3724–3731.
63. Weiland, P. 2003. Production and energetic use of biogas from energy crops and wastes in Germany. Appl. Biochem. Biotechnol. 109:263–274.
64. Wilkinson, S. 2000. "Gastrobots" - Benefits and challenges of microbial fuel cells in food powered robot applications. Autonomous Robots 9:99–111.
65. Willner, I., and E. Katz. 2000. Integration of layered redox proteins and conductive supports for bioelectronic applications. Angew. Chem. Int. Ed. 39:1180–1218.
66. Wilson, B. J. 1963. Experiments with a magnesium seawater cell incorporating a bacterial colonized cathode. NRL Report 5998.
67. Yagishita, T., S. Sawayama, K. I. Tsukahara, and T. Ogi. 1998. Performance of photosynthetic electrochemical cells using Immobilized *Anabaena variabilis* M-3 in discharge/culture cycles. J. Ferment. Bioeng. 85:546–549.
68. Yoshida, N., K. Yano, T. Morita, S. J. McNiven, H. Nakamura, and I. Karube. 2000. A mediator-type biosensor as a new approach to biochemical oxygen demand estimation. Analyst 125:2280–2284.

Index

L

M

N

P

R

S

T

V

W

Z

MIX
Papier aus verantwortungsvollen Quellen
Paper from responsible sources
FSC® C105338

If you have any concerns about our products,
you can contact us on
ProductSafety@springernature.com

In case Publisher is established outside the EU,
the EU authorized representative is:
Springer Nature Customer Service Center GmbH
Europaplatz 3, 69115 Heidelberg, Germany

Printed by Libri Plureos GmbH
in Hamburg, Germany